Handbook of Materials Management

P. GOPALAKRISHNAN
Formerly, Professor
Administrative Staff College of India
Hyderabad

Prentice-Hall of India Private Limited
New Delhi - 110 001
1996

Rs. 189.57

HANDBOOK OF MATERIALS MANAGEMENT
by P. Gopalakrishnan

PRENTICE-HALL INTERNATIONAL, INC., Englewood Cliffs.
PRENTICE-HALL INTERNATIONAL (UK) LIMITED, London.
PRENTICE-HALL OF AUSTRALIA PTY. LIMITED, Sydney.
PRENTICE-HALL CANADA, INC., Toronto.
PRENTICE-HALL HISPANOAMERICANA, S.A., Mexico.
PRENTICE-HALL OF JAPAN, INC., Tokyo.
SIMON & SCHUSTER ASIA PTE. LTD., Singapore.
EDITORA PRENTICE-HALL DO BRASIL, LTDA., Rio de Janeiro.

ISBN-81-203-0709-7

Second Printing . . . . . . January, 1996

Published by Prentice-Hall of India Private Limited, M-97, Connaught Circus, New Delhi-110001 and Printed by Syndicate Binders, B-167, Okhla Industrial Area, Phase I, New Delhi-110020.

To

the fond memory of my dearest son

G. GANAPATHY RAM

(The case study in this book is named after him)

Contents

Part II — COST REDUCTION METHODS

Part III — INVENTORY CONTROL

Preface

Today, Materials Management has gained acceptance in all developing economies. The materials department accounts for over 60 per cent of the cash outflow and 90 per cent of working capital of most of the organisations. With every Five Year Plan, the size of investments has been increasing, offering great challenges and opportunities to the materials professional.

I have had the privilege of intensive, indepth interaction with numerous senior executives engaged in materials, finance, production and warehousing management, inventory control, and so on, who attended the courses on Materials Management conducted by me in the Administrative Staff College of India, Hyderabad, as well as those conducted by leading private and public sector organisations. Experience gained in training, research and consultancy in Materials Management during the last three decades in India and abroad, including my role as a United Nations Advisor during 1979–1981, has convinced me the need for a Handbook on Materials Management; for, there is no single Handbook available in the country today, which covers all aspects of Materials Management.

This research-based Handbook on Materials Management is divided into 6 parts: Warehousing (Chs. 1–13), Cost Reduction (Chs. 14–29), Inventory Control (Chs. 30–42), Purchasing Strategies (Chs. 43–54), Materials Management Policy (Chs. 55–67), and an Integrated Policy Case (Ch. 68). The reader may note that the findings here could differ from organisation to organisation, depending on the product, organisational culture, and type of organisation (private or public sector or Govt. department/undertaking).

The book has certain unique features. Each chapter ends with 10 points to ponder, which also act as a summary. Another distinguishing feature is the provision of A to Z tips for improvement in Materials Management. The Integrated Policy case given at the end is aimed at improving the analytical skills of the reader. Each chapter is compact, comprehensive, concise and application oriented.

The text is primarily intended for students of Engineering and Business Management. It is also addressed to practising executives involved and interested in marketing, production, inventory, procurement, finance, materials function, logistics and inspection.

I wish to acknowledge the help rendered by a large number of organisations and senior executives, while preparing the manuscript, by answering the research survey questionnaires, which form the basis of this text. I thank the many colleagues, professionals and management students who encouraged and assisted me in writing this book. Also, the assistance provided by a large number of petrochemical organisations by giving data in the preparation of Ganapathy Ram Petrochemicals case is acknowledged. I also wish to thank Ms. Mahalaxmi Ramakrishna for her neat typing of the manuscript. Finally, I am indebted to my wife, Indira, who has significantly contributed to this book by suggesting various ideas and correcting the manuscript; besides, her patience and understanding during the preparation of the manuscript was remarkable. During my four decades of association with her, she has convinced me that the housewife is the best materials manager in the world!

P. Gopalakrishnan

PART I

WAREHOUSING MANAGEMENT AND LOGISTICS

1. The Warehousing Function

IMPORTANCE OF STORAGE

The concept of warehousing has been known since time immemorial. Kautilya's *Arthasastra* describes the storekeeper or warehouse manager, as follows:

> Praveno Vakpatu Dheeman Swami Bhaktscha Nischayaha !
> Pralubdaha Satyavadi Cha Bhandgari Sa Ishyate !!

This Sanskrit sloka suggests that a storekeeper or warehouse manager should be truthful, adept in the art of communication, intelligent, loyal to the organisation, and should suppress personal greed.

Supplies of essential commodities are often in excess when not needed, and scarce when required. To effectively combat the uncertainties in their availability, the management has to adopt better methods of storage. One of the earliest recorded narratives of organised warehousing is the story of Joseph in the Old Testament, giving an account of how he collected grain during the seven years of plenty to provide for the seven years of famine that followed.

Agricultural operation is always seasonal and man has to store cereals both for future consumption and for seed preservation. This habit of storage has permeated into industrial and trading organisations and every company has a storehouse or warehouse of materials for keeping regularly consumed items. Expansion of trade and commerce has enhanced the need for proper storage of materials. This has resulted in the emergence of various types of warehouses to facilitate receipt, storage, issue and disposal of all types of items in a systematic manner.

The concept of warehousing or stores function is not new in India. In many parts of the country, traditional castes such as Kothari, Kotwal, Bhandari and Bhandarkar have for generations been dealing in procuring, stocking, distributing goods and merchandise. However, though the concept of warehousing has been prevalent for over 2000 years, the warehouse has not yet obtained due recognition. In fact, it is considered as a dumping ground for unwanted people in quite a few organisations. To correct this erroneous impression and ensure optimal utilisation of warehouse, scientific principles of management need to be applied.

PROFIT CENTRE POTENTIAL

The Food Processing Ministry reports that out of the annual output of 80 lakh tonnes of fruits and vegetables, about 25 per cent or Rs. 3,000 crores are wasted due to lack of adequate post-harvest facilities, absence of linkages with markets, improper warehousing infrastructure, lack of cold storage facilities and spoilages in handling and transportation. Reduction in wastages through better infrastructure, warehousing and transportation facilities will automatically increase the surplus. Up to 30 million tonnes of food grains are stored annually and the damages caused by rats as well as rodents are well known. In the industrial sector, it is estimated that about one-third of bank deposits — about rupees 100,000 crores of money — locked in the form of inventories of raw material, machinery spares, and semi-finished and finished goods. India imports to the tune of about Rs. 40,000 crores per annum. The obsolete inventory stored in warehouses was estimated in 1974 by the then Finance Minister Mr. C. Subramaniam as Rs. 2,500 crores, which may well be around Rs. 10,000 crores now in view of the rapid industrialisation and increased inflation in India. Warehouse service charges are estimated at about 30 per cent per year of the average value of inventory.

RETURN ON INVESTMENT

One of the major corporate objectives in any organisation is to improve the ROI (Return On Investment). The return on investment is defined as return divided by investment. ROI can be improved either if the numerator, namely, return is increased or the denominator is decreased. Investment consists of fixed capital and working capital. If we want to decrease the denominator, then either the fixed capital and/or the working capital must be decreased. In an organisation, it is not possible to reduce the fixed capital, the capacity utilisation may be increased. So, if we want to improve the return on investment, we have to reduce the working capital. According to our research studies, 90 per cent of the working capital is blocked in the form of inventories of raw material, machinery spares, and semi-finished and finished goods. Hence, one important method of increasing the return is to reduce all kinds of inventories stocked in warehouses.

MATERIALS COST

Materials enter an organisation as raw materials, components, spare parts or as consumable items. Raw materials and components get converted into 'work-in-progress' and finally into finished goods that are sold. Spares and consumables are utilised for the smooth functioning of production machinery. In a majority of industries, these input materials contribute to over 60 per cent of the finished goods. Table 1.1 gives the materials cost as a per cent of total cost of finished goods in a representative sample of the Indian industry.

Table 1.1 Industrywise Input Percent Cost

Industry Groups	Percentage of Materials Cost
Asbestos, Cigarettes, Fertilisers, Ferro alloys, Newsprint, Newspaper, Steel, Transportation, R & D.	35–45
Cement, Chemicals, Electronics, Engineering, Explosives, Paper, Machine tools, Pharmaceuticals, Agriculture.	45–55
Bread, Cables, Textiles, Dyestuffs, Electricity generation and distribution, Tyres, Refrigeration, Ship-building, Heavy machinery.	55–65
Commercial vehicles, Cotton yarn, Furniture, Earth-moving equipment, Jute, Sugar, Scooters, Wool.	65–75
Electrodes, Projects, Civil construction, Fabrication, Petroleum refining.	More than 75

It can be seen that if the materials cost is as high as 60 per cent, wages, salaries, depreciation and taxes 35 per cent, and profit 5 per cent of the selling price of the product, then a 10 per cent reduction in materials cost increases profit by 120 per cent. If the materials cost is reduced by 10 per cent of the existing figure by scientific warehousing methods, then the new materials cost is only 54 per cent and profit increases by 6 per cent to a total of 11 per cent. This is equivalent to (11-5)/5 or 120 per cent increase in profits. This figure is the same as the benefit for more than doubling the marketing efforts or by a figure of 120 per cent, with other things remaining the same.

This shows the importance of economy in warehousing and the interrelated areas of purchasing and inventory control. Proper identification of how the materials have been consumed and an accurate updating of warehouse records on receipt, issue, price, balance, wastage, rejection losses, etc. are the initial steps involved in effecting economy.

NEED FOR WAREHOUSING

Identification of the needs of customer is a prime requisite. Once this activity has been completed, materials have to be procured. It takes considerable time — up to five years in some cases — before an identified need is converted into available materials ready for use and this time is called *lead time*. Materials have to be held in stock to meet consumption requirements of this period. In addition to this, there can be undue delays in procurement or unexpected increases in the rate of consumption. The stock of inventory maintained in the stores provides for this contingency. Materials are also stocked for speculative purposes, expecting price rise due to increased demand or shortages for supply. In India cold storages and railway wagons are extensively used by trade for the purpose of hoarding stocks of foodgrains, thereby causing an artificial shortage and rise in prices. In an industry, the warehouse manager holds stocks as protection against future drying up of supplies and other adverse factors. The storage and preserving of raw material and/or finished goods is regarded as a key element of marketing strategy.

Functionally, the goods that are held in stock enable a smooth adjustment of demand and supply. They act as buffers between factors that raise the demand and those that meet the requirements. Warehousing is not the simple act of storing goods, but rather a package of services that enables the smooth running of the wheels of industry, supplying raw materials to the operations executives and finished goods to the customs.

The warehouse should supply goods of the right quality at the right time in the right quantities to meet the pace of consumption. It has to forecast the demand to enable procurement, receive, handle, store and transport material without breakages or losses, and reduce wastages and obsolescence. In order to achieve this goal, it must have a clear picture of the nature of anticipated consumption, as well as the objectives and the environmental factors that influence the processes. Thus, the prime purpose of warehousing is strategic deployment of marketable goods in such a way as to minimise the total distribution costs and maximise customer service.

TEN POINTS TO PONDER

1. Describe the steps to improve warehousing function in India.
2. Identify the importance of warehousing function.
3. Trace the historical background of stores activity.
4. Discuss the profit potential of warehousing function.
5. Elucidate how stores can contribute to profitability of a company.
6. Explain the role of warehousing function in improving the ROI.
7. Articulate the need for raw material in warehouses.
8. Explain how warehousing activity improves customer satisfaction.
9. Amplify the service role played by warehousing activity.
10. Develop the idea that store is a profit centre in an organisation.

2. Organising the Warehouse

CONCEPT OF CUSTODIAN

Organising, planning and control are the basic functions of any decision-making executive, and this is more so in the case of the warehousing and inventory manager who is the custodian of stores. The fundamental requirement for success is proper organisation. Warehousing function should be so organised as to enable the personnel to effectively discharge their duties. As the custodian of materials, the inventory manager controls a sizeable portion of the total investment of the firm. The concept that warehouse is money should be appreciated by everyone in the organisation — for we fail to attach the same importance to one crore rupees of cash and one crore rupees worth of material. Issues should be done as withdrawals in banks by means of cheques in the form of indents and receipts should be properly documented so that the book balance and physical balance tally at any point of time. The manager should take prompt action on rejection, shortages, excess, scrap, surplus, slow moving, non-moving and obsolete items.

DUTIES OF A WAREHOUSE MANAGER

Warehousing is a buffer between the various consuming departments and the procurement cell. It has a comprehensive idea of the problems faced by both the consumer and the supplier. The strategy adopted should be a compromise between the seemingly unlimited resources of the procurement cell within the financial constraints imposed by the economic situation of the company.

The warehousing function has to gear up its activities of receipt, storage, handling, accounting, issues and disposal of materials to ensure timely procurement and reduce unnecessary stocks. Such a comprehensive approach enables the stores function to establish certain norms for storage — the labels of stocks to be held, the preservatives to be employed, etc., within the financial and environmental constraints so that the best possible service can be rendered to customers.

The warehouse manager is responsible for the security of the items. By proper identification of items and with the help of relevant documentation, he should service the customers/user departments promptly. He should ensure proper application of preservatives and identify slow moving, non-moving, surplus, obsolete, scrap separately to ensure proper inventory turnover.

OBJECTIVES OF WAREHOUSING

The primary objective of any storage system is to render service to the customer, which should include anyone who requires materials or stores. The defence services boast of one of the best examples of proper identification of objectives and development of this function. They have a well-knit supply corps under a quarter-master general. The army moves on its stomach, goes the saying, and the drying of the supply channels has been the Waterloo of many a force. This stresses the importance of warehousing.

The modern industrial or business scene presents many constraints and the stores function becomes an optimisation exercise wherein the limited available working capital resources have to be disbursed equitably.

Problem arises because materials that are held in stock incur expenditure in the form of capital cost, storage cost, handling cost, etc. This calls for striking a balance between storage costs and the level of service to be maintained. For this purpose, the warehouse has to update records accurately and minimise obsolescence by keeping the space tidy.

INTERFACE PROBLEMS

Rules and regulations must be clearly defined for receipts and issues of material in the stores by means of a stores manual. Proper distribution of power is a key factor for the effective functioning of stores. Delegation of powers should be such that procurement of high value items and capital machinery is done at the highest level and of other items by the purchasing department. This gives the latter a role in decision-making and prepares them for higher responsibilities. Physical organisation of stores is an important factor to be considered.

A common phenomenon in industries is the browbeating of the stores personnel by the consumers. This is often the result of the low position of the stores controller in the overall hierarchy. This anomaly must be rectified by taking into account the entire organisational structure. Within the warehouse, there is always interaction between receipt, issue and kardex sections. There is another type of interface between the central receiving and inspection departments. In the integrated materials set-up, the materials manager controls raw material store, inventory control and procurement activities. Machinery spare parts and tool room are controlled by the maintenance engineer, while the semi-finished goods inventory is looked after by the shop floor personnel. The finished goods stock — within the warehouse, transit, pipelines and with the dealers — is usually under the realm of marketing. A proper coordination of all the inventory is essential from the working capital control point of view. The data processing department and the top management are interested in the overall inventory function. Carriers, bankers, transporters, insurance companies, clearing agents, suppliers and ancillary units are the external agencies interested in the stocks.

Problems arising from internal and external sources should be sorted out in coordination meetings.

PRINCIPLES OF ORGANISATION

In order to successfully tackle the problems and serve the customer, the warehousing manager should organise the function using the following principles:

(a) Define and clearly spell out responsibilities to each manager according to the companys objectives and the manager's specialisation.

(b) Responsibility should be commensurate with authority.

(c) There should be no change in the scope and responsibility of a position without prior discussion.

(d) No employee should be subject to orders from more than a single source of authority so that unity of command is preserved.

(e) Orders should never be given to subordinates superseding a responsible executive.

(f) Criticism should be done privately and not in open meetings.

(g) All major/minor disputes between departments should be sorted out in coordination meetings.

(h) The immediate superior should approve promotion, disciplinary action, increments, etc.

(i) No employee should be the assistant as well as critic at the same time.

(j) The quality of the assistant's work should be periodically checked and mistakes pointed out so that corrective action can be taken at once.

(k) For carrying out the above tasks effectively, the span of control or number of employees reporting to an executive should not be too large.

(l) Adequate delegation of powers commensurate with the responsibility must be given.

(m) It is advantageous to develop an organisation warehousing manual describing all aspects of the storage function.

TEN POINTS TO PONDER

1. Discuss the concept of custodian in relevance to the stores manager.
2. Identify the principles of organisation involving warehousing.
3. Describe the duties of warehouse manager.
4. Enumerate the objectives of warehousing.
5. Elaborate on the interface problems relating to warehousing.
6. Amplify the need for a warehouse manual.
7. Describe the ideal warehousing organisation.
8. How do you resolve disputes with user departments.
9. Identify departments interested in warehousing activities.
10. Expand the following terms: authority, responsibility, delegation of powers, unity of command and span of control.

3. Location and Layout

LOCATION CONSIDERATIONS

The most important, but often neglected factor when establishing a warehouse, is locating it in the most appropriate place. The general principle is to minimise the total kg-km of transportation of materials within the constraints imposed by other environmental conditions. It is advantageous to have the warehouse nearest to the point of usage. Other considerations are the cost of investment expenses incurred for supplying inputs, manufacturing, and placing the product with the consumer. The environmental constraints are: availability of suitable land, water, power, manpower, approaches, source of raw material and markets. Any vacant site cannot be treated as the site of stores. Socio-economic factors, backward area classifications and availability of relevant facilities are the prime factors to be considered.

RAW MATERIAL STORAGE

The guiding factor in locating the store is minimising the material handling efforts while serving the user. In the broadest sense, this implies location of raw material near the production centre and location of finished good store and packing material store near the assembly shop.

The factory itself should be located near roads and railheads since the laying of additional roads, rail-links and railway sidings is a very expensive proposition. The location of steel plants offers a very good example. The major raw materials that have to be moved are coal and iron ore. Usually steel mills are located near coal fields. Steel plants at Durgapur, Rourkela, Bhilai, Bokaro and Jamshedpur are all in one economic area though located in different states.

FINISHED GOODS WAREHOUSE

Location of field warehouses can be decided by computing the total transportation costs involved by calculating the total cost of transporting all the inputs required from their respective sources and the cost of transporting outputs to the various locations.

When locating the warehouse, the territories which are good from the sales point of view may be inefficient from the distribution aspect. Determining the optimum number, location and size of warehouses for any given company is a highly complex task. Pooling the distribution needs of many products/departments/ companies makes it possible to eliminate pronounced peaks and valleys. The large-scale storage and handling operations permit economic utilisation of specialised facilities and personnel. The trend is to do away with regionalised warehouses altogether — except for very important customers or in large cities — and switch over to premium transportation of air or trucks to render a level of customer service, compared to that offered by an extensive network of warehouses. Private warehouses have also started renting storage space.

CENTRALISATION AND DECENTRALISATION

The number of every day transactions and the physical location of consumption centres play a leading role in matters pertaining to the location of warehouse. If consumption centres are located at the corners of a large plot, then decentralised location of stores may become inevitable. Similarly, as the number of transactions increases, there should be an increase in the number of service counters to serve customers without delay. A dysfunctional reaction to such a situation is the building up of surreptitious stores at various consumption centres, stocking small components, tools, hardware items and accessories.

Decentralised stores offer service where it is needed, resulting in the reduction of unnecessary consumption and prevent obsolescence and accumulation of surplus materials. However, a decentralised set-up costs money because the increased number of locations automatically results in increased stocks and personnel. Information handling also increases, thereby adding to the costs. These extra costs must be balanced against the advantages of the improved service being rendered.

Centralisation and decentralisation have their own advantages and disadvantages in multiproduct/multiunit companies. The stock should be nearest to the user in the form of a decentralised warehouse. However, the slow-moving and/or costly common items can be kept in a central place. The semi-finished goods lie on the shop floor. The finished goods, controlled by the marketing executive, may be in the central warehouse with depots serving important cities and major consuming centres. A judicious policy of central policy making with decentralised location, particularly of fast moving and routine items, serves the consumer better.

LAYOUT CONSIDERATIONS

The layout of a warehouse should be suited to the commodities that are being received, stored and issued. If the movements of commodities stored are analysed thoroughly, a fairly small percentage — around 10 per cent of the item — contributes to around 60 per cent of movements. This suggests that items which are handled frequently should be so located as to minimise the distance travelled.

Normally, the available floor space in a warehouse is split as receiving bay, inspection area, bulk storage area, activity area, gangway, outsides, smoking area and issue area. This is done in order to avoid confusion and overcrowding of bays.

The principle that layout means efficiency should be followed. Before designing the layout, the regulations enjoined by the Factories Act, 1872 must be taken into account. The layout should be such that entry of unauthorised persons must be prohibited. The premises should be kept clean by regular sweeping, use of disinfectants, and provision of adequate drainage and toilet facilities. There should be proper ventilation, and undue overcrowding of employees must be avoided. There should be sufficient lighting but glares should be prevented.

It has been calculated that for the same floor area, a square construction is cheaper than a rectangular construction. In both cases, the mean material movement distance does not vary significantly. Total cubic space available must be optimally utilised.

A TO Z OF LAYOUT PROCEDURES

In order to plan the layout of the warehouse properly, the following steps are useful:

(a) Procure the basic date on the number of items and their movements.
(b) Analyse and coordinate basic data with the plant layout.
(c) Calculate the mechanical handling equipment requirements.
(d) Select the general material handling plant.
(e) Make a rough sketch of the plot plan.
(f) Allow for flexibility and future expansions in the warehouse.
(g) Plan general flow pattern and short route movement of goods.
(h) Plan individual areas and workstations areas.
(i) Plan record keeping, issue, receipt and inspection areas.
(j) Select specific material handling equipment.
(k) Coordinate individual equipment with those available in shop floor.
(l) Plan the spare parts requirements of handling equipment.
(m) Coordinate individual operations.
(n) Calculate the area and cubic space required for warehousing.
(o) Calculate storage space area and stacking area.
(p) Construct flow diagrams of movement of materials.
(q) Allocate the different areas to the plot plan.
(r) Plan and locate the service areas.
(s) Construct the master layout.
(t) Check final layout.
(u) Make a trial run.
(v) Get official approval.

(w) Install the layout.

(x) Ensure coordination with substores and tool stores.

(y) Ensure adequate precautions for security against fire and pilferage and safety in handling.

(z) Thus the warehousing layout — ensuring the right material in the right place, and serving the user in the most effective manner through the shortest possible distance, in the least amount of time — must be developed.

A TO Z ASPECTS OF SECURITY AND SAFETY

Security and safety considerations which play a crucial role in the layout of stores are now discussed:

(a) Since stores is money (in stocks) and the warehouse manager is the custodian of the stores, he is responsible for its security.

(b) Security implies prevention of pilferage and theft by employees, vendors' men and carrier representatives.

(c) Security also means guarding the assets against theft, burglary and dacoity.

(d) The security force or the watch-and-ward staff keep vigil throughout the day/night, not only in stores but the entire premises of the organisation.

(e) The security force must be carefully selected and should have men of proven honesty and physical strength.

(f) Wherever necessary, the security men may be armed with guns.

(g) The security staff must be headed by a capable officer, usually a retired officer of the defence services.

(h) They should be trained in all aspects of safety and security checks.

(i) The appointment conditions should indicate that they must not be or become members of unions or indulge in undisciplined activities.

(j) Rotations of shift timings must be worked out.

(k) All warehouses are to have proper locking arrangements and periodic checks must be carried out.

(l) Everyone entering the warehouse must be informed that the security staff will check or search them while leaving.

(m) Unauthorised entry or exit of persons must be forbidden by the security staff.

(n) Receipt or disposal of materials should be checked by the security personnel for authorised documentation like invoice/gate pass, etc.

(o) The entire warehouse premises should be properly fenced or have high walls.

(p) Burglar alarms must be fitted wherever possible.

(q) Powerful lights must be provided in all the yards of warehouses, making provision for automatic switching over to battery lights in the event of power failure.

(r) Safety devices and firefighting equipment should be installed.

(s) Safe working loads should be exhibited on boards.

(t) Use of protective clothing, handgloves, helmets and gumboots should be ensured.

(u) Safety rules should be observed by all personnel.

(v) Excessive fatigue should be prevented by giving workers rest pauses, tea & lunch breaks.

(w) Dangerous chemicals, volatile substances, corrosive, and explosive materials should be segregated.

(x) Electrical wiring, overhead cranes and other handling equipment should be checked periodically. All machinery should be properly maintained.

(y) Adequate lighting, ventilation and moving space should be provided.

(z) All warehouse personnel should be trained in safety measures against fire, crisis, earthquake, flooding and calamities by audio visual demonstration.

TEN POINTS TO PONDER

1. Analyse the location consideration of a warehouse critically.
2. Explain the difference in raw material stores location and finished goods warehouse location.
3. Discuss the problems of centralised and decentralised stores.
4. Identify the advantages of decentralised warehouse.
5. Sum up the considerations involved in layout of a store.
6. Identify the procedures in evolving a good layout.
7. Enumerate the security considerations in warehouse.
8. Explain the safety aspects in stores.
9. Illustrate the role of location and layout in warehouse management.
10. Identify layout problems of raw material, semifinished and finished goods.

4. Physical Aspects of Warehousing

SCIENTIFIC WAREHOUSING

Unscientific method of warehousing has been one of the major factors that have been causing loss and worry in industries and agriculture. Physical aspects such as location, layout, type of material, shelf-life, handling equipment, storage methods, bins, racks, aisles and use of preservatives play a crucial role in the context of cost reduction. The buildings, racks, bins and aisles must be designed to give sufficient protection against hazards such as rain, fire, theft, vandalism and pilferage.

A TO Z OF PHYSICAL ASPECTS

(a) Utilisation of floor space is better in single product warehouses than in multi-product ones. This also depends upon the type of handling equipment employed and the number of items stored.

(b) The aisle width has to be designed on the basis of material handling equipment utilised in the stores. Godowns owned by the Central Warehousing Corporation have to be designed to utilise up to 80 per cent of the floor space, for their normal use is to hold foodgrains with the aid of manual labour.

(c) On the other hand, godowns using forklift, trucks and other mechanised handling equipment find it difficult to utilise more than 60 per cent of the floor space as they have to provide manufacturing area in addition to aisles.

Utilisation of Heights

(d) The performance of warehouse is measured by cubic space and hence utilisation of heights assumes importance. This storing height has to be decided on the basis of ease of storage, retrieval, type of packages, load characteristics, flooring, roofing, ladder facility and handling facilities. Stacking up to the roof without considering other factors is not feasible.

(e) Normally, floorings in warehouses are designed to hold up to 200 lb/sq ft of distributed load. If warehouses are located in upper floor buildings, the beams and columns have to be designed accordingly. In addition, they have to accommodate rolling loads due to the handling equipment and materials.

(f) Stacking should facilitate easy issue and receipt activities. The normal stacking height in grain godowns is 16–18 ft with about 20 bags. The latest trend in developed countries is to stack up to 30 ft, in view of the ever rising land costs. The upper limit of stacking heights depends upon the roof structure, lighting arrangements, exhaust arrangements, fire prevention facilities, use of sprinkler system, safety considerations, handling methods, etc.

(g) Normally, 3–4 ft clearance is allowed at the top and about six inches at the bottom, for free movement and prevention of dust accumulation.

(h) Adequate provisions must be made for maintenance of buildings, yards, pipelines, railway sidings, roadways, handling equipment, storage areas, wiring system, etc. in the warehouse.

Stacking Method

(i) The simplest stacking method that can be adopted is *block stacking*. In this method, unit loads are stacked one over the other, common examples being barrels, bales, drums, bags or other palletised goods. For temporary storage, *piling up stocks* in the same direction can be adopted. Alternatively, each layer can have *tiers* of lengthwise and breadthwise containers alternating to form a block. In this case, it is possible to issue out a part of the stack from top to bottom without disturbing the rest of the stack.

(j) Sometimes space can be provided in the centre to allow air flow to enable drying of wet materials. The advantages of the block stacking system are that no special storage equipment like bins or racks are needed and space utilisation is high. This is mainly used in the manual system of operation.

(k) A major disadvantage of this method is the difficulty in gaining access to materials. Quite obviously, the topmost layer has to be removed first. If random access is required, stack heights have to be reduced and locations reserved. Stability of the stacks has to be ensured to prevent crashing loads. This method is best suited for bulk storage of items in the form of bags, bales and other unit containers especially when there is less variety and more quantity.

(l) Retail issues are difficult in such a storage arrangement. Metals in the form of plates can be stored in this fashion. However, this method cannot be applied if a strict first-in-first-out system is to be followed for law shelf-life items.

Bins and Racks

(m) Another important approach used in industrial warehouses is the use of racks, bins, shelves, pigeon holes, almirahs and partitions made of slotted angle. These are basically wooden or metallic structures that divide storage space into compartments so that the items can be kept individually. Racks can be arranged either along the walls or back to back with sufficient moving space, as the case may be and can reach up to the roof. Bins should be properly numbered and a file indicating bin number and contents must be maintained by the warehouse manager.

(n) Moving racks fitted on cast or wheels are also used to make stores items more mobile and easy to use. The openings can be suitably arranged for receiving the particular size of items.

(o) Compartments can be multi-tier or single-tier to suit the needs. If necessary, bins can be partitioned to keep more items even though one bin must contain only one item. For small components, trays can be used. Closed or open bins or revolving racks can be used.

(p) At times, cupboard like structures are made to move on wheels to enable the supply of components at consumption centres. A variant of this is the oiling van used in machine shops.

(q) This method of storage allows direct access to every single item. Crashing load is not a problem as long as the compartments are able to withstand the loads. This arrangement is quite stable. However, the capital cost is higher because of the investment in structurals. The two-bin system of inventory control items in one bin being used for safety stock, while the items in the other bin are issued out and is also in use. Small costly items are locked in a cupboard placed near the warehouse manager.

(r) Storage of raw materials like roads, channels and sheets is accomplished by adopting both the above methods.

(s) If the volume warrants it, block stacking can be resorted to. But in other cases loose markers or stands, or even racks can be used.

Dry Storage System

(t) Some new types of storage have been evolved to suit Indian conditions. An extremely versatile and economical method of dry storage suitable for storing bagged type of materials is now available. The outer casing is made from synthetic rubber and steel mesh and is easily transportable. Its installation can be compared to putting up a tent.

(u) Bumper food crops have created storage problems, specially due to the paucity of concrete warehouses. One of the methods adopted by warehousing corporations is to provide Cover and Plinth (CAP) storage facilities. In this method, foodgrains are kept in the open on wooden crates under polythene covers. These protect the grains from sun and rain and are made rodent-proof by digging pits. However, severe cyclonic storms do uproot the polythene covers. Occasionally, school buildings are also requisitioned for storing foodgrains.

Loading and Unloading Areas

(v) To facilitate loading and unloading, trucks and wagons are normally parked against walls or docks. This is a critical area, often forgotten, though if this can give rise to bottlenecks. The dock height depends upon the type of vehicles used. Most transport companies use an iron framework and a wooden plank at an inclined plane. Care should be taken to have sufficient space for proper manoeuvering of vehicles. The distance between the edge of the loading bay and the nearest obstruction should be long enough to allow easy loading and unloading. A weighing scale must be kept at the unloading bay of the receipt section.

Structural Considerations

(w) Normal loading capacity for a warehouse is 200 lb/sq ft. If loading is to be more than this, then appropriate strengthening must be done. If the flooring is broken, there are frictional losses in handling, leading to wear and tear of handling equipment.

(x) Just as under designing leads to losses, over designing results in wastage of resources. Many organisations are tempted to locate their stores in bays meant for shops. They should remember that floorings in such bays are meant for heavier, dynamic and vibrational loads and should therefore avoid such wastage. Adequate infrastructure like a moving ladder, staircase, and handling equipment must be incorporated while planning structural facilities.

(y) Unless land cost is prohibitive or land is not available, use of first floor or higher floor warehouses is not an attractive proposition. Its advantage is the possibility of using gravity feeds for components, but this has to be compared with the extra work needed for storage and handling. Also doors should be as large as possible in order to facilitate handling. Columns should preferably be at least 30 feet apart. In the case of bulk commodities, columns can create plenty of wastage of space.

Lighting

(z) Regularly used stores areas should be properly lit in order to facilitate operations. Adequate natural lighting through the ceiling and windows must be provided. General illumination level for factories is around 100 lux, where lux measures lighting power in lumen per square metre. This general level of illumination is sufficient for block storages, but issue counters, inspection centres and areas where bins are used need twice or thrice this level. The scrap yard must be adequately lighted to prevent thefts. Storage areas where colour coded items and small components are stored require illumination levels up to 900 lux. Sufficient lighting arrangements as stipulated by the Factories Act, 1872, must be made to avoid accidents due to bad lighting.

TEN POINTS TO PONDER

1. Discuss the concept of physical aspects of warehousing.
2. Explain the utility of physical aspects of warehousing.
3. Discuss the relevance of area of square feet and volume or cubic feet warehousing.
4. Identify various stacking methods adopted in warehousing.

5. Describe the methods of storage in bins, racks and shelves.
6. Amplify the statement that all items cannot be stored in bins.
7. Elucidate the criteria for loading and unloading operations.
8. Enumerate the structural considerations involved in warehousing.
9. Discuss the problems of warehouses located in multistorey buildings.
10. Clarify the role of lighting in warehouse operations.

5. Safety Measures

FIRE HAZARDS — PRECAUTIONS

Protection against fire hazards is one of the primary safety measures to be undertaken in a warehouse.

1. Even in siting and construction stages, measures against fire hazards have to be considered.
2. Fire is a major menace in warehouses, because of the inflammable nature of boxed stores.
3. An extinguisher on hand, when fire breaks out, is better than a brigade coming 10 minutes later. So there should be provision for early detection of fire and a fire-fighting system which can swing into action at once.
4. Fire points have to be sited at strategic points and at the junction of gangways.
5. A bucket of sand and another with water must also be maintained, besides adequate number of fire extinguishers.
6. It must be mentioned that gas extinguishers are hazardous.
7. Fire exits and unobstructed passages must be provided as per the requirements of law.
8. First aid and medical kits must be made available at all points.
9. Adequate number of fire alarms must be placed throughout the storehouse.
10. Automatic alarm and sprinkler systems may be installed in the warehouse.
11. All fire equipment must be inspected regularly, and the potency of the chemicals used in extinguishers must be ensured.
12. Fire fighting teams of volunteers must be recruited and trained to effectively control the spread of fire.
13. While water is the fire extinguisher generally thought of, it cannot be used in all cases, especially where chemical, oil or electrical fires occur. Instead of extinguishing, water spreads such fires and can even lead to explosions, causing irreparable damage to equipment. To overcome such fires, other agents like carbon monoxide gas, sand and chemicals must be used.
14. Metallic chairs, rings, hooks, shackles and swivels must be regularly annealed. Such materials should not be used directly after they are lengthened or repaired or altered. They should be properly tested before use.

15. It is necessary that the maximum safe working loads are marked on the hoists and lifts.

16. It is also desirable to use pallets or specially designed platforms for the stacking of goods.

PILFERAGE PROTECTION

One of the perennial problems faced in storage of materials is pilferage, particularly in today's context of deteriorating moral values and poor standards of living. The dictum — prevention is better than cure — aptly indicates the strategy that should be adopted. Costly items that are small enough to be carried out without provoking detection are likely to be pilferred. Stationery items also fall in this category.

Caged enclosures or closed cribs with locks for available items may be arranged. This limits access. Care has to be taken to restrict the number of persons who have keys so that the responsibilities cannot be shirked.

The check-out systems should be well-established and foolproof. When a large workforce is present, a thorough random inspection must be adopted and anybody caught must be summarily punished. The warehouse manager usually executes a fidelity bond and some organisations go to the extent of deducting salary for negligence in such matters.

If precious metals are stored, deployment of professional guards becomes necessary. Security patrol is required for outdoor storage. The cost of protection agents must be matched against the expected pilferage losses. Detailing more protection agents to high-value items may be a working solution.

In this context, the railway protection force (RPF) is cynically termed by the industry as the railway pilferage force! Hence, suitable steps must be taken to ensure that the security staff do not collude. Appealing to the moral values and ensuring that the wages of the workers are comparable to other organisations may help reduce pilferage.

A TO Z PRESERVATIVES IN STORAGE

If materials are to be stored without depreciation of their physical properties, they have to be guarded against the effects of dampness, rain, dryness, heat, cold, humidity, temperature variations, dust, dirt, attack by insects, fungus formation and other biological aspects.

(a) Depending upon their individual physical characteristics, the materials have their own shelf-lives. Special arrangements like cold storage or irradiation do help in extending the shelf-life, but the costs involved must be comparable with the savings or the cost of materials themselves. These protective measures can alter the properties and care should be taken to keep these alterations within limits. It is also desirable to develop a manual on preservatives as a part of the warehouse manual.

Chemical Items

(b) Storage of corrosive chemicals must be carried out with systematic care. All containers must be labelled and storage areas demarcated by signs.

(c) The storage place containing chemicals must be well-ventilated and far removed from active fire hazards. Incompatible chemicals must never be stored together. For instance, never store acetone next to concentrated nitric acid or sulphuric acid mixtures, or copper near acetylene or hydrogen peroxide.

(d) Isolate fluorine from everything. Never store sulphuric acid near chlorates, perchlorates or permanganates.

(e) Protective clothing should be used by the staff when dust, mists, fumes, gases or vapours are produced in harmful quantities. Control must be maintained by exhaust ventilation.

(f) Acids must be kept away from paper, wood and other items that are likely to corrode. They must be handled with great care and by properly attired personnel. Cylinders holding gases like oxygen, hydrogen, acetylene are usually kept padded with coir covers in a separate godown, away from sources of heat.

(g) Storage of large quantities of carbide requires special permission. It should be kept free of moisture and away from flames. Similarly, it is essential to keep phosphorous in kerosene. Chemicals that are normally supplied in gunny bags must be protected against moisture as dampness solidifies them — cement is a typical example.

(h) Oil is stored in bulk in tanks under dust and dirt-proof conditions. Leakages and evaporation should be minimised. A quick turnover of oils is essential for maintaining properties like viscosity. Petroleum is explosive and the inspector of explosives has to certify the safety of the storage. Vegetable oils should be protected from air, light and moisture. Paints must be guarded against moisture. As they age, the top surfaces form into a crust. The drums should be inverted to prevent settling. It is preferable to use them within a year. Varnishes do not deteriorate, but they contain alcohol and hence are inflammable.

Agricultural products

(i) Wood is susceptible to attack by white ants. It should not become damp. A coating with paint or coal tar can be used and there must be sufficient ventilation. On the other hand, bamboo must be kept submerged under water.

(j) Foodgrain bags usually contain moisture and up to 16–18 per cent would not pose much problem. However, it is advantageous to have a sound ventilation system. Foodgrain can be stored either in silos or in bags. While this method of storage is capital intensive, fumigation costs are about 50 per cent cheaper around Rs. 1.50 per ton per fumigation.

(k) Fumigants have to be used in order to protect grains against pest attacks. Some of the commonly used fumigants are (i) DDVP, a 100 per cent emulsion, can be used in weekly sprays in concentration of .05 per cent in water; (ii) Pyrethrum emulsion can be used at 2.5 per cent concentration level on bags and .025 per cent level on bulk as often as required; (iii) Methyl bromide can be used as a fumigant for an exposure of 48 hours at a concentration of 32–40 mg per litre for exposure periods of one week; (iv) Aluminium phosphate can be used at 3–6 gm per ton of free exposure for periods up to seven days.

(l) It should be noted that when chemicals are used, their residues in grams must be within limits or else the grains will be unfit for human consumption. The tolerance limits for some chemicals are given in Table 5.1.

Table 5.1 Tolerance Limits

Chemicals	Maximum Tolerance (mg/kg)	Chemicals	Maximum Tolerance (mg/kg)
Aldrum	0.01	Dichlores	1.00
Carbonyls	1.50	Heprachlores	0.01
Dialdrin	0.01	Hydrogen phosphide	0.05
Diazinons	0.05	Pyrethrins	1.50

Manufactured Items

(m) Brushes must be protected from moisture and insects. It is preferable to keep them in original cartons, sprayed with insecticides. Their shelf-life is low, especially in dry hot conditions, where the bristles become brittle. Under humid conditions growth of moths can spoil their properties.

(n) Another manufactured item with low shelf-life is battery cells, especially under damp and humid conditions. Hence, their stock must be under control. Electrodes suffer under humid conditions and have to be warmed before use. Rough handling can damage their outer coatings; hence they should be constantly aired and wiped.

(o) Fire bricks should be protected from rain and inundation as they will disintegrate if they absorb water. Cloth and paper should be protected from moisture, water, rats, white ants, etc. Woollen goods must be protected from moths. Insecticides must be regularly used.

Rubber Items

(p) Rubber goods need special treatment. Factors affecting them include light, sun rays, air, heat, moisture, ozone, oils, dust and dirt. Their shelf-life is low and it is advisable not to hold items for periods longer than six months.

(q) Tyres should not be placed in piles as this practice tends to kink the bead wires and distort their form. They should be carefully covered with a tarpaulin or some reasonably heavy woven fabric for protection against light, air and dirt. Tyres should not be stacked on dirty floors without first laying a foundation of wood strips. They should not be exposed to light from skylights, windows or other openings. They should not be piled near radiators or other sources of heat. Storage arrangements should be planned so as to rotate the stocks and move the oldest tyres first, thus avoiding accumulation of old stocks. Under no circumstances should tyres of large size and greater weight be piled on smaller varieties.

(r) Glass sheets should not be stacked one over the other or horizontally. It is preferable to keep them upright and protect them from sunlight. Window glasses should be painted over with dark paint and the openings kept closed so that minimum light enters the warehouse. Electrical discharges can occur as a result of the operation of electric motors, arc welding, generators, switches and other electrical devices. These discharges generate ozone and greatly accelerate oxidation that spoil items like tyres.

(s) Atmospheric temperatures in the warehouse should be kept as low as possible, since high temperature tends to increase the rate of deterioration. The preferred temperature is about 80°F. Air currents circulating in the warehouse, keep replenishing the supply of oxygen. The more the oxygen present, the more rapid will be the deterioration of some rubber items. The storehouse should be as free from moisture as possible.

Pallets

(t) Where it is desirable to utilize more vertical space than is permitted by conventional piling, a pallet type crate can be used in combination with lift trucks to avoid handling losses. The pallet should be constructed with sufficient strength to withstand handling and support the load imposed upon it in the pile without damage to the tyres. Two or more sizes should not be piled in the same pile unless a shortage of space makes this practice necessary. There is a wide range of pallets of different sizes, made from steel, wood, disposable fibre, etc.

Metal Items

(u) Castings improve with age in atmospheric conditions. But rusting of iron castings must be prevented. Steel castings can remain in the open if they have a coating of paint. All castings, specially non-ferrous ones, have to be protected from pilferage. Machined components, especially the threaded and mating portions should be protected from rusting by greasing and pluffing. Pipes must be given a coating of anticorrosives. Their ends should be plugged and the threads must be protected. If they are stacked vertically in the open they should be protected from the accumulation of rain water.

(v) Steel must be protected from rusting by using oils or anticorrosive coatings on the surface. Abrasives can be used to remove the rust and pittings. Kerosene is an useful agent for this purpose. Cutting tools must have their surfaces protected by a coating of grease or by dipping in paraffin solution. Files must be greased with mineral jelly or lighter oils and should be wrapped in paper.

Machinery Spare Parts

(w) Bearings should be greased and packed grease proof with a thin coat of 80 per cent mineral jelly and 20 per cent bees wax melted together for long term preservation. Cylinders and barrels should be plugged.

(x) Expensive and delicate instruments should be kept in their boxes with small 2 oz cotton bags of dehydrated silica gel, which should be checked every three months for discoloration and reactivated by heating: if necessary, vapour phase barrier paper can be used for wrapping delicate metal components. Hydraulic cylinders, if kept filled with hydraulic oil, must be exercised or operated 10 to 15 times each, at least twice a year, as otherwise the hydraulic packings will perish. Sintered bush bearings must be soaked in warm oil for 24 hours. Grinding wheels must be stored vertically in individual portions so that they do not rub against each other.

(y) In conclusion, it should be stressed that adequate stock turnover and issuing on a first-in-first-out basis will reduce the incidence of losses in warehouses. It may be noted that heat accelerates chemical breakdown, and rubber, leather, textiles and a host of goods need protection. In temperate climates, the items should come under cover as quickly as possible to protect them against rain. In the tropics the problem of humid heat has to be tackled. Heat builds up unless it is allowed to escape; clean air circulation is necessary.

(z) Air cooling and airconditioning may be necessary in the case of electronic items. Humidity is a condition in which moisture is dissolved in the air. Fungus attacks anything which provides any kind of nutrient. Bacteria and viruses of large varieties can etch glass and render expensive lenses useless. Insects thrive in hot, humid and dusty conditions and spread diseases. Cleanliness inspection, proper ventilation, and use of preservatives may reduce the wastage.

TEN POINTS TO PONDER

1. Identify the precautions to be taken in the warehouse when fire occurs.
2. Enumerate the factors to be considered in siting and building stages against occurrence of fire.
3. Describe the steps to be taken against pilferage.
4. Discuss the methods of storing high value, small items.
5. Explain the role of preservatives in material utilisation.
6. Examine the precautions to be taken for chemical stores.
7. Identify the steps needed to store agro-based commodities.
8. Enumerate steps recommended for storing tyres.
9. Mention the methodology for storing metal items.
10. List out methods of preserving machinery spare parts in stores.

6. Warehouse Operating Systems

FACTORS IN STORES SYSTEM

In order to satisfy the customer/consumer/user, it is essential for the stores function to systematise its working. Proper control of the throughput has to be maintained so that there is no overstocking or wastages. However, in a growing organisation, the rate of growth of material needs is exponential. Hence, the system that is designed should not only be capable of matching requirements with existing supplies, but also be resilient enough to take care of growth without buckling under pressure. This suggests the necessity and importance of feedback in the system. The factors that have to be considered while designing a storage system are the type of material, volume of transaction, physical facilities, receipt, issue, documentation, flow of materials, and the warehousing manual. Let us now consider how the flow of materials influences the working of stores and the steps taken to regulate the flow.

TYPES OF MATERIALS

Materials held in the warehouse can be classified according to the use they are put into. The advantages of such a classification are that each broad category has its own characteristics and hence needs its own system for proper control. In a manufacturing establishment, the material can be classified as raw materials, semi-finished goods, bought out components, spares, consumables and packing materials. In an engineering industry, work in progress and bought out components play an important role, whereas in a process industry, machinery spare parts fall in the critical category. Even in a non-industrial situation, such as departmental stores or provision stores, the items are grouped according to their use, e.g. clothing, utensils, provisions, textiles, stationery, vegetables, and fruits. Depending on the number of items in a major class, it can be split into subclasses to facilitate better control.

RAW MATERIAL STORES

Proper control of raw material is one of the major factors that influences the profitability of a company. Typical raw materials include M S steel, alloy steel, stainless steel, plates, pipes, bars, non-ferrous items, condensor tubes, conductors, chemicals, bulk items, heavy items, direct delivered items, production items, etc. These, being the major input into the organisation, get converted into output.

A stockout in the raw materials is an unusual phenomenon and is immediately reflected by an idle manufacturing line. External factors in the supply such as availability, lead time, seasonality, staggered deliveries and credit conditions all combine to influence the stocking of raw material. As the organisation grows in size, it is advisable to have a separate raw material stores. Items like castings can be stored in the open in sun or rain whereas packing cases have to be protected from rain.

Quite expectedly, transactions are done more in raw materials, necessitating tight control on these. Stock levels and consumption levels should be clearly fixed, with regular follow-up. To reduce handling, stores should be located as near the consumption centres as possible.

COMPONENTS STORES

In engineering industries, the finished products require components of the final assembly. These are either manufactured in feeder shops or procured from contractors. They include items such as castings, forgings, instruments, valves, subcontracted items, etc. These are stored in a component store from where the assembly lines are fed. The stockout of a single components can make the entire assembly line idle. Hence, components must be monitored closely and adequate steps must be taken to avoid stockouts. Selective control techniques may be used to ensure that stockouts do not occur due to non-availability of low value components.

R & D INSTRUMENTS

Storage of instruments and laboratory tested items is necessary for activities like research, development, design, inspection in national scientific institutions, in-house R & D, and quality control. These are similar to tool cribs. Here, in addition to general precautions of a tool crib, the physical storage conditions must be conducive to the needs of storing the delicate instruments. Moisture, dust, heat, light, intense cold, and vibrations are some of the factors that can impair their working, and hence adequate steps are necessary for their protection. Breakages, handling losses, poor maintenance and wrong calibrations must be avoided. The person in charge of these specialised stores must be technically knowledgeable. An often repeated complaint of a harassed user is that the storekeeper does not know what he is holding and expects to be educated. On the other hand, the storekeeper's complaint is that the scientists are unable to forecast their requirements adequately and abandon their respective projects midway, leading to obsolescence in stores. These problems have to be sorted out by better planning of the requirements of research activities.

CONSUMABLES

Consumable items like cotton waste, grease, lubricating oils, etc. enable the smooth running of the wheels of any industry. Generally, these items do not cause any supply problems, but occasionally some of them can become scarce. Petroleum products can be cited as example particularly in stores located in remote areas. Packing materials are also consumables, and a general characteristic is that they are voluminous. Depending on the demand and availability of storage space, their stocking policy should be controlled. They have to be protected against moisture and excess heat, for if they lose their appearance they cannot be used for marketing. Offices and banks handle large volumes of stationery entailing proper storage facilities.

MACHINERY SPARES

Machinery spares are normally controlled by maintenance engineers as the pattern of consumption is rather unpredictable. This obsolete inventory in most cases comprises only the spare parts. As such, proper coordination, planning, monitoring and control have to be done in collaboration with the plant engineering function. Regular collection, collation and analysis of consumption data reduce the problem of a stockout. A common failing is the inadequacy of existing nomenclature which leads to multistocking by using the suppliers, part number instead of internal codification. The damages due to non-availability of a part is several times more than the cost of the part itself, leading to hoarding of machinery spares stock.

TOOL CRIBS

Tool cribs are specialities of most industries where the necessary tools and fixtures are stored. These can be cutting tools, farm tools, jigs, fixtures, die sets, hand tools, accessories, etc. The requirements are determined by the facilities plan. If a tool is not available, costly machinery will lie idle. Therefore, tools must be kept in good condition and stored properly. Blunting of edges and rusting should be prevented. Tools must be handled with care. Tools are borrowed and returned to the tool cribs. Replenishment of the tools is done periodically through the central stores. Location of the tool-crib, tool-life, standardisation of tools, accounting of tools, charging off tools, etc., are some of the aspects that need to be considered. The tool crib is often under the control of the operating/maintenance department. Proper liaisoning and control by the warehouse manager are necessary.

WORK IN PROGRESS

The concept of semi-finished goods is largely found in engineering- and assembly-oriented companies where the 'work-in-progress' is lying on the shop floor. In a process industry, semi-finished goods are in the process line itself and hence external storage is not needed. The complexity of the problem increases with the complexity of products. An effective solution has to be techno-economic in nature. The normal practice is to let the production, planning and control have the responsibility of such materials. But in many industries, where the size or the nature of manufacture does not warrant this allocation, the accounting responsibility and control of these materials are added to the stores section. Another common problem area is the accusation of the shop floor people that the 'work-in-progress' accumulates as matching components have not been made available by the stores. The scrap generated by the departments — sometimes used again in the process and more often disposed of — is another category of items stored in the yards.

FINISHED GOODS

Distribution and trading organisations are mainly concerned with finished goods. The marketing department usually controls the finished goods. The peculiarities of the finished goods depend upon the type of industry. In a consumer product or a process industry, the input to the stock is as per the production schedule, while the output depends on market conditions. A slack market leads to stockpiling and a buoyant market leads to scarcities. Shelf-life is an important consideration in the case of finished goods. Storage space should be adequate to meet these extreme conditions. The system of control should be oriented towards coordination not only within an organisation but also with the competition and consumer profile.

BONDED WAREHOUSE

A common practice existing in many organisations is the use of bonded warehouse, where goods are kept under the control of customs or excise authorities. The bonding or licensing requirement is little more than a revenue producing measure. The posting of a bond by the warehouse means that inspection is undertaken by the state authorities. The items are drawn after paying the necessary levies to the authorities — the levies fixed by the central state governments vary with the type of commodity stored. The location may be inside the factory or in a general warehouse attached to the shipyard/ airport.

Quite often imports are in quantities too large for current use, due to various reasons. If these have to be cleared at once, a sizeable amount of duty has to be paid. Instead, the goods are kept in a bonded warehouse without payment of duty. A bond is raised and the cost of maintaining the warehouse is borne by the importer. As and when the materials are needed, they are released for use. It should be noted that this cost can be incurred so long as it is less than the demurrage rates and the interest rate on import duty. Export zones can be considered as the enlarged version of bonded warehouses.

Products are liable for excise duty, once they cross the production area. But there is always a time lag between production and the actual despatch, depending upon the market environment, availability of suitable mode of transport, and packaging activities. It is not advisable to block large amounts of money as excise duty paid when products are lying as finished goods inventory. Here too, they can be stored in a bonded warehouse operated under proper sanction from the excise authorities. According to the rules, packaging should also be done in the bonded area, and the goods can leave the bonded area on payment of duty and with a gate pass. In these cases, MODVAT can also be considered so that optimum benefits to the company are obtained.

The concept of decentralised stores or departmental stores or substores, is implemented to meet the requirements of consumers and to avoid hoarding in unauthorised surreptitious substores. Materials should be as near the consumer as possible in order to reduce transportation time, cost, double handling, etc. Periodic meetings with the users help identify problem areas and increase the service facility of stores. The guiding factor in all cases must be the cost of capital and benefits arising therefrom.

TEN POINTS TO PONDER

1. Explain the concept of warehouse operating system.
2. Identify the type of material stores in an organisation.
3. Describe the concept of raw material stores.
4. Elucidate the features of component stores.
5. Discuss the problems associated with instrument stores.
6. Amplify the term 'consumable stores'.
7. Enumerate the problems of machinery spare stores.
8. What are the materials kept in tool cribs.
9. Articulate on the need for WIP.
10. Examine the necessity of the finished good stores and the relevance of bonded warehouses.

7. Management of Receipts in Stores

CRITICALITY OF RECEIPT MANAGEMENT

The popular English saying "a battle well begun is half-won" aptly emphasises the management of receipts or inputs in the warehouse. Input management in the warehouse includes the following activities:

(a) Requirement determination
(b) Raising of purchase requisitions
(c) Chasing supplies through the purchase section
(d) Request for expediting suppliers
(e) Receiving the materials
(f) General visual inspection
(g) Quantity inspection
(h) Coordination with quality inspection
(i) Taking stock of the accepted materials
(j) Taking care of rejected materials
(k) Completing documentation
(l) Making book entries
(m) Endorsing suppliers bills

Unless the warehousing department systematises its working by better planning, it will not be able to complete the input formalities. This is one of the critical areas where peaks of workload often occur as the time of deliveries is uncontrollable and depends upon the suppliers' transporters'/carriers' convenience. Lorries and wagons from outstations cannot be considered as extension of warehouses, and have to be cleared on a priority basis to avoid demurrages. Coordinating quality control and inspection personnel can also lead to difficulties. In many organisations, the central receipt warehouse overflows with quarantined goods while the department is starved of the same item, leading to production stoppages. To overcome these drawbacks, it is necessary, as a first step, to have a clearly laid out procedure and a warehousing manual.

RECEIPT DETERMINATION

Even though determination of requirements does not fall under the purview of warehousing function, the customer/user approaches only the stores to get his requirements. The inventory controllers help the personnel in most cases.

However, for some repetitive low value stock items, the warehousing manager is asked to develop the norms. In fixing the levels, it is necessary to seek the cooperation of not only the inventory controllers, but also the concerned users. Past consumption, anticipated forecasts, lead time, exceptional known cases, are relevant to determine the requirements in the case of regularly consumed items.

In the project and construction stages when the organisation is still nascent, stores is one of the first sections to be established. In engineering industries and construction projects, the starting point is the engineering drawing which includes a bill of materials (BOM). The bill of materials usually lists the requirements of all items and material specifications.

Scheduling the arrival is an important factor in these cases. The aim should be to interrelate the project activities on a time scale and procure the materials accordingly. The requirement determination, and procurement and manufacturing activities can be interlinked through network analysis techniques. Warehouse manager must be able to discern the requirements; otherwise he will be blamed for not having what is wanted while being saddled with surplus/non-moving items.

PURCHASE REQUISITIONS

The procurement section has to be informed about the materials requirements so that it can take necessary purchase actions. This is accomplished by raising purchase requisitions. The purchase requisition must contain the material description, code, specifications, required quantity, alternatives, brand names and probable sources of supply. Table 7.1 gives a specimen of the required format.

Table 7.1 Purchase Requisition

To: The purchase department — Number: ________

From: The stores — Date: ________

Prepared by: ________

Sl. No.	Item Code	Description	Cost Centre	Quantity	Probable Source	Alternatives	Required by

This format enables the procurement department to easily locate the supplier and procure the required materials. For new items it is essential to consult the user before releasing the requisition. Such a step ensures corrected, updated entries and prevents accumulation of obsolete items.

On the basis of the purchase requisitions, the purchase department issues purchase orders, a copy of which is forward to the stores for facilitating receipts. This enables the stores to prepare for receiving the materials and correlate with the suppliers' invoice sent with the material. The receiving section should constantly chase the follow-up wing of the procurement section so that materials arrive in time.

RECEIVING PROCESS/SOURCES

The schedule of receiving activities depends upon the supplier, mode of transport and terms of supply. Supplies can be effected through the following means: (a) company's transport, (b) supplier's transport, (c) public lorries, (d) railways, (e) inland waterways, (f) airfreight, (g) shipping through port trusts, and (h) pipeline system. In most cases supplies are released through documents of the suppliers sent in advance. Depending upon the terms of supplies, the documents are either sent directly or through a bank. In such cases, the documents have to be retrieved from the bank after paying the necessary dues and then presented to the carrier who subsequently releases the documents.

The external carriers charge demurrage if the consignment stays in their premises. Demurrage rates, particularly in the case of railway wagons, are steep as organisations tend to use wagons as warehouses. Freight charges could have been paid by the supplier or he might send them on to-pay basis. Full loads on lorry are usually delivered at the door. This reduces handling.

But smaller packages generally have to be collected from the depots. Railways door delivers containers only. All other packages have to be collected from the goods marshalling yard.

Formalities such as octroi, duties, levies, tariffs, excise and customs have to be paid at the time of clearance. Each one of these require specific documents, procedures and rates to be paid. The sums involved in these dues need to be decided at the time of placing the order. The warehouse is often loaded with the function of these clearances. Dealing with governmental agencies needs plenty of tact and ability to communicate. Another problem area is the behaviour of transport contractors and lorry drivers. To overcome these problems, many organisations employ clearing agents who are specialised in this field.

ACTUAL RECEIPT

The central goods receiving bay should be clearly demarcated and cordoned off to facilitate unloading activities. Entries of unauthorised persons must be restricted to minimise the chances of traffic jam and pilferages. The packages are accompanied by the invoices/despatch challans of the supplier, indicating description, specification, purchase order number, quantity and rate. These need to be checked with the purchase order after allowing for minor mistakes in the documents. Usually the packages have a packing slip declaring the contents.

The supplier requires a document/challan/copy of the invoice, wherein the receipt of the goods has been accepted. Either one copy of the challan is signed and returned as a token of acceptance or a provisional receipt is issued.

SOURCES OF RECEIPT

Normally, the bulk of receipts come from external vendors/suppliers. The receipt section has to regularise miscellaneous receipts such as samples and cash purchases by raising receipt notes. In these cases, there is no inspection as they are already accepted. Stores also receive components from the feeder shops. This is an internal transaction and as such there are no receipt notes other than inter-departmental transfer notes in vogue in the organisation. Receipts are issued for directly delivered production material, bulk items and safe-custody items from different departments. Stores receive scrap for disposals and returnable tools and equipment. These need to be accounted on a priority basis. The factories material obtained from subcontractors after carrying out the relevant operations, also have to be accounted separately.

INSPECTION INFRASTRUCTURE

In all organisations, the warehousing manager is accountable for the physical count of the receipt, whereas the quality assessment is invariably carried out by the inspection and quality control department. A portion of the receiving bay is earmarked as the place of carrying out inspection operations. The final incoming material cannot be taken into stock until it passes through the prescribed incoming quality checks. Technically the procurement action is not complete until the inspection is over. This implies that this part of lead time is dependent upon the availability of inspection staff and the testing infrastructure facilities.

General visual inspection should not pose any problem. Common items such as consumables, brooms, cotton waste, detergent, etc., can be accepted by general visual inspection. The supplier's credentials also play a role in this regard. For doing this effectively, the stores receipt section should be provided with adequate measuring instruments like weighing scales, measuring vessels for volumes, ratio balances for counting etc. Regular maintenance and calibration of measuring instruments must be ensured to avoid unnecessary litigation with the suppliers regarding shortages on excesses. Coordination between inspection and quality control personnel is necessary to pass judgment on quality. As the materials are received, they should be indexed and the concerned inspector contacted for conducting the required tests before taking them to stock.

GOODS RECEIPT NOTE

Once the quality of materials received has been assured, they can be taken into stock. The document used to regularise receipt of materials is the goods receipt note or GRN. GRN indicates the suppliers, purchase order, job number, item code, quantity received, quantity accepted, reasons for the quantity rejected, and remarks such as excess or short supply, etc.

Specimen copy of the goods receipt note is given in table 7.2.

Table 7.2 Goods Receipt Note

Sl. No: ________ Date: ________

Sl. No.	Supplier	Item Code	Description	Purchase Order Number	Quantity Received	Quantity Accepted	Remarks

Goods received by: ________

Inspected by: ________

Warehouse Manager: ________

Copies of goods receipt note are sent to the purchase section and the accounts section for further transactions.

REJECTED MATERIALS

Arrangements should be made to request the suppliers to remove rejected materials immediately from the stores receiving bay. Accumulation of rejected material in the receiving section might occupy too much space and this may become a big problem. Sometimes the suppliers purposely do not lift the rejected materials because they know that, when faced with shortages, the organisation will definitely use them. Accepting substandard material when badly needed and rejecting good material when there is plenty of stock lead the suppliers to come to the above conclusion. By adopting a consistent policy and scientific sampling schemes for acceptance/rejection, the above situation can be avoided.

The storekeeper who has received the materials has to keep rejection notes in the case of rejected materials. This note contains details like the reasons for rejection, the quantity rejected, etc. A specimen of rejection note is given in table 7.3. Copies of this note have to be sent to the purchase section for necessary follow-up and the accounts section for adjusting the payments.

Table 7.3 Rejection Note

From: | To
XYZ Company Ltd. | Supplier
ABC Company Ltd.

Your Ref: Consignment Note No. . . . . . .
Our Ref: Purchase Order No. . . . .

Please note that we have rejected the following items received as per your above consignment. We request you to take appropriate action for replacement by items as per specifications.

Sl. No.	Description	Quantity Invoiced	Quantity Received	Quantity Rejected	Reasons for Rejection

Purchase Manager

The procedure to be followed for sending back the rejected material to the supplier depends upon the terms of supply and the company policy as stated in the manual. If there are numerous rejections of supplies from a single source, it becomes necessary to develop another source.

SHORTAGES AND EXCESSES

Often, the warehouse manager has to tackle the problem of shortages and excesses, according to the operating conditions in the market. When shortages arise, a shortage note needs to be presented, giving the necessary details. There is an urgency in this regard, because claims with carriers like lorry transporters or railways can be time barred if they are delayed. Handling of excess supply is more tricky and depends upon the corporate policy, nature of the item and the relationship with vendors. Some organisations have specified time limits up to which they accept any excess supply. Copies of these notes should also be sent to the purchase section for necessary follow-up and adjustments. Usually, only one side of the GRN document is used and the other side can be used for excess/shortage/rejection, etc. This reduces the number of documents in use.

ENDORSEMENT OF BILLS

The organisation's credibility among its suppliers depends to a large extent on its promptness in paying its bills. It has been a common experience that suppliers enhance the price to take care of excess delays in payments. Sometimes payments are delayed due to lack of all supporting documents of supply. The person incharge of receiving can remove this irritant and contribute towards better buyer-seller relations. Since he receives a copy of the purchase order, he needs only to bunch them together along with the supplier's invoice challan and the goods receipt note for necessary action. Problem arises when the supplier sends a single invoice for a set of purchase orders or supplies in part. In such cases, the warehouse manager can maintain a supplier and commoditywise history to enable easy settlement of bills. As soon as the material is accepted after inspection, the warehouse manager should promptly forward the invoice, bills, goods receipt note, rejection note and other documents for settling bills.

TEN POINTS TO PONDER

1. Discuss the role of receiving in warehousing management.
2. Identify the sources from whom the material will be received.
3. What is the role of goods receipt note in warehousing?
4. Describe the manner in which shortages/excesses/rejects are to be tackled.
5. Identify the importance of bill settling at the appropriate time.
6. Discuss the activities of receiving in the stores.
7. Enumerate the steps to be taken if the supplier refuses to remove the rejected goods.
8. Examine the role of inspection in receiving.
9. How does the receiving incharge control directly delivered bulk items to the department?
10. How does the receiving section regularise receipts of miscellaneous items?

8. Management of Inventory Issues

IMPORTANCE OF OUTPUT MANAGEMENT

The performance of the warehousing function, which is primarily a service function, is largely evaluated on how it satisfies the users/customers, by meeting their demands. Hence it is important for the warehousing personnel to streamline their output/issue/sales/customer order processing procedures and prevent unnecessary delays and locking of scarce resources.

ISSUE TO MANY AGENCIES

Issues must be made to the right persons. Regular issues to the different consumer departments constitute the bulk of issuing activity of warehouse. Issues also have to cater to temporary requirements such as equipments, tools, kits, projects, subcontractors, imprest issues, sample issues, returnable issues empties to manufacturers, etc. Obviously, separate procedures have to be developed for these issues and the method of accounting should be specified and effectively implemented.

A TO Z ISSUE SYSTEMS

(a) Issue procedures must be unambiguously spelt out to streamline the output of materials from the warehouse. Clear instructions regarding all categories must be given to the stores staff.

(b) Issues must be made on the basis of valid authorisation. Authorisation can be in the form of a material requisition voucher as shown in Table 8.1.

Table 8.1 Material Issue Voucher

Department

Sl. No.	Item Code	Unit	Quantity Requested	Quantity Issued	Cost Centre Job Order	Remarks

Authorised by: __________ Storekeeper: __________

Received by: __________ Date: __________

(c) This voucher should indicate the name of the requisitioning consumption centre, material code, quantity required, job order, cost centre, budgeted quantity and the signature of the competent authority.

(d) Requisition voucher can be either for a single commodity or for batches intended for special projects.

(e) The person who is authorized to sign is either the executive in-charge of the consumer department or the production executive.

(f) In organisations, where computerisation has been effectively implemented, requisition voucher is printed out by the computer.

(g) In the case of production items which are issued regularly, it is advantageous to have pre-printed requisitions, giving the code, unit of issue etc.

(h) The consumer fills in the quantity column and affixes the authorised signature.

Checks and Balances

(i) Once the warehouse manager receives the requisitions, he checks the availability, identifies the authenticity, correlates the control limits and issues the materials.

(j) He makes the necessary account adjustment for the quantity after having obtained the signature of the person authorised to collect the materials.

(k) The requisition on which an issue has been effected is called a cashed requisition.

(l) Copies of the cashed requisition must be sent to the concerned departments and the costing section for further processing of data.

(m) Procedures for other types of issues are only minor variations of this basic consumption issue. Consumables like soaps, lubricants, cotton waste, small tools, stationery, etc. are usually issued periodically to other departments according to the quantities fixed without any issue voucher.

(n) This facilitates easy material, accounting and smooth flow of consumables.

(o) Similarly, low value C class items can be directly charged off as overheads to reduce paperwork. Many private sector organisations treat bazaar purchase items as directly charged off overheads.

(p) One-time issues for specific purposes to some departments are directly charged off as overheads or to respective accounts. Examples are bath tub to the Director's house or an airconditioner for another.

(q) Issues of capital equipments and machineries are invariably of one-time nature occurring mostly in project stages. A detailed register must be maintained for this purpose.

Loan Issues

(r) Loan issues are made in the case of items like equipment, tools and instruments that are stored centrally in the stores. The signed requisition slip can be used as a sort of pledge at the store and returned when the user returns the equipment. The movement of such equipments must be recorded in a register to keep a check.

(s) Items deposited in the stores for safe custody should be treated in a similar fashion.

(t) Issues to subcontractors will be authorised by the subcontracting or planning department. As the material has to move out of the premises, gate passes be prepared. For issues to outstation firms, transportation formalities have to be taken care of. These records must be passed on to the receipt section in order to tally with the quantities received from the subcontractors.

(u) Issues to sister concerns, substores and departmental stores will either be on an imprest basis or on a returnable basis. If the issues are on a returnable basis, then the procedure to be followed will be similar to that of loan issues.

(v) If the issues are on an imprest basis, the free issues will be effected only when consumption vouchers are presented to the main stores.

(w) The fundamental factor in the management of issues is that issues should be authorised and accounted for accurately and posted with the least possible time delay.

Problems of Shortages

(x) Material issue vouchers are normally released only on an assurance that the materials are physically available in the stores. However, there can be instances when sufficient quantities are not available, especially in the case of batch issues for a production run.

(y) In multiplant organisations, in the case of common materials, items meant for one section are grabbed by another, creating artificial shortages. Such practices should be avoided by talking to the people concerned, if necessary with the help of higher authorities. The warehouse manager has to ensure replenishment from external sources or feeder shops.

(z) Some organisations have adopted the procedure of issuing *owing in* slips where in the materials owed by the stores are indicated and are supplied as soon as they are available. It must be ensured that important jobs are not kept pending for want of materials and stores.

PHYSICAL ISSUE DIFFICULTIES

The common place of delivery is the stores windows, but that is possible only if the items are small enough to be handled across the counter. As the materials become heavier and larger, handling has to be specialised. This has given rise to the practice of delivering on the shop floor at the place of requirement. The handling men are either contract labour or skilled personnel hired exclusively for this purpose by the stores. This reduces breakages and wastages during handling. Special materials like paper rolls, glass, acids etc. are best handled by trained men.

Another strategy that is often adopted is to unload bulk material such as coal, pig iron, mineral ores, etc. nearest to the point of use. In these cases, care should be taken to complete the receipt, stocking and issuing formalities, as soon as the direct delivery is effected. At times, especially in the case of bulk issues, materials may be transferred from one consumption centre to another. The procedure that should be followed for this should be well defined and such transfers should take place only after the sanction of the warehouse executive. The transfer vouchers should indicate the item code, the present location, the place where it is required, quantity, reasons and the authorisation. In this case also, the respective data must be updated on a priority basis so that accounting can be accurate and effective.

Table 8.2 gives a specimen of the Stores Transfer Voucher.

Table 8.2 Stores Transfer Voucher

Sl. No.	Item Code	Unit	Quantity Transferred	From Whom	To Whom	Under What Account	Authorisation

Prepared by: __________ Checked by: __________

FIFO ISSUE SYSTEM

To enable a proper rotation of stocks, a *first in-first-out* (FIFO) principle must be adopted in the issue of materials. As the name implies, items which are received first must be issued first. This is influenced by the methods of storage adopted. For heavy plate items stocked one over another, this may be difficult. If random storage is done, the older material cannot be located easily, resulting in issue of late arrivals. Hence the storage system must be organised so that the first in materials are handy enough to be issued out first. Here, the role of kirana merchant or dispensing chemists serves as a good example.

FIFO can be effected by having a material preparation cell and fixing the times of delivery. The preparation cell can physically arrange the materials so that FIFO can be adopted. Various consumer sections can be given different times so that a build-up of queue at the service counter is avoided. The role of the preparation cell increases for sheet metal and other bulk commodity issues, where the requirement is for small quantities but the supplies are in large sizes. Savings effected by a material preparation section is enormous, especially in engineering industries.

EMERGENCY ISSUES

Whatever system of issues may be proposed and implemented, there are occasions when some materials are required on an urgent basis, and the normal procedure cannot be followed. A common example is the requirement of a machinery spare part in the night shift, when the store is closed. In such cases, persons who are normally authorised to issue the materials are not available to make a prompt issue. The executives controlling the operations should then have the sanction to open the stores and take whatever is needed. The security personnel can supervise to prevent misuse of these powers. If a pattern can be established on emergency consumptions, items can be issued in advance to minimise emergency issues. The procedure to be followed must be clearly indicated in the warehouse manuals.

RETURN TO STORES

Material issued from the stores are tested items and are for consumption as per the requisitions. Hence one does not expect such items to be returned to the stores under normal conditions. There can be instances when shop personnel are forced to return the materials taken earlier. One reason can be that they have received in excess of what they can normally store or consume. Another reason may be that wrong material had been issued. In such cases, the consumer must be given due credit for the returns and the material should be reissued as and when the needs arise.

Another reason for the return can be that, though the materials have passed inspection, they do not match the production specifications or prove to be of inferior quality and unworkable. Now, not only must these be replaced to keep the production going, but also necessary steps like returning it to the supplier for replacement and informing purchase and accounts for follow-up action must be taken. A specimen of the material return slip is given in Table 8.3.

Table 8.3 Material Return Slip

Department . . . .

Sl. No.	Item Code	Unit	Quantity Returned	Cost Centre Job Order	Remarks

Foreman . . . .

Storekeeper . . . .

TEN POINTS TO PONDER

1. Discuss the importance of Issue Management in stores.
2. Identify the agencies to whom issues are made.
3. Discuss the issue systems and the documents used therein.
4. What are the checks and balances to be provided in issues?
5. Describe the problems associated with loan issues.
6. Enumerate the difficulties encountered in issue management.
7. How do you solve the difficulties of issue management?
8. Describe the FIFO issue system in detail.
9. How do you tackle emergency issues in stores?
10. Explain the procedure to be adopted in the case of material returned to store.

9. Warehousing Costs

STORAGE COSTS

The Warehouse manager must be aware of the fact that materials cost the organisation in the form of the intrinsic price paid to the vendor, freight charges, handling expenses and inspection charges. Overstocking charges, stockout costs, order processing costs or ordering charges, and inventory charges are usually included in the overheads. It is necessary to identify them for better control.

1. Stocking items in the form of raw materials, machinery spares, semi-finished items and finished goods cost the organisation a substantial sum. This cost of having inventory is known as warehousing charges, storing charges, stock holding charges, inventory holding cost, inventory servicing cost, inventory carrying cost, etc. Materials stored in the warehouse cost the organisation in terms of men, materials, time, space, insurance, risk and so on.

2. This inventory carrying cost is about 30 per cent per annum on the average value of materials stocked.

Cost of Funds

3. The major component of the inventory is the cost of the money invested in stocks which can be equated to the present average bank lending rate which is around 18–20 per cent even though the base lending rate is only 15 per cent. This figure should be increased to include the cost of getting the loan sanctioned, financial executive time, documentation, transportation to the bank, obliging the loaning authorities etc.

4. Some companies, particularly those with foreign collaboration, to whom stocks have been primarily mortgaged, do not borrow from commercial banks for their working capital needs.

5. In the absence of borrowed funds, we can assume the weighted average cost of capital or an opportunity cost for funds invested in lieu of bank rate.

6. If the company uses its internal resources for financing inventories, the opportunity cost concept which will be around 18 per cent, may be used.

7. For public utilities like electricity boards who borrow through debentures at 12 per cent, this figure may be low. But in the interest of overall economy, the opportunity cost should be computed at the prevailing bank lending rate.

Cost of Storage Space

8. For storing items, a firm needs space. This includes the space for documents, manpower, office furniture, etc. The cost incurred on the physical storage facilities may be in the form of rental value or opportunity cost basis.

9. For work-in-progress inventory stored on the shop floor, this component is likely to be higher due to increased foundation cost of the factory floor space.

10. For finished goods warehouse, and items stored in downtown areas nearest to the points of consumer, the cost of floor space is very high.

Salary and Wages

11. The salary and statutory payment to all the stores staff is another important component in inventory charges.

12. The salary component should also include security staff, and stores audit staff for the time spent by them in connection with warehousing activity.

13. Depreciation charges of bins, racks, handling equipment furniture, office equipment and other devices used in stores is another component determined on the basis of depreciation policies.

Obsolete Items

14. Items stored with low shelf life deteriorate over time and have to be disposed of.

15. Losses in evaporation of volatile substances and chemicals are also included in inventory carrying costs.

16. Breakages in handling and in transporting within store premises can be substantial in the case of glass items.

17. The cost of application of preservatives intended to prolong the life of items, particularly machinery spare parts, must be included in warehousing charges.

18. The above items 14, 15, 16, 17 included under categories could result in non-moving obsolete items which have to be sold at throw-away prices. The rate of obsolescence is higher in high-technology spare parts — intensive, maintenance-oriented organisations.

19. Losses due to theft and pilferage are included in the inventory cost.

20. Insurance cost against fire, theft, burglary and other risks is about 0.5–1 per cent of the value of goods stored.

21. Wastages due to spillage on the floor sometimes occur, particularly in the case of fuel and liquid substances.

22. The cost of all types of documents used in stores like bincard, kardex, GRN, indent, issue voucher, gate pass etc. is yet another component of inventory charges.

23. The cost of use of electronic data processing department relevant to warehousing activity is another component.

24. The total inventory carrying charges are usually denoted by C_c. It comprises all the elements discussed above and is about 30 per cent on the average value of inventory. If figures are available, we can ascertain the fixed and variable portions. In the best organisations, this has been reduced to about 28 per cent, while there are organisations which have higher figures of around 32 per cent. The warehouse controller has to manage the stores efficiently by reducing this cost without affecting the service level to the consumer.

OVERSTOCKING COST

The concept of overstocking is an extension of inventory carrying cost in the sense that it signifies the investment in inventory for a longer duration than necessary. Hence the opportunity of alternative investment is lost. If all items are ultimately used, this cost can be approximated to the inventory carrying charges. On the other hand, for items that cannot be used after a certain period, this cost will be the difference between the cost of the item and its salvage value. To this amount, the cost incurred for ordering it and storing it till it is salvaged should be added for estimating the overstocking cost. In the case of one-time purchases, overstocking cost would be (purchase price-scrap value). The overstocking cost is usually represented by K_o. Many organisations particularly the spares-intensive times — overstock machinery spares to avoid risk.

STOCKOUT COST

Executives from all disciplines are afraid of stockout cost or understocking cost, arising due to non-availability of an item at the required time. In the finished goods area, non-availability of an item implies that the customer will go to another vendor or substitute the brand with another. This results not only in loss of profit on sales, but also affects the image of the organisation. In the case of semi-finished goods, non-availability of a matching component will increase the quantum of the semi-finished goods on the shop floor. In the case of raw material, stockout results in idling of a production line, the cost of lost production during the period of stockout and the extra cost per unit which might have to be paid in an emergency purchase. There are other intangible elements, like the machine operator or maintenance fitter 'letting off steam' on the stores executive, inertia of workers, etc. which cannot be measured directly. However, many organisations consider only the profit lost due to loss of production, as the stockout cost. Stockout cost is represented by K_u and is always greater than the cost of the item. Hence there is a tendency to overstock, particularly machinery spare parts.

ORDER PROCESSING CHARGES

The warehouse manager should be aware that the very act of procuring or getting items costs the organisation. To the householder, it may be in terms of time spent in the market, whereas to the company, it is in terms of money spent in organising and maintaining a purchase department.

1. The cost of ordering is represented by C_o and includes the following elements:

2. Salary and statutory payments of purchase staff and tender committee members who process the order

3. Rent or opportunity cost of office space

4. Depreciation of office furniture and office equipments

5. Cost of documentation, receipts of sale and tender documents

6. Advertisement in newspapers and trade journals

7. Follow-up costs for expediting, including travel costs

8. Communication costs of post, telephone, telex, telegraph, fax, satellite communication and VHF costs relevant to acquiring the material

9. Cost of source developments in providing material/financial/technical assistance and accepting substandard material initially

10. Cost of electronic data processing relevant to procurement

11. Cost of entertainment

12. The total amount spent by the purchasing division divided by the number of orders gives the average cost per order. If data are available for individual elements, then we can classify them as fixed and variable portions for better control.

13. The average cost depends upon several factors such as lead time, procedures, delegation of powers, follow-up system, imported item etc. But an average of Rs. 600 per order is a reasonable figure in Indian conditions.

14. The largest central purchase agency in India, namely the Directorate General of Supply and Disposals, charges the indentors 0.5–1 per cent on the value of procurement.

15. Hence, another way of looking at the ordering charges would be that it should be less than 1 per cent of the value of procurement.

16. If staggered deliveries are adopted, cost of additional documentation, additional follow up etc. must be added to get the total cost incurred.

Set-up Cost

17. Corresponding to the ordering cost, we have the set-up cost of different jobs due to job changes in production shops.

18. Numerous jobs are set up, dismantled and set up again, resulting in unnecessary additional expenses.

The set-up may be broken down for a new job and a week later the original job is back on the machine. This involves stoppages and includes charges of jigs, fixtures, tools, scheduling paper work, scrap, inspection, accessories and cleaning. If there are too many set-up changes, the set-up cost goes up.

19. On the other hand, if we have the same set-up continued for a long time, the inventory of semi-finished goods will pile up.

20. It should be noted that these two costs — set-up or ordering cost and inventory carrying charges — go in the opposite directions requiring an optimum result.

21. In the case of finished goods, this is known as customer order processing cost.

22. All the elements of ordering charges appear in this as well, and the marketing executive has to determine the priority allocation for despatches.

23. It is advisable to work out broad norms for these costs in order to effectively control it after examining the procedures, powers of delegation, etc. by means of an organisation and method study.

SERVICE LEVEL

The efficiency of a warehouse is normally measured on a negative scale, i.e. in terms of stockouts. On the other hand, the achievement of the utopian goal of zero stockout will require a very high idle investment and even then cent per cent service cannot be assured. One approach to this problem is to fix the service level as percentage. Thus for a critical spare part, we can fix a 99.5 per cent service level. This implies that if there are 200 demands requisitions, they will be met on 199 occasions in the long run. As the criticality of the item decreases, the service level can be reduced.

We can arrive at the service level by considering the costs involved. There are two opportunity costs that have to be considered for computing the required service level. The first is the overstocking cost, which basically measures the opportunity cost due to the investment in the stocks for longer time than necessary. For items that will be ultimately used, this can be equated to the carrying charges. However, for items that cannot be used after a certain time, the cost due to scrapping must be considered.

The opposing opportunity cost is the understocking cost which measures the damage due to non-availability of material. Mathematically, these two costs are related to the service level as

$$\text{Service level} = K_u/(K_u + K_o)$$

For example, K_o is Rs. 1000 and K_u is Rs. 10,000, then we have 1000/11,000 or 1/11 or 9%.

It can be deduced at once that as the understocking cost increases, service level has to be increased for major critical items. On the contrary, if the overstocking cost is comparable to understocking cost, then it would be advisable to have lower service level. Whenever service level is fixed on other considerations, understocking cost can be imputed since overstocking cost is fairly easy to compute. These factors can be advantageously used by the warehouse manager to fix the stock levels of various items.

TEN POINTS TO PONDER

1. Identify the relevant costs for warehousing operations.
2. Explain the term inventory carrying cost.
3. Describe the components of stock holding charges.
4. Discuss the term overstocking cost.
5. Enumerate the steps involved in calculation of ordering charges.
6. Discuss the concept of stockout cost.
7. Identify the mechanism for finding out set-up costs.
8. Explain the procedure of computation of order processing in the area of marketing.
9. What are the norms for holding and carrying costs?
10. Elucidate the concept of service level in warehouse operations.

10. Linear Programming for Warehousing

LP APPLICATION AREAS

One of the common problems faced by warehouse managers is to identify which factories should supply to a particular store in a multi-warehouse/multi-factory situation. The aim of the warehouse manager is to reduce the cost of transport, at the same time, meeting the warehouse demands from the factory production. One of the earliest applications of linear programming has been the solution of such transportation problems. One product can be transported from any of the several alternative sources to each of many destinations. The transportation rates for all possible source-destination combinations are known. Manufacturing capacity of each source and the demand volumes for each destination, are also available. The problem is to determine which destinations are supplied by which sources so that the total transportation cost is minimised.

Mathematical skill and intelligence must be exercised in building linear programming models. The steps involved are formulation of problems, development of alternative strategies, identification of the best alternative by analysing the computed results, implementation of the solution and review of the performance. Linear programming method is useful in managerial functions like (a) allocation of scarce resources, (b) scheduling production facilities and resources, (c) optimising capital investment decision, (d) inventory control, (e) routing oil tankers, aircrafts, salesmen (f) allocation of maintenance crew or toll collectors over a period of time, and (g) structural design. Computer facilities are necessary to solve linear programming applications to transportation problems. By comparing different costs for the various alternative strategies, the concept can be extended to location of warehouses by simulation method. Further sophistications of linear programming are non-linear programming, dynamic programming, etc.

PROBLEM FORMULATION

In the most general case we can assume a set of *m* sources/supply centres located at various places feeding a set of *n* *sinks*/consumption centres/warehouses at different locations. Let A_i – [$i = 1 \ldots m$] be the availability in some units of measurement at each source. Let B_j – [$j = 1 \ldots n$] be the demand in the same units at each sink respectively. We can assume that the total quantity available is equal to the total demand. Otherwise the same methodology can still be used by considering dummy sources or sinks as the case may be. Let R_{ij} be the cost of transportation of a unit from source i to sink j. Let Z be the quantity transferred from source i to sink j. Hence the total cost incurred is obtained by summing over all i and j the product $R_{ij}Z_{ij}$.

Another objective is to choose the various Z_{ir}'s so that this total cost is minimised subject to the restrictions placed by the availability conditions at the sources and the requirement at the sinks. Here the equations are assumed to be linear ones. Such a problem is basically a linear programming problem and can be solved by the use of simplex algorithm or an iterative approach with the help of the computer software package. It can be illustrated as follows.

Illustration

Consider a company with three manufacturing units or sources, one each at Kakinada, Wardha and Cochin. The warehouses are located at Bilaspur, Bombay, Shahabad and Madurai. Table 10.1 gives, in the form of a matrix tabulation, the production and storage capacities at the sources and the sinks in the fourth column and fourth row, respectively. The cost of transport for one truckload in some rupee units from a source to a sink is also given. Our aim is to allocate the truckloads from the sources to the sinks so as to meet the supply and demand conditions with minimum cost of transportation.

Table 10.1 Cost of Transportation

	Bilaspur	Bombay	Shahabad	Madurai	Production Capacity in Truckloads
Kakinada	5	1	2	8	24
Wardha	1	4	1	8	31
Cochin	10	6	4	2	24
Storage capacity in truckloads	16	16	21	26	79

Initial Feasible Solution

The first step in the linear programming approach is to find an arbitrary allocation that meets the supply and demand restrictions. This is known as the initial feasible solution and is given in Table 10.2. We can either follow the *north-west corner rule* or *the least cost method*. In the north-west corner rule method, we start the allocation at the left hand top corner and then proceed step by step fulfilling the requirements. In the least cost method, we allocate first to the least cost cells and proceed to other cells with high costs. But the main idea is to obtain an initial solution that meets the constraints. The cost of allocation in Table 10.2 is as follows:

$$(6 \times 5) + (16 \times 1) + (2 \times 8) + (10 \times 1) + (21 \times 1) + (24 \times 2) = \text{Rs. } 141$$

Table 10.2 Initial Feasible Allocations

	Bilaspur	Bombay	Shahabad	Madurai	Total Capacity
Kakinada	6	16	–	2	14
Wardha	10	–	21	–	31
Cochin	–	–	–	24	24
Total requirement	16	16	21	26	79

Improved Solution

Now, we have to find out whether this is the optional solution or whether we can improve upon it. To improve the solution, we have to find out where to reduce and where to increase the allocations. These are called leaving and entering variables. To find this, we define a unit cost of transportation and calculate as follows:

$$[K - X + Y]$$

where X and Y are evaluated for various cells. Again we start with an arbitrary value of $X = 0$ and derive the set of values. X's are for the sources and Y's for the sinks. In Table 10.3, we start with the value X to be zero for Kirloskarwadi. Based on this and K_i value of the cells, where allocations have been made, the Y's of sinks are computed. Then we go back to obtain the value of X's for the other sources, by a similar process, i.e. $K_j = X + Y$ if an allocation has been made in the ijth cell. In the case of empty cells or where no allocation has been made, the corresponding $E_{ij} = K_{ij} - [X + Y]$ value is entered as the imputed economic cost. This E_{ij} gives the marginal economic cost of altering the allocation to that particular cell.

In the given example, 'X' for Kirloskarwadi has been assumed to be zero. This fixes the Y values for Bombay, Shahabad and Madurai on 5, 1, and 8 respectively. Using these values we fix the X values for Wardha and Cochin as – 4 and – 6 respectively. Now we compute the E_{ij} values for the empty cells. For example, for Kirloskarwadi and Shahabad cell, $X = 0$ and $Y = 5$ and

$$E_{ij} = K_{ij} - (X + Y) = 2 - (0 + 5) = -3.$$

This has been entered within brackets in Table 14.3. From linear programming theory it can be shown that when E_{ij} is negative, their reduction in cost is possible to that cell. Therefore, we choose that cell with the most negative E_{ir} as the entering variable and the allocation with the maximum K_{ij} as the leaving variable. We also correspondingly alter the figures in the columns to meet the constraints. This is done in order to reduce the number of iterations or, in other words, to reach optimality faster than by trial and error methods.

Table 10.3 Evaluation of Economic Cost

	Bilaspur	Bombay	Shahabad	Madurai	Total Capacity	*X*
Kirloskarwadi	6	16	(–3)	2	24	0
Wardha	10	(+7)	21	(+4)	31	– 4
Cochin	(+11)	(+11)	(+5)	24	24	– 6
Total requirements	16	16	21	26	79	–
Y	5	1	5	8	–	

Second Allocation

In Table 10.4, we have shown the second allocation. We find that Kirloskarwadi to Shahabad shows a saving and hence we allocate six truckloads from Kirloskarwadi to Shahabad. Now to meet other constraints, Shahabad gets only 15 truckloads from Wardha. On the other hand, Bilaspur gets its entire requirement of 16 truckloads from Wardha and nothing from Kirloskarwadi.

Table 10.4 Second Allocation

	Bilaspur	Bombay	Shahabad	Madurai	Total
Kirloskarwadi	–	16	6	2	24
Wardha	16	–	15	–	31
Cochin	–	–	–	24	24
Total	16	16	21	26	79

We again proceed to calculate the new *X*'s and *Y*'s. This is shown in Table 10.5. We find that all E_{ij}'s computed are positive and hence there can be no improvement in the allocation.

Table 10.5 Economic Cost Evaluation

	Bilaspur	Bombay	Shahabad	Madurai	Total Capacity	*X*
Kirloskarwadi	(+3)	16	6	2	24	0
Wardha	16	(+4)	15	(+1)	31	–1
Cochin	(+14)	(+11)	(+8)	24	24	– 6
Total requirements	16	16	21	26	79	–
Y	2	1	2	8	–	

Hence the optimum allocation that minimises the cost is as given in Table 10.4. The cost incurred is

$$(16 \times 1) + (6 \times 2) + (2 \times 8) + (16 \times 1) + (15 \times 1) + (24 \times 2) = 123$$

It should be noted that this solution is an interactive process and in the given example, we have reached it in two steps. The optimum has been reached by the positive sign of the imputed economic cost i.e. E_{ij}'s. By using this method and proper simulation methods, it is possible to plan the location of warehouses for optimising the cost, provided the capacity, requirement and transportation costs are known..

TEN POINTS TO PONDER

1. Explain the concept of linear programming method.
2. Enumerate the advantages of LP models.
3. Describe how LP can be used for optimum location of warehouses.
4. Discuss how LP can be used for optimum transportation.
5. Identify the steps in formulating the LP model.
6. What is meant by initial feasible solution?
7. How do you proceed from initial feasible solution?
8. Amplify the mechanism to locate the optimum feasible solution.
9. Discuss the role of computers in LP models.
10. Describe the difficulties in application of linear programming method.

11. Stock Verification

VERIFICATION THEME

The physical stock and the book balance of an item in any company, never tally exactly at a point of time. The warehouse manager is responsible for the safe custody of the largest current asset of the balance sheet, namely the raw material, machinery spares, semi-finished and finished goods. But the checking process of correlating the number for each item in the book balance with that of physical balance is usually confined to the raw material stores and stock verification is very rarely done on the work in progress, even though the value may be accumulating on the shop floor. The finished goods inventory under the control of the marketing executive is also rarely checked for the tallying of physical and book balance.

In the case of raw material stores, the warehouse manager must avoid discrepancy in stocks. For this purpose, he has to constantly review his stocks, update the posting, and reconcile the physical and book balance. He is in the unenviable position of having not only to be honest, but also to exhibit his honesty. The magnitude of the problem will increase as the number of transactions increases because errors may be committed in each transaction, leading to major discrepancies. The cost of stock keeping may be excessive, especially in a large warehouse. But, the main objective of verification process is to identify the weaknesses in the operations of the warehousing system. Besides, the stock verification process ensures that inputs, outputs and stock on hand have been accounted for accurately. The process also enables proper location of the item in the respective bins. Stock verification procedure, if conducted thoroughly, enables identification of slow moving and obsolete items. The verification process should not be construed as a witch-hunting exercise. Stock verification is done to detect the chances of collusion amongst the various staff members in the company. The purpose thus is to set up records as the basis for control and secondly to pass on accountability from one department to another.

VERIFICATION PROCESS

Stock taking or stock verification can be defined as the physical counting, weighing, or measuring the stock materials held and making a record of these figures. In most organisations, physical stock verification is usually an annual exercise coinciding with the closing of accounts. In this period, the stores will be closed and staff from the accounts department carry out the verification process. Many shops dealing with finished goods also adopt this practice of closing for a couple of days for the process of verification. Since there is always a pressure to complete this exercise quickly, this may be ineffective and nullify the advantages.

To overcome the above drawbacks, organisations adopt continuous monitoring or perpetual inventory taking. In this case, materials are verified throughout the year by verifying a few items daily. The perpetual inventory control is so staggered that all the materials are verified and covered in the course of the year. Obviously, it will be advantageous to have the high value items verified more often than the other items. Spot checks can be undertaken by senior officers for verifying the items. The warehouse manager can arrange verification of high value items at the time of issue of receipt of these items, in order to avoid accumulation of errors towards the end of the year. The frequency of stocktaking of the balance items should be related to their movement and cost.

The primary purpose of verification is an assurance that inputs, outputs and the stock on hand have been accounted for accurately. The verification process, whether continuous or periodic, should be thorough and well organised and needs careful planning and execution. The basic requirement is that materials should be properly measured and recorded. This measurement can either be in numbers, or weight or volume, or length or area, as the case may be; what has to be ensured is the use of a consistent basis and proper standards of measurement. In this context, electric motors are specified by weight in USSR and horse power by other countries. Indians measure height by FPS system and weight by metric system. We are also familiar how land is measured by different parameters in India, causing confusion. Measuring instruments must be calibrated so that they do not introduce errors in measurement. Care should be taken to avoid double counting or omitting some items.

A TO Z REASONS OF FAILURE

The primary purpose of stock verification is to check that the physical stock agrees with book balance. If these fail to match, as often is the case, then it could be due to the following reasons:

(a) Failure of the system because the paper system is in error.

(b) The item may be incorrectly labelled due to mistakes in packing or inspection as the wrong issue could lead to less stock.

(c) Stock could have been binned in the wrong location.

(d) Poor warehousing as boxes containing items may be in gangways.

(e) Items may have been mislaid or lodged between shelves and dislocated later on.

(f) Regular theft of valuable items in collusion with the stores personnel.

Documentation Discrepancies

(g) Bad handwriting and figures copied incorrectly account for a majority of errors.

(h) Loss of documentation and elimination of the number of copies to be made.

(i) Deteriorating accuracy of records.

(j) Transferring the paper work from first shift to the second shift has not been proper.

(k) Receiving by weight and issuing by number, in the case of items like nuts, bolts, screw etc. could cause discrepancies.

(l) Inaccuracy of weighing scales at receiving or issuing or both. If the scales indicate 1 kg less every time and if there are 100 issues in a period, but with one receipt only — then the differences will be (100 – 1) = 99 kg in the actual quantity.

(m) Issuing machinery spares without authorisation or indents or issue vouchers when the mechanics claim that the machines are idle.

(n) Urgent issues made when the store is closed, without posting entries.

(o) Delays in inspection, posting the goods receipt note and other receipt documents.

(p) Using rejected material during emergency and not accounting properly.

(q) Improper communication with user/indentor, inspection, external vendors etc.

(r) Each bin containing more than one item leading to mixing and confusion at issue.

(s) Indentor using suppliers part numbers and not internal codification, leading to confusion.

(t) Stores personnel unable to identify the item and issuing wrong material.

(u) Too many users coming to take material particularly in the morning creating confusion at the issue counter.

(v) Issuing without documentation as all the users are in a hurry.

(w) Direct delivered items and bulk items are received and removed without adequate documentation.

(x) Understaffed stores and presence of unskilled stores personnel.

(y) Improper records on the transfer from one store to another, lead to discrepancies.

(z) Lorries with materials are received when the store is closed, resulting in delays in posting.

In spite of the above difficulties, the warehouse manager should ensure the proper receipt, issue and accounting of the items adequately by placing the items in their proper location.

VERIFICATION AGENCY

Verification agency should be impartial. The person handling the documents should not do verification as the roles of the judge, prosecutor and culprit cannot be combined into one. However, for his own internal control purpose, the warehouse manager can do this particularly at the time of issue of each item. In large organisations, stock-taking is the responsibility of some senior officer from the material accounting or finance department. In government organisations, there is a statutory requirement for stock-taking, to be carried out to satisfy the propriety of public accountability. Auditors and bankers sometimes undertake verification of stocks on a sample basis ensure authentication of the correctness of the corporate performance. Sometimes the purchase and inventory control officers check the items to acquaint themselves with the identification details. There is also the classic example of the Managing Director, casually dropping a hundred rupee note on the shop floor, and the foreman running to pick the note and return it only to be told that the semi-finished goods are worth several millions more than the hundred rupee note.

ROLE OF WAREHOUSE PERSONNEL

Success of the verification process depends on the skills of the personnel involved. These people should be honest and adept in the art of numerical calculations. They should have a thorough knowledge of the location of various materials and must possess the sixth sense to dig up the right thing from wrong places. If the organisation is able to motivate its own stores personnel by a proper development of the environment, culture and value systems, then it can have the best stock verifiers. A suitable incentive scheme may also be thought of. Warehouse personnel have an intimate knowledge of what is where and are best suited to do the job. Then stock-taking ceases to be witch-hunting exercise and the organisation gains immensely by enabling its employees to be their own judges and motivators. Manufacturing organisations operate on two primary documents, namely *receipt* and *issue*. Hence every effort must be made by the warehouse manager to keep these two statements accurate and up-to-date to minimise the year-end imbalances. The verification agency should collate the entire data on each item and present it, emphasising the areas needing control. Physical verification is of no avail, if no follow-up action is planned for future and hence the agency should suggest the follow-up action like writing off, imposing penalties, tightening security, etc.

TEN POINTS TO PONDER

1. Identify the stock taking objectives of raw material, WIP and finished items.
2. Discuss the details of verification process of raw material, WIP and finished goods.

3. Illustrate the various agencies associated with verification process.
4. Describe the statutory requirements of stock verification.
5. Explain the possible discrepancies in issuing.
6. Locate the possible weakness in receipt management.
7. Enumerate the differences due to documentation.
8. What are the possible failures due to physical store keeping?
9. Who should be responsible for the differences in physical and book balances?
10. Amplify the preventive and corrective measures in verification.

12. Legal Aspects of Storage

LEGAL ISSUES

No business can exist in isolation. In an organisation environmental factors influence decision-making at all levels and warehousing is no exception. Tax formalities such as sales tax, entry tax, octroi, excise and customs influence the movement of goods to the warehouse from external suppliers or to the external customers from the finished goods godowns. The basic rules and regulations remain the same, but the rates, tariff and materials affected change from year to year. The warehouse manager must be aware of them in order to function effectively, though sometimes these issues are handled by clearing agents and consultants. In this chapter, we shall discuss customs, excise, bonded warehouse, sales tax and octroi formalities. Customs duties, levies and sales tax account for a major portion of Government revenues.

CUSTOMS DUTIES

Customs duties are of very ancient origin and were introduced by the Moghuls. The law relating to customs is embodied in the Customs Act 1962 and the Customs Tariff Act, 1975. The power to levy duties of customs is provided by Section 12 of the Customs Act. The effective notified rates are applicable on imported goods and exported items, subject to the fulfilment of conditions in the notification. The commodities are assigned a six digit code — the first two indicate the chapter, followed by the heading and subheading. Once the goods are correctly classified and the rates applicable in the tariff read with prescribed notifications, the duties can be determined. The duty on imported goods is on — Cost + Insurance + Freight (CIF) value; for this value of goods at the time and place of import has to be determined.

The classification of goods is based on BTN — Business Trade Number (BTN) and valuation on GATT — General Agreement on Trade and Tariff. This value for assessment includes the cost of transportation of goods up to import, loading, unloading, handling and insurance, but this total shall not exceed 16 per cent of FOB — Free On Board value. This cost will be 21 per cent of import by air. In case of disputes, payment under protest or lower rate with execution of security bond can be made, and the disputes can be sorted out with the assistant collector or appeal to the customs collector.

A TO Z PROCEDURE OF CLEARANCE

(a) The importer of his agent files the bill of entry

(b) Bills are numbered and noted in the relevant manifest

(c) The date of filing the bill of entry is important since the rate of duty will be calculated from that date

(d) The following are the essential documents that should be made available for filing with the customs:

Documentation

(e) Bill of lading/airway bill issued by the carrier shipline or airways

(f) Invoice in duplicate

(g) Packing list issued by the suppliers

(h) Broker's note

(i) Catalogue/literature or a write up of the product given by the supplier

(j) Contract purchase orders, with acceptance and amendments

(k) Price lists

(l) Correspondence with the supplier if any

(m) Import licence and its annexures

(n) Net or gross value

(o) Freight note and insurance cover

(p) A certificate of origin as imports from certain countries enjoy concessional duty or the item may be banned

(q) The clearance will be delayed if the above documents are not filed

(r) In case the goods are to be tested to determine the classification, samples are drawn for testing

(s) The appraisal often takes time since the rules have to be interpreted. It must be noted that the licensing authority is under the Ministry of industries, while the customs works under the Ministry of Finance

(t) All the necessary formalities are to be adhered to and the goods must be cleared in time or else the landed cargo incurs demurrage

(u) A strategy that can be adopted is to have the papers processed before the arrival of the cargo and the bill of entry is duly adjusted. This is possible only if the innumerable queries of the appraisers are answered

(v) Unless the organisation has regular frequent import transactions, it will be advantageous to employ clearing agents, who are specialised and have a knack of public relations as well as a knowledge of the regulations

(w) The catalogue should bring out the exact nature of the cargo, elucidating the equipment on which it is to be used, its location, chemical composition, and the performance of the plant. This will enable the appraiser to correctly classify and assess the cargo

(x) But, when 'open Inspection' is ordered due to non-availability of correct description, there is unnecessary delay

(y) The appraiser determines the aggregate assessable value and levies the custom duty, auxiliary duty and additional duty as per the Customs Tariff Act. If the price is fixed in terms of foreign currency, it is converted into Indian rupee with the exchange rate prevailing on the entry date. The duty is to be paid in order to clear the goods

(z) Imported goods can be warehoused in approved warehouses subject to execution of bonds. These can be cleared under Section 68 of the Customs Act by presenting an ex-bond bill of entry which is assessed again. The rate of duty prevailing on the date of removal from warehouse is applicable to the goods.

EXCISE FORMALITIES

The warehouse is concerned with excise, only if it is dealing with the despatches of finished goods/products. Almost all manufactured products come under one of the many tariffs of the central and the state excise laws. The states levy excise duty on alcohol; whereas the central excise duties cover the manufactured items as announced in the annual budget. Excise duty has to be paid when the manufactured product leaves the premises of manufacture. To facilitate such collection, the self-removal procedure has been instituted, by which deposit is made to the credit of the excise account. This is known as the *personal ledger account*. The excise duty is reduced from this amount as the manufactured products leave the warehouse. A separate account must be maintained for each tariff of excise duty. Petroleum products, steel items and cement account for about 40 per cent of the excise duty in India.

REMOVAL OF GOODS

For each removal, an excise gate pass must be prepared, wherein the value of goods removed and the duty paid must be indicated. A removal goods register must be maintained for each tariff, wherein daily production figures and stock figures are indicated. The entries in these books and the gate pass must be signed by authorised persons. Registers and gate pass books must bear the office seal and these should be initialled at the excise office. The enforcement wing of the excise authorities makers periodic visits to check whether the rules and regulations are being followed. In the case of exports, rebate is allowed. The goods can be cleared under the normal procedure on payment of duty and then the necessary documents can be filed with the department. After scrutiny, refund will be ordered, even though it is a time-consuming process. Another approach is to execute a bond account, where a bond is raised for a given value under sufficient guarantees.

MODVAT SCHEME

The MODVAT scheme was introduced on 1.3.1986 as part of the central excise law to mitigate the cascading effect of taxation by granting duty relief already paid on inputs used in the manufacturing of further products. This scheme provides for taking instant credit of duty paid on inputs received and used in or in relation to the manufacture of final product. One important condition for availing MODVAT is that the final products must not be wholly exempt from duty or charged to nil rate of duty. In order to give encouragement to the products of small scale industry, MODVAT credit to user industries is available on normal rates of duty and not at the subsidized rates paid by the small scale sector.

PROCEDURE FOR CREDIT

The manufacturer should file a declaration to the assistant collector in the prescribed form showing the details of inputs intended to be utilised and the final product to be manufactured and obtain acknowledgment in order to avail the credit. Credit of duty can be taken on inputs only if they are received under cover of duty paying documents, such as central excise gate passes or bill of entry. The manufacturer should maintain the prescribed MODVAT account and file monthly returns indicating the credit taken and the credit utilised in respect of inputs utilised. Penalty is levied for wrongly availing or utilising credit without proper care and the goods may be confiscated.

SALES TAX FORMALITIES

The movement of materials is affected by sales tax regulations. Unless the carrier has the invoice wherein the sales tax dues are recorded and the 'C' forms, if a lower rate of tax has been claimed, the materials being moved are liable for confiscation. Hence the warehouse manager/purchase executives must send the 'C' forms to the supplier to facilitate his despatches. Similarly, while making despatches, these formalities must be cleared with the customer.

Sales within a state are covered by the State sales tax, while inter-state transactions are covered by the Central sales tax. To be eligible for a lower rate of tax, especially in inter-state sales, the organisation must be registered with the concerned authorities. These registration numbers must be obtained separately for intra-state and inter-state sales transactions. Only then, the C form can be issued. Quoting the registration number is an important factor that should not be forgotten. It is worthwhile to follow these regulations as the resultant reduction in tax rate is up to 10 per cent.

OCTROI FORMALITIES

Octroi duty is levied by municipal Corporation/Panchayat Unions, and similar local self-government authorities like Blocks, in order to boost their revenue.

This duty is levied on the goods unloaded for trade purposes in the jurisdictional area of the concerned authority. Besides this, some states have started levying entry tax. Goods that pass in transit, are not charged. Goods that have to enter such an area in order to reach a port or railhead are also exempted from this levy.

Implementation of the octroi is in the hands of the tax collector at the border check posts. The rates of levy and goods that are liable for levy are decided by the concerned authority. The value of products is obtained from the declared invoice value. Decision-making on the value, rate of tax, interpretation of rules for exemption, etc. is in the hands of the people manning the checkpost. Personal understanding on the part of the transport undertakings, often quickens the flow of traffic at the checkposts. In view of the long queues of trucks at the innumerable octroi checkposts and the consequential national waste — including corruption at many checkposts, a consensus is emerging at all levels that octroi collection should be abolished in India.

TEN POINTS TO PONDER

1. Identify the legal issues in warehousing.
2. Describe the customs duty procedures.
3. Elaborate on customs clearance procedures.
4. Mention the documentation needed for clearance.
5. Discuss, the excise formalities and procedures.
6. State the steps for removal of excisable goods.
7. Explain the MODVAT scheme.
8. Illustrate the procedure of obtaining credit under MODVAT scheme.
9. Amplify the utility of sales tax levels.
10. Illustrate the problems of octroi levies in transportation of goods in India.

13. Business Logistics Management

CONCEPT OF LOGISTICS

Business logistics is also referred as rhocrematics, physical supply management, physical distribution management, distribution management control etc. It aims at delivering material in production areas by better freight management.

Business logistics comprises a number of activities dealing with control of incoming and outgoing material. These activities cover functions of incoming material, transportation, warehousing, materials handling, stock control, order processing, supply scheduling and despatch of finished goods. Business logistics embraces the management of all activities which facilitate movement and the coordination of supply and demand, in the creation of time and place of utility of goods. Thus, business logistics is connected with the smooth functioning of procurement and physical distribution. Hence it contributes substantially to the final product cost. An efficient logistics system is one which handles the flows of inputs of raw materials into the production shops and the flow of the finished products to the customer according to plan with benefits greater than cost.

PHYSICAL DISTRIBUTION

There are many situations where the warehouse manager has to undertake the distribution management of finished goods. Unless a business is able to place these finished items in the hands of the ultimate consumer, it will not be able to reap a sizable return on its investment. Distribution can be undertaken either by the manufacturer himself by having his own channels or he can establish a distribution network consisting of stockists, distributors, wholesalers, and retailers. Distribution depends on the type of product. For instance, consumer products have to be made available to the purchaser when he is in a position to buy. Here again, different strategies have to be adopted for regular consumables like eatables, food products and toiletries on the one hand, and for durables such as furniture, radios, vessels and clothing on the other. Pharmaceuticals have their own seasonal markets. Strategies adopted for industrial products are entirely different as the methods of selling are different. However, the use of containeration has facilitated the logistics problem to a great extent.

COMMON PROBLEMS IN DISTRIBUTION

The distribution of all finished products has the following factors in common:

They have to be suitably packed for customer acceptance and for handling, transporting and delivering to the customer. Storage automatically assumes importance because of time lags between completion of manufacturing activity and packaging, between the packed state and loading into the carrier and between unloading at the destination and final delivery to the consumer. Added to this, the products may be detailed, while waiting for connecting transport arrangements *en route* to their destinations.

Efficient management requires that the time lag be reduced to as minimum a level as possible. The problems are more acute in food products, pharmaceuticals and other such items, whose shelf-lives are low. Thus, the warehouse manager is often confronted by the problems of keeping the distribution channels full and at the same time minimising distribution costs. Suboptimal policies, like a mere reduction of transportation cost, often result in the drying up of supplies at consumption centres. For instance, in a few cases, it may be even advisable for the organisation to have its own vehicle for delivering finished goods at depots. The inventory carried in the pipeline and warehouses costs about 30 per cent to the company. The transportation cost can be minimised by using linear programming methods for multi units. Commission to dealers, and discount to individual consumers have to be included in appropriate areas. A total solution, considering all the relevant factors, is needed in relation to the following areas: (a) Space cost economy in warehouses; (b) Labour economy in handling and in warehouses; (c) Transportation efficiency; (d) Reduced working capital investment; (e) Accurate forecast and control of costs; and (f) Marketing efficiency and customer satisfaction.

STEPS AND TECHNIQUES OF LOGISTICS

Business logistics views the movement of goods not as an activity by one firm, but by and for all firms in a channel of logistics. It analyses all costs resulting from the use of the methods to accomplish a logistics task. The system involves designing of men, machines and information in such a way that the parts are closely integrated to create greater productivity in the whole system. The elements include the following: (a) The objective of the logistics function; (b) quality of management; (c) inputs like products design, choice of carrier; (d) outputs or products of the firm; and (e) the linkages through transportation network. Putting all logistics function under one project manager and staffing with experts is the first major step in the logistics system.

Business logistics calls for the appraisal of all costs of transportation, storage, handling, processing, documentation, inventory carrying, etc. in order to optimise the total costs, by questioning the nature and extent of each activity. The detailed distribution plan must then be worked out and implemented. This should take into account the type of items, geographical pattern of demand, location of warehouse, transportation lead time, distribution network, etc. An overriding factor in the case of the finished goods inventory is the shelf-life of the product. Beyond the shelf-life, there is a drastic reduction in the intrinsic value. Ultimately, the value left may be equal to whatever can be salvaged out of them. In certain cases, the value can even be negative if the products have to be physically destroyed to prevent any misuse. For instance, in the pharmaceutical industry, the value of products destroyed may go up to 10 per cent.

WAREHOUSING AS A BUSINESS

Warehousing is the activity of storing goods pending consumption. The manufacturer has the option of building his own warehouse or of buying stores premises or leasing a branch of warehouse for a specific duration.

Any entrepreneur can offer the package of services of handling, transport, safety, security, distribution and documentation to facilitate the flow of materials from the producer to the supplier in an organised manner and run the warehouse as a business. In India, the concept of public warehousing has been of recent origin. The Royal Commission on Agriculture in 1928, the Reserve Bank of India in 1944, and the Rural Banking Enquiry Committee in 1950 have emphasised the need for establishing a public warehousing system. But the development has not been appreciable. The reasons for the slow growth in public warehousing system in India may be due to (a) low rate of return on investment, (b) loads of peak and valley on requirement of space, (c) unique individual specifications, and (d) legal complications and liaison problems between the two parties.

CENTRAL WAREHOUSING CORPORATION

The above phenomenon has necessitated the development of public sector warehouses with the establishment of the Central Warehousing Corporation by an Act of Parliament. Individual states also have their state warehousing corporation. The broad objectives of public state warehouses are as follows:

(a) Creation of a negotiable paper to provide for expansion of credit through commercial banks for the benefit of producers, dealers and others.

(b) Adding to the Nation's wealth by reducing waste and losses in storage by developing warehousing and scientific storage facilities.

(c) Assistance in orderly marketing by introduction of standard grades, specifications and the warehouse receipt.

(d) Training of personnel for management of the warehouses.

SERVICES BY WAREHOUSES

Scientific warehousing of the stocks deposited has been the motto of warehouses. Besides the conventional storage godowns, warehouses also have air condition and cold storage facilities for preservation of perishable goods like fruits, vegetables and dairy products. Storage facilities are also provided for preservation of hygroscopic delicate commodities like shellac, jaggery, tobacco, coffee beans, spices etc. Other commodities include foodgrains, pulses, oil seeds, spices, fibre, fertilisers, milled product, oil cakes, sugar, seeds and 75 notified commodities. Foodgrains account for nearly 60 per cent of the total quantity of items.

Goods deposited at the warehouse are properly sampled, weighed, analysed and graded. They are insured against the risks of fire, flood, theft and burglary. For the goods deposited, a valid warehouse receipt, which gives the description, weight and value of goods, is issued. The depositor may obtain a negotiable or a non-negotiable warehouse receipt for items deposited. A negotiable warehouse receipt can be pledged with a bank for drawing loan and can be used for sale/purchase without physical movement of goods. The competitive tariff rate is about one rupee per standard bag per month and varies from time to time as well as depending on the type of items stored.

The warehousing corporation undertakes handling and transport arrangements at the warehouses at the request of the bulk depositor. It extends disinfection services to the stored items and undertakes preshipment fumigation of exportable commodities. It also operates bonded warehouse in major cities. The corporation serves as a model for central warehousing, storage and handling.

PUBLIC DISTRIBUTION

The need for a well-organised public distribution system, in ameliorating the suffering of masses, with more than 50 per cent of people living below the poverty line, has been appreciated in India. The major activities of public distribution include procurement of essential commodities at fixed prices, storing them, and transporting and distributing them through over five lakh retail ration shops. Distribution to the ultimate consumer is largely by ration card in most of the areas. Control of stock at ration shop, quality of goods stored and service rendered by the ration shopkeepers in the public distribution system are important parameters. The peak load will be immediately after the salary days. Common complaints by consumers are inferior quality of goods, short weighments, increased foreign matter etc. Surprise checks by senior officers of the Civil Supplies Department will reduce the incidence of malpractices.

SUPER MARKETS

"Consumer is the king" is the motto of any business. This motto places a tremendous responsibility on the person incharge of the storage function; for, the non-availability of any consumer request will be directed to him. Forecasting and identifying consumer requirements is essential for proper management of stores. All the same, such stores should guard against the onslaught by innumerable, aggressive salesmen.

Each counter head must prepare a daily account of the quantities moved of the items handed in that particular counter. An average consumption pattern can be worked out from this after integrating corrections for growth trends. There should be periodic analysis of movement of the items, monitoring on high value items, identification of slow moving items, recognition of style changes, stock clearance sales etc. Merchandising and displaying the goods for sale is an important factor to be considered in the super market (super bazar). Regular storage racks, storing in mezanine floors, use of slotted angles etc., can be considered for storage space.

The number of counters depends upon the flow of customers. The average time spent at a counter should be less than the rate at which the customers arrive to prevent the build-up of a queue. Shop lifting and pilferage have to be controlled by security personnel, who should be on constant vigil by making rounds. Closed-circuit TV systems have been installed in many places as a control measure. Stock verification and auditing should be done on a routine basis to prevent pilferages and malpractices. The stock management in stores which issue mail order business and door delivery is slightly easier, since the customer is not physically present. This also enables such organisation to hold stocks at lower levels.

Super bazaars (markets), both private and public, are concentrated only in metropolitan areas and commercial towns. With the accent on rural development, the super bazaars will have to extend their services to the interior and remote villages. This will throw new challenges and opportunities for warehouse managers of these organisations in the area of logistics management.

TEN POINTS TO PONDER

1. Explain the concept of business logistics.
2. Discuss the relevance and advantages of logistics management.
3. Identify the physical distribution parameters.
4. Name the problems in the area of physical distribution.
5. Enumerate the steps and techniques used in business logistics.
6. Amplify the scope of warehousing as a business.
7. Examine the necessity of Central Warehousing Corporation.
8. Explain the services rendered by Central Warehousing Corporation.
9. Describe the concept of public distribution management.
10. Articulate on the role and utility of super bazaars in India.

PART II

COST REDUCTION METHODS

14. Statistical Aids for Demand Analysis

ROLE OF PROBABILITY

The warehouse manager is always in search of scientific techniques to improve the working in the stores. Statistical decision-making techniques help him, particularly when risk or uncertainty exists in the case of demand/supply/issue or anticipated consumption. Besides, the statistical techniques help him in sampling, calculating the fluctuations of demand, inventory control, price forecasting, quantity forecasting, averaging the seasonal factors, correlation analysis, projection by trend, testing various hypotheses accepting the incoming material on sample basis, fixing norms for consumption of items, as well as in other areas. All the statistical principles used to measure risk are based on the theory of probability. The term probability refers to the chance of happening of an event. For example if there are two outcomes in a random experiment of tossing the coin, then the probability of each outcome in a large number of trials is 0.5. Similarly, the outcome of a particular face, in the six faced dice, when thrown a large number of occasions, is 0.1667. Thus, the probability of occurrence of an event is equal to the number of possible outcomes of the experiment or proportion of occurrence of the event, assuming all outcomes are equally likely.

NEED FOR CONSUMPTION CONTROL

Financial institutions have already introduced norms for working capital, which include the norms of different components of inventory — namely raw material, machinery spares, semi-finished goods and finished goods. Similarly, it is desirable to have norms for consumption/sales in order to plan for smooth working.

The role of consumption control is not one of inhibiting consumption. Rather it is a measure of management control that aims at better utilisation of materials. Since the materials reach the consumer from the issue counter of store, this is the best place for implementing control. Controls can be effected in many ways and must be tailor-made to suit the needs of the organisation. Usually, quantity limits can be prescribed for issues. These are fixed on the basis of past consumption, the average consumption rate and the maximum variance from the average.

Experience has shown that in large undertakings, while dealing with commodities and components, which are often liable to supply shortages, the consumer divisions adopt selfish policies. They tend to corner all available stocks and hoard such items as a hedge against anticipated shortages. At the same time, other departments requiring the same item are starved as they are not available in the central stores. The storekeeper should avoid such situations by taking adequate precautions through better coordination and control of issues. He should promote a mutual give and take philosophy amongst the consumers.

CONSUMPTION PATTERN ANALYSIS

Let us now consider the monthly consumption of a regularly consuming fast-moving raw material in some units, for a duration of 25 months. These figures are tabulated in Table 14.1.

Table 14.1 Consumption Pattern

Month	1	2	3	4	5	6	7	8	9	10	11	12	
Consumption	17	18	22	19	17	21	18	20	19	20	21	23	
Month	13	14	15	16	17	18	19	20	21	22	23	24	25
Consumption	20	19	16	22	20	18	16	24	20	21	22	19	23

This data can be grouped to indicate the frequencies of consumption of 16, 17, 18, 19, 20, 21, 22, 23 and 24 as 2, 2, 3, 4, 5, 3, 3, 2, 1 or with a relative frequency or proportion of .08, .08, .012, .16, .20, .12, and .004, respectively. These can be plotted in the form of histogram or frequency polygon, a series of rectangles, each proportional in width to the individual values on the horizontal axis, and height proportional to the frequency.

NORMAL DISTRIBUTION

If we join the mid-points of the frequency polygon, then we get a bell-shaped symmetric pattern known as the normal distribution. The normal curve is the most widely used probability distribution, for most of the regularly consumable items or for any continuous variable. Instead of 25 representative observations, if we take consumption for longer duration, then also we get the normal distribution curve which is uniquely defined by its mean and standard deviation. Once the mean and standard deviation are calculated, we can calculate the proportion of observations lying beyond a distance of any value. The data obtained are given in Table 14.2.

MEAN/MODE/MEDIAN

The most common average used in the industry is the arithmetic *mean*. This is obtained by summing or totalling of individual observations and dividing by the number of observations. In the previous example, the total of the 25 values is 495 and the average is given as (495/25) = 19.8. If there are a large number of observations, then they can be grouped into classes and the arithmetic mean obtained. This arithmetic mean gives an idea of the average consumption. The average can be in decimals. In the example, 20 is the mode as it occurs five times. Mode represents the typical or common consumption figure. One can observe how this concept is applied in choosing the typical people's representative in multi-cornered elections.

Median is the central value and in order to arrive at it, we have to arrange the observations in an ascending or descending order as follows: 16, 16, 17, 17, 18, 18, 19, 19, 9, 20, 20, 20, 20, 20, 21, 21, 21, 22, 22, 22, 23, 23, and 24. The central value which has twelve observations below and twelve observations above is 20, and this is called the median. In the case of the symmetric normal distribution the mean, median and mode coincide at the average. Another kind of average is the geometric mean used in compound interest calculations and investment decisions. The arithmetic mean, obtained by averaging the reciprocals of individual observation is used in rates and ratios.

STANDARD DEVIATION

Range of (maximum–minimum) or (24–16), or 8, gives an idea of fluctuations, but is affected by extreme values. The standard deviation, as the name indicates, is the only measure to estimate the unavoidable fluctuations or variations. It is the standard method of calculating the dispersion. It is defined as the RMS value or root mean square of the deviations of the observations from the mean.

(a) The differences of the individual values from the mean are obtained as (2.8, 1.8, 2.2, etc.)

(b) These differences are squared and then summed up as
(7.84 + 3.24 + 4.84 + . . . +)

(c) The above total is 112.40.

(d) Divide this total by (25 – 1) or 24.

(e) (112.40)/24.

(f) Then obtain the square root of

$$\sqrt{112.40/24} = 2.2.$$

Use of Standard Deviation

For regularly consumable items resulting in normal frequency distribution, we have the following use. An interval containing (Mean ± 1 standard deviation) includes 68 per cent of values. (Mean ± 2 standard deviation) includes 95 per cent and (Mean ± 3 sD) includes all the values. Hence in the given example, we have [19.8 ± 3(2.2)] or (19.8 ± 6.6), which includes all values as per statistical theory.

The interval 13.2 – 26.4 indicates the consumption norms under the existing circumstances. If the inventory controller feels that this interval is too much, then he should make appropriate investigations to reduce the variability observed in the process. Standard deviation is also helpful in arriving at safety stock, estimated rejection in incoming material, as well as for measuring the forecasting accuracy.

Table 14.2 Area under Normal Curve

t	Proportion	*t*	Proportion	*t*	Proportion
– 4.0	.99997	– 1.3	.9032	1.4	.0808
– 3.4	.99995	– 1.2	.8849	1.5	.0668
– 3.8	.99993	– 1.1	.8643	1.6	.0548
– 3.7	.99990	– 1.0	.8413	1.7	.0446
– 3.6	.99984	– 0.9	.8159	1.8	.0359
– 3.5	.99977	– 0.8	.7881	1.9	.0287
– 3.4	.99966	– 0.7	.7580	2.0	.0228
– 3.3	.99952	– 0.6	.7257	2.1	.0179
– 3.2	.99930	– 0.5	.6915	2.2	.0139
– 3.1	.99900	– 0.4	.6554	2.3	.0107
– 3.0	.9987	– 0.3	.6179	2.4	.0082
– 2.9	.9981	– 0.2	.5793	2.5	.0062
– 2.8	.9974	– 0.1	.5398	2.6	.0047
– 2.7	.9963	–	.5000	2.7	.0035
– 2.6	.9953	0.1	.4602	2.8	.0026
– 2.5	.9938	0.2	.4207	2.9	.0019
– 2.4	.9918	0.3	.3821	3.0	.0013
– 2.3	.9893	0.4	.3446	3.1	.0010
– 2.2	.9861	0.5	.3085	3.2	.00069
– 2.1	.9821	0.6	.2743	3.3	.00048
– 2.0	.9772	0.7	.2420	3.4	.00034
– 1.9	.9713	0.8	.2119	3.5	.00023
– 1.8	.9641	0.9	.1841	3.6	.00016
– 1.7	.9554	1.0	.1587	3.7	.00011
– 1.6	.9452	1.0	.1357	3.8	.00007
– 1.5	.9332	1.2	.1151	3.9	.00005
– 1.4	.9192	1.3	.0968	4.0	.00003

Proportion of observation lying above '*ts*' from the mean. (*s* is the standard deviation and '*t*' is the variable distance on the horizontal axis).

POISSON DISTRIBUTION

While the normal distribution helps to determine the frequencies of regularly consumable items, the Poisson distribution is helpful in analysing the frequencies of spare parts. The failure pattern of spare parts follows the highly skewed, non-symmetric, Poisson pattern. This Poisson distribution is uniquely determined only by its average. For instance, if a part on an average fails six times in a year, the actual future observation or failure in future may vary from 0 to 16. The Poisson distribution is given in Table 14.3 and is useful for forecasting individual failures. Thus past consumption data not only provides the basis for consumption norm but also future forecasts by providing the probability of occurrence. This information enables the inventory controller to determine the safety stock levels.

Table 14.3 Poisson Distribution

Number to be consumed in Future	Appropriate Proportion Given the Average Consumption Value *m*. 0	1	2	3	4	5	6	7
0	.368	.135	.050	.018	.007	.002	.002	–
1	.368	.271	.149	.073	.034	.015	.006	.003
2	.184	.271	.224	.147	.084	.045	.022	.011
3	.061	.181	.224	.195	.140	.089	.052	.028
4	.095	.090	.168	.196	.176	.134	.091	.057
5	.003	.036	.101	.166	.176	.161	.128	.092
6	.001	.012	.050	.104	.146	.161	.149	.122
7	–	.003	.028	.060	.105	.138	.149	.140
8	–	.001	.022	.030	.065	.103	.130	.140
9	–	–	.003	.013	.036	.069	.101	.124
10	–	–	.001	.005	.018	.041	.071	.099
11	–	–	–	.002	.008	.023	.045	.072
12	–	–	–	.001	.003	.011	.027	.048
13	–	–	–	–	.001	.005	.014	.030
14	–	–	–	–	.001	.002	.007	.017
15	–	–	–	–	–	.001	.002	.009
16	–	–	–	–	–	–	.001	.005
17	–	–	–	–	–	–	–	.002
18	–	–	–	–	–	–	–	.001
19	–	–	–	–	–	–	–	–
Total	1.0	1.0	1.0	1.0	1.0	1.0		1.0

For higher average values of *m*, the formula for upper limit is $[m + k\sqrt{m}]$, where *k* is the service level factor. For 95% service level, $k = 2.3$. Suppose the average is 16. Then the maximum consumption norm is given by $(16 + 2.3\sqrt{16}) = (16 + 9.2)$ or 25.2 or 26 items.

CONTROL BY CONSUMPTION NORMS

For proper delegation of powers, selective control techniques and MUSIC-3D principles can be applied. The rules can be specified so that it is obligatory to obtain permission from the higher levels of the hierarchy for any deviation from the norms of consumption for high value items. Lower consumption levels, if frequently occurring, will lead to obsolescence. Alternatively, monetary limits can be fixed on the value of issues on monthly basis to individual departments. It should be noted that mere control measures will be of no avail since methods can always be adopted to sidestep any system. Therefore, the consumption patterns will have to be analysed regularly and the control limit updated. In this context fixing the limits of consumption is helpful not only in tackling the above-mentioned problems, but also in planning for future.

TEN POINTS TO PONDER

1. Discuss the meaning of the term *probability*.
2. Identify the need for consumption control or issue control in a warehouse.
3. Describe the relevance of mean/media/mode in warehousing.
4. Examine the need for consumption pattern analysis.
5. Explain the meaning of standard deviation.
6. Specify the steps for calculating standard deviation.
7. Discuss the use of standard deviation in warehousing.
8. Elaborate the role of normal distribution in inventory control and formulation of consumption limits.
9. Discuss the role of Poisson distribution in warehousing.
10. Examine the scope of consumption control in raw material stores/finished goods warehouse.

15. Forecasting Principles

USES OF FORECASTS

The business activities undertaken today must be based on yesterday's plans and tomorrow's expectations. Future plans cannot be made without forecasting the events and their relationships. Leadership requires ability to recognise a problem before it becomes an emergency. Up-to-date and accurate forecasts can facilitate the identification of problems. The inventory manager, materials manager and the warehouse manager have to forecast their future activities and make their plans.

Sales forecast must be done first. It should then be converted into material requirements. The top management is interested in long range planning, diversification and expansion. The finance manager is interested in the forecasts of operating budgets, capital budgets, cash inflows and cash outflows. The marketing executive is interested in evolving sales strategies to meet the future target. The personnel officer is interested in manpower planning and promotion aspects of the executives in future. The production manager is interested in scheduling the production on the basis of the sales forecasts. The materials manager is interested in the forecasts of prices and quantities as well as in materials planning. The problem is *not* whether to forecast or not but how to improve the forecasting accuracy.

ACCURACY OF FORECASTS

Robust common-sense, intrinsic analytical skills and high IQ are prerequisites for any forecast. Forecasts are rarely cent per cent accurate. The accuracy of forecasts depends on the method used, factors considered, and the period for which the forecast is made. For instance, the forecast for tomorrow or next month will be more accurate than the forecast for the next year. Obviously, the short term forecasts must fit in the long term forecasts. Successful forecasts are based on subjective experience, knowledge, skills, expertise and judgment of the executive. Therefore, it is advantageous to use more than one method to improve the accuracy.

The factors influencing forecasts, include social, economical, political, religious, ethnic, fiscal, governmental, natural, macro, micro, and technological aspects. Some of these factors are tangible while others are not. The macro factors comprise influencing business forecasts policy aspects such as credit regulations, taxation policies, budgetary policies, export/import policies, industrial licencing policies, deficit financing, and inflation. The micro level policies cover the market share, promotion expenses, advertising, research/development aspects, cost/pricing aspects, motivation, management information system, and criticality — availability — reliability of an item. These factors can be classified as controllable/uncontrollable or as tangible and intangible factors.

SWOT ANALYSIS AND 7S APPROACH

Before forecasting is attempted, it is essential to do a SWOT — Strength, Weakness, Opportunity and Threat — analyis. Strength and weakness are analysed with respect to the factors internal to the organisation. The opportunities and threat are the effects external to the company.

These can be analysed by the seven 'S' approach of the Harvard-Boston group. This approach looks at every aspect from the following viewpoints: (a) super ordinate goals, (b) strategies, (c) structure, (d) systems, (e) skills, (f) staffing, and (g) styles of management. The corporate planning group in the organisation must carry out such an analysis before forecasting. The course of action is to improve the organisations strength and reduce the weakness so that the opportunities can be capitalised and threats eliminated. For example, an effective forecasting ability can be used to reduce the threat of fluctuating prices. Ideally, SWOT analysis has to be done separately for the department, company, industry, nation and international scene, both for the key raw material and the important finished goods.

METHODS OF FORECASTING

A large number of forecasting techniques with varying degrees of complexity have been devised in the recent past. The advent of the computer and easy availability of computer programs have facilitated the task of the forecaster. The methods can be classified as subjective depending upon individual judgments and objective methods arrived at after analysing the quantitative information. Let us discuss the subjective method initially and then proceed to the analytical methods of forecasting.

Visionary Forecast

The top management either as an individual or as a team fixes the forecast on an *ad hoc* basis, usually at 10 per cent more than the last year. Obviously, the management would have considered all the factors before coming to the final figures. This team is called the jury of executive opinion. The forecast uses personal insights, judgments, thumb rules and, if possible, works out different future scenarios. The forecast is characterised by subjective guesswork and imagination and the methods are unscientific. These forecasts are particularly useful for long range, new product sales and forecasts of margins. If more than one person is involved, a set of possible scenarios about the future are prepared by the experts in the light of past events. This method adopted in conservative firms presents forecasts in the quickest possible time. Admittedly, luck plays a crucial role in any forecast. Research survey confirms that many organisations follow this method only.

Panel Consensus

Panel consensus is based on the assumption that several experts can arrive at a better forecast than one person. There is no secrecy, and communication is encouraged amongst the panelists. The forecasts are obviously influenced by the social background of the individuals. The panelists analyse the available data in the quickest possible time, before giving their opinion. In the marketing field, this method is known as sales force composite, where the individual salesmen send the sales forecast of each product to the regional headquarters. The regional incharge collects all such information and develops a forecast for each item for the region as a whole. This process is repeated till the forecast for the entire organisation is evolved at the corporate headquarters.

Market Research

This process involves the collection of information from a representative sample of possible consumers through a carefully structured questionnaire, on the utilisation of the company's products. Considerable amount of information on market data is needed before coming to a conclusion on the forecasts. Systematic, formal and conscious methodology of obtaining the subjective opinion of customers is commonly adopted for new product sales forecasts long range plans, efficacy of after-sales service and forecasts of margins. Increased number of respondents, cost of conducting the survey, and non-cooperation of the respondents are the common problems encountered in this field. A large number of consulting and professional organisations have started undertaking market research surveys and render service on a commercial basis. A case in point is the opinion polls conducted during the pre-elections period.

Delphi Method

In order to minimise the biases in the subjective methods, the Delphi method has been found quite useful. In this method, a panel of experts is interrogated by a sequence of questionnaires, in which the response to one questionnaire is used to produce the next questionnaire. Any set of information available to some experts and not others is passed on to the others, enabling the experts to have access to all the information for forecasting. This technique is an iterative process and eliminates the bandwagon effect of majority opinion. The coordinator issues the sequence of questionnaires and consolidates the responses after editing. This method, though it involves precious executive time, is very useful for long range planning and new product sales forecasts.

Time Series Analysis

A simple plot of the past information on consumption or sales on the *y*-axis and the time dimension on the *x*-axis, often results in well-known patterns. The philosophy behind this method, also known as economic rhythm method, is as follows: "History is the wisest counsellor of all and past patterns tend to repeat in future". The term *time series* is used to refer to any group of information collected at regular intervals. This technique splits the past information known as the time series into: (a) long term growth trend, (b) seasonal variation, (c) business cyclical fluctuations, and (d) irregular component. With a minimum of three years data and used with special analytical skills, this is perhaps the best method for medium range forecasts, allowing one to predict the turning points and to time special events.

Trend and Cyclical Effects

This technique projects by means of a trend line fitted to a mathematical equation. The steady increase in the cost of living, recorded by the consumer price index is an example of long term trend. Seasonal changes within a year, which tend to be repeated from year to year, are the second kind of fluctuations. Increased sales of umbrellas during monsoon season or sales of fertilisers before the sowing seasons are classic examples of seasonality. The seasonal index, where it can be constructed, is an adequate forecast of seasonal variations. The most common type of cyclical fluctuations is the concept of business cycles with booms/peaks and troughs/valleys. Over a period of time, there are years when the business cycle has a peak above the trend line and times when the business activity is likely to slump, touching a low point below the trend line. In this context, Ravi Batra's prediction of the great depression during the end of 1990s is well known. The time lag between touching peaks or falling to low points is generally seven to ten years. Irregular variation are usually due to random fluctuations. These variations are generally considered as interacting in a multiple manner of $T \times S \times C \times I$ to produce the observed values of the overall time series. The forecast of the actual value is obtained by the extension of the trend line, seasonal index and the position of extended cyclical pattern. Usually computer programs are available for applying this method. Sometimes, a leading or lag-indicator time series is also used. A time series of an economic activity whose movement in a given direction provides the movement of some other time series in the same direction, is a leading indicator.

Moving Average Method

Moving average, exponential smoothing and time series analysis are the three methods under this category of economic rhythmic method. The computation of moving average is straightforward and simple. This method can be considered as an artificially constructed time series, in which each periodic figure is replaced by the mean value of that period. In other words, each point of a moving average of a time series is the arithmetic weighted average of a number of consecutive points of the series. Equal weights are given to all the observations. Here the number of data points is chosen so that the effects of seasonality or irregularity or both are eliminated. Usually a minimum of 25 month's data on sales/consumption is chosen in order to locate the seasonality and smoothen the same. An average of a period of 4 or 5 months is taken. Suppose we have a set of figures on consumption as 54, 45, 40, 35 and 56. Then the five-monthly total yield is 230, and the average is 16, which is taken as the forecast for the sixth month. Suppose the sixth month actual is 44 units, then the total obtained by omitting the first, but including the sixth month, is 220, yielding a forecast of 44 for the seventh month. In this way, we proceed to get the forecast based on moving averages.

Exponential Smoothing

The exponential smoothing technique is similar to the moving average, except that more weightage is given to the more recent observations. Descriptively, the new forecast is the sum of old forecast plus some proportion of the previous forecasting error. Computer programs are available for the exponential smoothing method.

The forecast for time $(t + 1)$ made at time t is given below:

$$\hat{F}_t = dF_t + (1 - d)\, \hat{F}_{(t-1)}$$

Here $\hat{F}_t$ is the forecast of $(t + 1)$th period made at time t,

F_t is the actual performance in time t,

$F_{(t-1)}$ is the forecast for time t, made in $(t - 1)$th period, and d is the smoothing constant lying between zero to one.

The term *adaptive forecasting* is somewhat the same as above, except that seasonal effects are also computed. There are many variations of the exponential smoothing method and they are used for production, inventory control and forecasting. Exponential smoothing is a special case of Box-Jenkins technique. The time series is fitted with a mathematical model which is optimal in that it assigns smaller errors to past data. This is the most accurate statistical method, but it is also one of the most costly and time consuming methods.

Input-Output/Diffusion Index

The input-output model gives a method of analysis concerned with the inter-industry or inter-departmental flow of goods/services in the economy or a company and its markets. It shows what flows of inputs must occur to obtain certain output. These models are more useful at macro levels. Considerable effort must be expanded to use these models properly. Extra information specific to a particular business on product and service flows for which an input-output analysis is desired is needed for forecasting in an individual company.

In this context, econometric models and input-output models are sometimes combined for forecasting. The input-output model is used to provide long term trends and is used to stabilise the econometric model. This is particularly useful for forecasting the company sales for industrial sectors and subsectors. Diffusion index gives the percentage of a group of economic indicators that are going up or down. The diffusion index has been used by the National Bureau of Economic Research in the United States. The movement of the diffusion index is generally related to that of general business activity and shows that the aggregate economy starts expanding only when the diffusion index has crossed the 50 per cent line from below.

Correlation Analysis

The forecaster has to identify the influencing variables like disposable income and sales on the forecasted variable. The correlation analysis measures the strength of the relationship between two or more variables. The value of correlation coefficient lies between –1 and +.1, the positive values indicating that there is a tendency for the second variable to increase with the increases in the first. The scatter diagram, or a plot of the two variables in a graph sheet, gives some idea of the nature of relationship to be considered. This may be a straight line or a curve and the extent may be positive or negative.

Regression Model

This regression functionally relates sales to other economic, competitive or internal variables. It estimates an equation using the least square technique of minimising the total square of deviations of the observed and expected value. The relationships are primarily analysed statistically although any relationship should be selected for testing on a rational ground. The usual equation is $y = a + bx$, y is the dependent variable or the forecast, x is the influencing independent variable. $(a = \bar{y} - b\bar{x})$ is the intercept on the y-axis and $b = (\Sigma xy - n\bar{x}\bar{y})/(\Sigma x_i - n\bar{x}^2)$. Computer programs are available for complex situations involving more influential variables or non-linear equations to forecast by the regression method.

Thus, the analytical techniques are applied to indicate the forecast for the total business as well as for the individual item, by identifying any pecularities or sudden changes in trends or patterns. This information is then incorporated into the details of the item forecasts with adjustments, considering the smoothing mechanisms, seasonals, executive judgment and other influencing factors.

TEN POINTS TO PONDER

1. Identify the utility of forecasts in warehousing and inventory management.
2. Articulate on the accuracy of forecasts.
3. Explain the significance of the SWOT analysis in forecasting.
4. Examine in detail the 7S concepts and other influencing factors.
5. Discuss the merits and demerits of the jury of executive opinion and the role of the Delphi method.
6. Define the term panel consensus.
7. Detail the procedure for market research for a consumer item.
8. Explain the terms moving average, exponential smoothing and Input-output analysis in the context of forecasting.
9. Indicate the steps needed for a Time series analysis.
10. Describe the significance of correlation-regression analysis in the context of forecasting.

16. Codification

A TO Z ADVANTAGES OF CODIFICATION

The role of the warehousing and inventory controller is to serve the user promptly.

(a) Unless the warehouse manager is able to identify instantaneously the user's requirement, he will not be able to render good service.

(b) Delays lead to the building of queues at the service stations and consequent delays in the production line.

(c) In the case of finished goods warehouse, the customer will go to another company, thereby affecting the image and profits of the organisation.

(d) The starting point is the consumer, who has to unambiguously identify his requirements, through proper nomenclature.

(e) The procured item may have a trade name of its own, and the stores executive has to correlate the two names, with the problem going out of hand if the number of items is very large.

(f) Codification helps to uniquely identify the item by having one-to-one correspondence.

(g) The advantages of unique identification and disadvantages of ambiguous nomenclature can never be overstated.

(h) Long description by tongue twisting long names must be avoided.

(i) Systematic grouping of similar items is facilitated by codification.

(j) Codification automatically leads to the process of standardisation.

(k) Ordering becomes more economical as like items are ordered together.

(l) Price advantage can be availed out of the resultant bulk ordering with discounts.

(m) Location problems of items in bins are reduced by codification.

(n) Service levels can be improved for the same level of investment.

(o) Servicing time is less due to easy identification.

(p) Stock levels are less as duplication has been eliminated by codification.

(q) Obsolescence is less as the number of items is less.

(r) Codification facilitates computerisation.

(s) The records, entries and transactions are less as less number of items are handled per person.

(t) The overall efficiency of warehouse staff is much better with higher level of motivation.

(u) Logical grouping of items facilitates variety reduction easily.

(v) Procurement operations become easy as the purchase people, even if they are nontechnicals, can understand the requirements easily.

(w) Communication between different departments is smoother by process of codification as they know what they are talking about.

(x) The simplicity of codification enables even unskilled workers to distinguish one item from another and trace where the wanted item is kept.

(y) Codification introduces discipline and reduces inventory by removing nonstock items and reducing items on temporary codes. After all we live in a coded world consisting of permanent account number of income tax, telephone number, postal index number, ration card number etc.

(z) Minimising the number of items today by means of codification, enables everyone in the organisation to meet future challenges created by expansion, diversification and technological upgradations. Codification aims at maximum service to users with minimum input resources, after identifying items uniquely and classifying them into major class, subclass group, subgroup and minor groups.

Names of a Screw

It is interesting to note that in an electric company, a few decades ago, a metal screw of 6" length and 3/8" diameter was known by 111 names.

These names were given by the respective users in their indents according to the particular types of usage. Some of them had common names as pin, locating peg, stud, rod, spindle, collar pin and screw etc. with prefixes attached to them. The details of all the names are given below:

(1) Pin, (2) Pin hinge, (3) Carrier link pin, (4) Value cottor pin, (5) Clutch toggle pin, (6) Swing bolt pin, (7) Main spindle locating pin, (8) Pin towing eye, (9) Roller pin, (10) Pin link, (11) Locating pin, (12) Operating pin, (13) Trip arm pin, (14) Special pin, (15) Eye bolt pin, (16) Bearing pin, (17) Water pump driving pin, (18) Plain pin, (19) Shaft pin, (20) Coupling pin, (21) Jaw pin, (22) Piston rod pin, (23) Kicket catch pin, (24) Thrust collar pin, (25) Resec frame shop pin, (26) Drive pin, (27) Pin anchor, (28) Pin pivot, (29) Pin crank, (30) Dowel pin, (31) Wire pin, (32) Bond pin, (33) Driving pin, (34) Jaw pin, (35) Rod, (36) Hinge rod, (37) Lock rod, (38) Push rod, (39) Valve rod, (40) Plain rod, (41) Dash pot rod, (42) Connector link rod, (43) Canopy rod, (44) Dowel, (45) Wheel stud dowel, (46) Peg, (47) Locating peg, (48) Volve cotter, (49) Roller shaft, (50) Pivot, (51) Brush, (52) Plug, (53) Spindle, (54) Stop, (55) Roller, (56) Plunger, (57) Connector, (58) Traunion, (59) Electrode, (60) Sealing sleeve, (61) Intercooler, fitter dowel, (62) Starting handle, (63) Distributor drive, (64) Governor drive, (65) Man handling bar knuckle, (66) Pump drive, (67) Centre drive peg, (68) Slave spindle, (69) Clutcher roller, (70) Counter shaft, (71) Rest spindle, (72) Drain tap boss, (73) Fuel pump tupper, (74) Pin push lever, (75) Jacket leg plug, (76) Davel from main bearing, (77) Breather hole plug, (78) Selector bar plug, (79) Governor weight spindle,

(80) Roller spindle, (81) Inlet rocket bracket, (82) Oilway plug, (83) Armature cradle bearing rod, (84) Connector moving oscillator coil, (85) Latch, (86) Button, (87) Stud, (88) Rivet, (89) Spacer, (90) Bar, (91) Value stem, (92) Flag pivot, (93) Cran river, (94) Post, (95) Post mileage, (96) Pawl spindle, (97) Driving bar, (98) Lever spindle, (99) Armature stud, (100) Pincon spindle, (101) Water pump dowel, (102) Chain anchor dowel, (103) Oil relief ball stop, (104) Contact lever axle, (105) Pin clamp lever axle, (106) Float lever stud, (107) Pin mould holding, (108) Motor drive spindle, (109) Sanding valve roller, (110) Woodscrew, and (111) Rock drill.

Neither the storekeeper nor the users thought that they were referring to the same item by different names resulting in 111 locations in stores, 111 sets of documentation, high stock for each item, stockouts if one name was not available, etc. Introduction of codification removed the 110 items in one stroke, with reduced documentation, less inventories and more space.

BRISCH AND KODAK SYSTEMS

The *Brisch system of codification*, developed by a British industrial engineer, is one of the earliest developed codification systems. It is a comprehensive system, consisting of *seven* digits and is applied in three phases. The seven digits split into groups such that similar items are brought together. The preliminary set of categories are categories of assemblies, subassemblies and bought out items. The classification is done after duly considering materials, sizes, function and uses.

Another established system is the *Kodak system* developed by Eastman Kodak company of New York. It consists of *ten* digits divided into groups of three, four and three. In addition to the classification based on materials, function and use of the item, importance is given to the mode of procurement of the item by the first two digits. When arranging within a group, a precedence system based on alphabets is used. Thus within a group, the numbers used are 00 to 99. If nuts and bolts are to be coded, then bolts can have the number 28 and nuts 29, because *b* is ahead of *n* in the alphabet.

CHECK DIGIT FOR CATALOGUING

In addition to the digits that build up the codes, we have to introduce a check digit to take care of any mistake that may arise during transfer of data. When processing is done by the computer or any other data processing machine, computations are done on the code numbers themselves to verify the accuracy of codification. The check digit is introduced as an additional digit and is usually added at the right end. Its value is calculated by a 'modules 11' procedure. Here the digits of the original code are multiplied from the right by two, three, four and the sum of these products is divided by eleven. The check digit is given by the remainder. It is zero if the remainder happens to be zero or ten. While processing the data, the same computations are repeated, and if the figures do not tally, an error is indicated. The system of using check digit has been found to be very effective for computerisation. The computer gives error signals if there is any ambiguity in codification process.

DESIGN OF CODIFICATION DIGITS

The codification system should be so designed as to form a logical method of unique identification for each and every item in the raw material warehouse. Sometimes the system includes semi-finished assemblies and final products. The final product is usually given a part number at the sellers end. However, the user must develop internal codification and avoid part number, particularly for the spare parts as the same item can be supplied by more than one vendor. If necessary, a digit of the codification can be allocated to represent the suppliers, as rarely the organisation will have more than ten suppliers for a specific item. For this purpose, the system must be logical, unique and simple so that cyphering expert is not needed to decode the items every time. One item should be represented by one code and one code must indicate only one item.

All organisations grow in size and level of activities. Hence the design of codification system should be flexible enough to allow for expansions. Codification in multi-plant organisation is more complex. It is desirable to develop a comprehensive codification to facilitate inter-plant transfers by fixing one place to include the location of the plant as well.

CODIFICATION STRUCTURE

Alphabets or numerals or a combination of both can be used for codification purposes. An example of the alpha numerical system is the permanent account number of the income tax department. Names are earliest examples of alphabetical codes. Telephone system uses numbers as an example of numerical devices. Colours can also be used as codes and this system is being extensively used in electronic components and steel rods. The three colours red, orange and green are used all over the world as a means of traffic communication.

In practice, codification is done by a group of industrial engineers, who consult all the concerned people from stores, purchase, inventory control, maintenance, production, design, marketing and accounts. The first decision to be taken is whether to use alphabets or numerals or a combination of both. A place in the alphabetical system can cover A, B, . . ., Z, or 26 characteristcs, while a digital system can cover only ten varieties, namely, 0, 1, 2, . . ., 9. However, it is difficult to interpret and handle alphabets in practice. These are normally used in one or two situations for ease of use or including some special characteristics, like income tax permanent account member.

SIGNIFICANCE OF DIGITS

The digits can have some significant intelligence built in all of them or only in some of them while the rest are in chronologically ordered codes. A critical decision that needs to be taken on codification is the number of digits or number of places. This is basically a compromise between various opposing factors such as (a) ease of remembrance, (b) error-free usage, (c) building up of intelligence, (d) number of items, (e) logical ordering, (f) special requirements such as supplies etc. For easy remembrance and reduction of errors in usage, the fewer the digits, the better. On the other hand, for increasing the intelligence, for including various varieties and for minute classification, the opposite codification of as many digits as possible is better. A seven to ten digit system is used in a majority of organisations. Once this decision is taken, the process of filling in the digits must be planned. It would be better to design the code in such a way as not only to clearly identify all items, but also indicate the use and location of each item.

The digits are grouped together so that each group can possess intelligence about a particular type of quality. Thus we can have separate groups to indicate the material, sizes, suppliers, use, bin location and any other groupings as needed by the organisation. The different items within a group are denoted by different codes. Normally zeroes are avoided in codes to reduce mistakes. For example the first digit 1 may indicate raw material, 11 may be metals, 112 non-ferrous metal, 1123 aluminium, 11234 may be aluminium plates, then sizes, check digit follow. Sometimes all the places will not be required by the items. Uniformity of numbering will mean unnecessarily expanding the digits, where they are not actually required. This problem has to be tackled at the designing stage itself.

PROBLEM AREAS

Codification is not beneficial until and unless all the concerned personnel make use of it advantageously. In a large number of organisations, temporary codes are given for new items and the practice is continued indefinitely. The maintenance staff invariably use supplier's part number while indenting. This forces the warehouse manager to be at the mercy of a single supplier, with his stipulated price, quality and lead time. Use of the suppliers part number can be avoided by indicating a place for vendors in the internal codification. The users avoid codifications by resorting to non-stock items. Non-stock items are those that are procured only upon specific requirements and are issued at once. Some organisations show more than 50 per cent of items as non-stock items, which is very dangerous from control and service point of view. Rarely required non-repetitive non-stock items should account only to about 10 per cent of the total number of items. These should be segregated and the remaining items must be included in the codification process. The codification system must clearly identify the raw material, spares, semi-finished items and finished goods. It has to group like items together for better management and control through acceptance of all those involved.

NATIONAL CODIFICATION

So far, the codification process — in a department, in an organisation and multi-plant units — has been stressed. In the long run, the advantages of a proper codification system far outweigh the problems in implementation. It is necessary to expand the horizons of codification to the national level by the following stages : (a) individual, (b) departments, (c) company, (d) inter-plant multi-units, (e) industry, (f) national and (g) international level. The Universal decimal classification for library is a case in point. Of course the suppliers, as a whole, act as a base for national codification but with different identification details.

A comprehensive national codification has to be developed by the users, vendors and research agencies with the assistance of professional associations, canalised agencies, government departments, industrial associations and chamber of commerce. National codification will help avoid duplication of resources, efforts, materials and working capital at all levels in the economy resulting in better utilisation of national resources.

TEN POINTS TO PONDER

1. Explain the relevance of codification in warehousing.
2. Identify the benefits of codification to the stores manager.
3. Expand Brisch system of codification.
4. Elucidate the Kodak system of codification.
5. Amplify the relevance of check digit in codes.
6. Describe the designing of codification process.
7. Discuss the structure of codification.
8. Enumerate the problem areas in codification.
9. Articulate the differences between part number and code.
10. Illustrate the steps to be taken for national codification.

17. Standardisation

A TO Z BENEFITS OF STANDARDISATION

The warehousing and inventory manager must have an unambiguous nomenclature for identifying the items, in order to serve the customer promptly.

(a) In this process, standardisation which reduces the variety plays a vital role by reducing the number of varieties of the same item held in the warehouse.

(b) More than any other executive, the stores-incharge comes into direct contact with a variety of materials. He can gain knowledge on the uses of the material in the store and can contribute significantly to achieve standardisation.

(c) Standardisation enables one to concentrate on large quantity of fewer items.

(d) It is possible to place economic order quantity or orders of staggered deliveries with bulk discount.

(e) Economic lot size at manufacturing will not be a problem as items can be manufactured with the same set up timings for standard items.

(f) Since the inventory is likely to be less, the working capital commitment of the organisation will also be less.

(g) The items can be easily identified by all persons in stores as the bin location can also be standardised.

(h) Standardisation reduces the time involved in negotiation process with the suppliers as communication is better with prompt delivery schedules.

(i) The procurement lead time gets reduced as everyone knows the item clearly.

(j) The buyer-seller relations can be improved as all dimensions including price analysis, specifications lead time etc., are standardised without scope for disputes.

(k) Standardisation promotes healthy competition amongst vendors as standard items can be bought by several companies.

(l) Standardisation minimises obsolescence.

(m) It is possible to have rate/running contracts for standard items.

Cost Reduction

(n) Standardisation reduces design time by concentrating on fewer parts.

(o) It minimises drafting time through repetitive use of standard drawings.

(p) It expands the engineers list of known and proven items.

(q) It reduces specification writing by repetitive use of standard specifications.

(r) Interchangeability of parts is assured through standardisation.

(s) Lower cost of production through continuous manufacturing run with less material handling problems and increased capacity utilisation is achieved by standardisation.

(t) The workers at all levels and departments become more efficient by repeated handling of items and it is possible to utilise less skilled workers for standard operations.

(u) Standardisation enables reduced inspection and quality control burden.

(v) Since methods are standardised, it ensures safety to all.

(w) Standardisation directly reduces the variety and improves the information system of the warehouse manager.

(x) Standardisation enables reduced number of maintenance tools and improves maintenance practices.

(y) Standardisation aims at maximum variety of finished products with minimum categories of assemblies, subassemblies and components.

(z) In general, standardisation results in simple operations, minimum paper work, improved inter departmental coordination, ease of computer application and pinpoints areas involving inefficiency. In short, standardisation is a means to faster and better understood communication, with fewer mistakes and hence smoother life.

STANDARDISATION AND SIMPLIFICATION

A standard is defined as a model or general agreement or a rule established by authority, consensus, or custom, created and used by various levels of interest. Standardisation is the orderly and systematic formulation, adoption, application and review of industrial standard which leads to simplification or variety reduction. This implies reducing unnecessary varieties and standardising to the most economical sizes, grades, shapes, colours or types of parts. The word "standard" is multi-valued as it may mean a physical standard, prototye of mass or a technical document. Industrial standard is the generally accepted statement of the requirement.

Standardisation eliminates the rule of thumb method as the width of the thumb varies widely from one individual to another. When Noah built the Ark, measurements were made in cubits, cubit being the length of a man's forearm. Thus, if Noah had been a smaller person or a hulking giant of a man, the size of the Ark of 300 cubits could have varied considerably.

In this context, simplification is the process of reducing the variety of items. It reduces the number of types and sizes of items to a minimum, consistent with the needs of an organisation. Simplification in company operations is an essential process of standardisation.

Specification is defined as the concise statement of requirement to be satisfied by a product, a material or a process indicating the procedure by means of which it may be determined whether the given requirements are satisfied. Specification changes according to the purpose of the use of material. Thus Aluminium used for cooking utensils should be at least 99 per cent pure, where 95 per cent may be sufficient for an electric conductor. An important factor is to avoid both overspecification and underspecification as both these strategies will lead to increased costs. Overspecification increases the cost due to the extra effort needed to reach higher standards.

Standardisation is the process of formulating and applying rules for a beneficial and orderly approach to a specific activity. It is based on consolidated efforts of science, technology and experience, reflecting in: (a) units of measurement, (b) terminology, (c) products, (d) process, and (e) safety of persons.

Procedure of Standardisation

The task force, responsible for standardisation must follow a procedure to get the items standardised. They should get all relevant facts from the concerned executive. They should develop standards which are easy to understand and to implement. There should be a constant reviewing, updating and monitoring committee so that best benefits accrue to the company.

The process of standardisation can be conceived on a three dimensional plane, to include factors such as levels, aspects, subjects and their interrelations. The levels include individual, department, company, industry, nation, international, universe, while the aspects may include forms, ethics, code of conduct procedures, contracts, inspection, testing sampling, agreements or reporting, notes etc. The third dimension, namely the subjects, encompasses industry, education, agriculture, communication, transport, mining, trade, commerce, energy, family planning, construction and administration.

The standardisation process may take the form of a document containing a set of conditions to be fulfilled, a fundamental unit or physical constant or an object for physical comparison. The apex body of standards in India is the Bureau of Indian Standards, formerly known as the Indian Standards Institution (ISI). It has formulated 15,000 standards, after consulting the involved people. At the International level, International Standard Organisation is based in Geneva.

Items to be Standardised

The Warehouse and inventory manager should involve executives from design, maintenance, inspection, operations and costing department in order to ensure that the recommended standards are implemented. This team should update its knowledge on the latest available technological developments to effectively implement the process of standardisation. If the standard is approved as a company standard, the standard department should make necessary reproductions and distribute copies as mandatory to all concerned.

We have the kilogram as the standard of weight, metre as the standard of length. Keeping to the left as the standard of the road, motors are made only in certain ratings, bulbs only in specific voltages, etc. Procedures, formats, reporting systems, sampling inspection plans, letter heads, operating manual, standard costing methods, etc. are the examples one comes across in the field of standardisation. It is easy to standardise items for which published standards by the Indian Standards Institution are available. It is advantageous to tackle non-critical high value, easily available items for standardisation, as otherwise there will be opposition from design departments. MRO — Maintenance, Repair and Operation-items like fasteners, hardware items, maintenance, etc. are a good starting point to impose standardisation.

METHODS OF STANDARDISATION

Whenever published standards are available from outside sources like Indian Standards Institution, the same can be applied. The frequency distribution of demand for each type and size of an item in terms of quantity of materials purchased/used/or products sold, would be an excellent starting point for standardisation and variety reduction in an organisation. For instance, consider the example where the diameters in multimeters of pipes are 5.0, 5.1, 5.2, 5.3, 5.4, and the past pattern shows that only two sizes 5.1 and 5.4 are often used. Then the orders for the remaining three dimensions can be gradually discontinued if these items are not critical from performance and safety angles.

The preferred number series — a series of standard sizes in suitable steps used in bulbs and currency — developed by Renaud and known as R-5, R-10, R-29, R-40 etc. is another way of starting the standardisation process. The values of the series are available in the Indian Standards 1076 published by the Indian Standards Institution.

ROAD BLOCKS

Standardisation should not lead to monotony and prohibit innovation. For instance, it is not possible to introduce standardisation of the end product in job shop, but assemblies/parts constituting the final product may be standardised. Similarly, it is possible to standardise items like shoes in terms of seven or eight sizes. Standardisation must not lead to rigidity in design or restrict technical development. Standardisation is also based on previous practices. For instance, Russians specify motors by weight and not by horse power. Indians usually measure weight in kilogram but height in British units as six feet and not centimetres. These practical approaches have to be borne in mind while standardisation process is developed as it should not make life complex.

TEN POINTS TO PONDER

1. Identify the relevance of standardisation in warehousing and Inventory Management.
2. Discuss the benefits of standardisation to different departments in the organisation.
3. Explain how you will identify the items for standardisation.
4. Describe the procedure for standardisation.
5. Enumerate the pitfalls of standardisation.
6. Elucidate the application of the concept of frequency distribution to standardisation.
7. Clarify the concept of preferred number series in relation to standardisation.
8. Specify the interrelationship between standardisation, simplification and specification.
9. Explain the role of Indian Standards Institution in development of standards.
10. Examine the scope of standardisation in job shop, office and administration.

18. The Learning Curve

THE LEARNING PROCESS

The concept of 'learning' is familiar to every one, as every individual learns from his previous mistakes. The supplier learns at the expense of the buyer over a period of time. The concept of learning is based on the fact that a person improves his skills with practice. Every worker tries to gain experience over a period and tries to reduce the amount of time for each job, particularly in labour intensive jobs.

LEARNING THEORY

The learning theory examines critically the following areas:

(a) What the learner brings to the transactions

(b) What the group leader or supervisor brings to the transactions

(c) The learning background in terms of technical, technological and human skills

(d) Man-machine interactions

(e) Conditions necessary for improvement

(f) Utilisation of learning in the worklife of the learner

(g) The establishment of the process of continued learning.

THE LEARNING FACTOR

The learning factor is based on the principle that all jobs are performed more efficiently as greater experience is gained. This is particularly witnessed in the labour intensive jobs. The learning factor depends on the type of industry and past experience.

The learning factor has been estimated in American industries as 80% for the aircraft industry, 85% for the manufacture of engines, 90% for electrical industry and 95% for machinery industry. In the above industries, the labour breakdown for the labour intensive assembly time has been found to decrease from 75% to 25%, whereas the machine time increases from 25% to 75% respectively. The ratio of machine time to assembly time varies from one industry to another. As the machining time goes up, the rate of learning declines, resulting in a flat curve. Research abroad has indicated that the learning factor varies from about 70% for the labour intensive jobs to about 90% for the less labour intensive activities. We also conclude that if the learning is zero, the learning factor is 100%, as every job will repeat up to hundred only.

Some people use the complement as the learning factor, namely zero to indicate zero learning and unity or 100% to indicate perfect learning. Hence due care is necessary for understanding this. The Stanform Research Institute, about four decades ago, found that to build the 1000th B-29 bomber aircraft took only 3% of the time used on the first plane, based on 80% learning factor. This was done after trying a series of learning factors and finally arriving at the 80% factor. In other words, if the first job takes 100 hours, then the cumulative time for the second takes (100 × 80); the cumulative time for the third is (100 × 80 × 80); the cumulative time for the fourth is (100 × 80 × 80 × 80); and so on. We can tabulate cumulative time against the cumulative number of units.

LEARNING CURVE

Learning curve is obtained by plotting the cumulative number of units on the horizontal *X*-axis and the cumulative time — of the values obtained as above — on the vertical *Y*-axis. The curve gives a decreasing value in hours in the form of the curvilinear *L* — decreasing in shape as the number of units is increased. This curve, when plotted in the form of log-log graph, by taking logarithm of both values on cumulative units and cumulative time, gets converted into a straight line. A horizontal straight line of 100% learning factor curve indicates that no learning is possible. On the other extreme lies the 50% learning straight line, lying between the *X*- and *Y*-axis from the top indicates that when the production quantity is doubled the second half is produced at no cost, which is an impractical situation. Hence useful learning factors are based between 70% to 90%. Thus the labour time required to produce an item will decrease each time the job is done, but the amount of decrease will go on reducing each time.

Mathematical Approach

Instead of drawing the curve or calculating every time, we can work out the calculations on the basis of a mathematical formula. This is represented as follows:

$$T(q) = Hq^{1-h}$$

Here, $T(q)$ represents the cumulative time for q units, h is the learning factor lying between 0 to 1 — usually between 70% to 90% as mentioned earlier — and H represents the time for manufacturing the first unit which has been mentioned as 100% earlier. The learning factor (h) depends on the work involved in the manufacturing of products.

Use of Learning Curve

In the present high inflationary spiral situation, the supplier always demands price increase for his products due to increased cost of material and labour. This learning curve method is very useful for negotiating the price increases by isolating material costs and applying the learning factor for labour intensive jobs. This method is particularly useful for negotiating works contract and labour contract with subcontractors. The logic is that the subcontractor is gaining experience at the expense of the purchaser. It is therefore natural for the buyer to benefit from the learning process.

Sometimes scientific-minded vendors also plot their own learning curves for labour intensive activities and use them in negotiation process after giving the breakdown of materials cost, labour cost, overheads, and profits. The learning curves of some labour intensive jobs are more advantageous than the quantity discounts offered. Before accepting quantity discounts, it is necessary to compare the vendor's quantity discounts with the benefits to be obtained by using the learning curve. By repetitive application of learning curve, the quality of the item improves.

The learning curve originated in the American aircraft industry is now being applied to other labour-intensive, assembly-oriented operations. Because the curve shows the relationship between the number of units produced and the labour hours needed to produce them, it can be used to forecast future labour costs and hence the cost of bought out items.

In the American context, learning curve has also been used in forecasting labour hours for new products. The concept of improvement occurring in any repetitive situation can be successfully used for negotiating price with the vendors as well as for production scheduling in vendor's premises and for make-buy decisions. Some American firms have claimed a benefit of cost reduction of up to 30 per cent by using the learning curve. The application of learning curve philosophy results in the simplification of complex problems.

Limitations in Use

The learning curve should be used when (a) proprietary factors limit competition, (b) tooling is expensive, (c) specialised know-how is involved, and (d) direct labour is a substantial portion of the product cost. The learning curve enables the buyer to get a complete breakdown of the cost structure, which is necessary for negotiation purposes.

It is not desirable to apply the learning curve to an estimated price as the errors will be magnified in the original estimate. It is dangerous to apply one curve universally for all items. Application of learning curve is recommended only for non-standard items. The first purchased standard items cannot be produced first, and hence cannot be given the 100% time value, as many similar identical parts have been made and sold to other users. Thus, the learning curve is an important tool not only for cost reduction in labour intensive products, but also for improving buyer-seller relationships as facts are known to both parties.

TEN POINTS TO PONDER

1. Describe the learning process in any industrial situation.
2. Define the concept of theory of learning.
3. Explain the significance of the learning factor.
4. Identify the process of obtaining a numerical value for the learning factor.
5. Elucidate the mathematical equation of learning curve.
6. Clarify the methodology of drawing learning curve.
7. Illustrate the uses of learning curve.
8. Examine the scope of learning curve as a powerful tool for cost reduction;
9. Amplify the misuses of learning curve.
10. Compare and contrast the benefits of quantity discounts with the advantages of learning curve.

19. Value Management

SCOPE OF VALUE ANALYSIS

The quest for value has continuously found expression in new methodologies and techniques. According to the society of Indian Value Management, value analysis and value engineering, practised in Indian organisations earlier have graduated into the value management concept today. The concept of value analysis due to L.P. Miles of General Electric in America has gained ground as a useful device to identify the requirements of a product in the form of well-defined functions and assess the worth of product vis-a-vis cost. Through this effort, the desired performance is obtained to the satisfaction of the user at a cost that he is willing to pay.

VALUE ANALYSIS

Value analysis is defined as an organised creative approach aimed at identifying and eliminating unnecessary costs, without affecting quality, safety, performance, guarantee and functional utility. Value analysis is concerned with ascertaining whether the material or item purchased is good value for the money for the required purpose and end use. *Value engineering* is the application of the value analysis cost reduction technique at the design stage itself.

Value analysis investigates whether the buyer is obtaining the most economical material or equipment suitable for the end product or end use. The approach is concerned less with price negotiating and bargaining, than with the right quality, design, specifications, standards and methods of manufacture. It naturally involves the substitution of material or component of better quality, or obtainable at lesser price more easily.

The concept of value analysis is not new. It is used most unconsciously by all shrewd purchasers in transactions. An example in this context is the housewife who buys tomatoes instead of green peas for making soup as the price of tomatoes has decreased due to the hailstorm the previous day. This is value analysis in practice. In India, many companies are using value analysis approach as it offers excellent scope for cost reduction and accommodates import substitution as well.

Cost Price and Value

Often the terms *cost, price* and *value* are used interchangeably. The fundamental principle of finance management stipulates that *cost* is an *actual fact*, where *price* is *an option*, formulated on the basis of market conditions. Cost consists of fixed and variable costs, including material, labour, overhead, etc. The difference between price and cost is the profit/loss element. Price is what you pay and cost signifies that it costs to an organisation. But, price and cost signify the same at the user's end, as price is the cost the buyer has to pay to get a product.

Value lies in the satisfaction one receives from the product's performance, quality or appearance. Value management constantly seeks to identify and isolate areas of poor value or high cost and low satisfaction and improve them subsequently. The functional orientation — what the customer requires of the goods — seeks to give him the maximum value for his money. Yet for the producer, it helps to quickly identify unnecessary costs which are disproportionate to their functional worth and seeks to find the least cost combination of designs, materials and processes, which can readily perform the needed functions to the consumers' satisfaction.

OBJECTIVE OF VALUE MANAGEMENT

The concept of value is usually divided into the following classification for better control:

(a) Use or *functional value* signifies the properties and qualities which primarily accomplish the use, work or service.

(b) The properties, features or attractiveness which create a desire to own are known as *esteem value.*

(c) *Cost value* is the sum of labour, materials overheads, taxes and other costs required to produce it.

(d) *Exchange value* implies the properties or qualities which enable us to exchange it for something else we want.

(e) Value is a relative term and changes with place or time.

(f) Hence it can be measured in comparison with other products.

(g) Value relates cost to the function by the equation.

Value = (Function/Cost)

Hence value management asks the question: Is the cost proportional to the functional worth?

(h) It increases the worth by reducing the cost after identifying the unnecessary costs in the system.

TEAM TASK

Value management is at best combination of many proven tools and techniques in management process. It is a systematised and organised common sense approach — but common sense is uncommon. While it can be initiated by individual managers and value specialists, fruitful results are achieved by teams of managers, from different disciplines. The process is through various phases of the value management methodology. These phases are: (a) familiarisation phase, (b) information phase, (c) creation of alternatives, (d) evaluation of alternatives, (e) implementation phase, (f) review and feedback phase.

Value management teams usually work part time and occasionally full time for pressing problems. The team members are drawn from the functions of marketing, design, operations, finance, purchase and warehousing. Many progressive organisations in India are integrating their value management efforts into their annual planning. The key to success in effectively utilising value management technique lies in (a) team work, (b) clearly identifying the function, (c) establishing the target cost for each needed function, and (d) striving towards that target with every creative means known to man. The team makers should have robust commonsense and an intimate knowledge of the process and technology. They should not be like the Central Minister who asked the farmers to grow red chillies instead of green chillies to get more profits.

ITEMS FOR VALUE ANALYSIS

In practice it will be advisable to start with simpler items and progress to more complicated ones. The thinking has to change from performance-oriented to value-oriented philosophy. High value non-critical items, overdesigned items, packaging items, items with least contribution to the company and products with decreasing return on investment are ideal for value analysis. For a number of items, cost reduction by value analysis may not be possible, but with proper indoctrination amongst the different departments, more and more items have to be selected to yield profitable results.

The matrix method of linking emphasis coefficients with emphasis factors is another method used in the industry.

The alternative items and their functional utility with suitable weightages arranged in the form of a matrix by competent personnel in order to arrive at the best contributions is the matrix method.

IQ Questions

The IQ questions to be administered on the value analysis team can be as follows:

(a) Does this item add anything desirable to the end-product for operation.

(b) Is this item the only one suitable for the end use?

(c) Is this item standard as regards specifications, size, shape, type, economic batch, quantity etc.?

(d) Is the cost of the item reasonable in terms of its contribution to the end product?

(e) Does the item meet all the design specifications?

(f) Is the manufacturing method too expensive?

(g) Can the item be profitably manufactured in our workshop?

(h) How can we reduce the cost of material, conversion cost, profits, and the price quoted?

(i) Is the present source of supply the best and who else do they supply?

(j) What does our competitor do for this item?

(k) Can a local manufacturer supply so that transportation cost is saved?

(l) What are the primary, secondary and tertiary functions of the item?

(m) Do I spend like this, if it is my own money?

(n) The other questions consist of what, when, where, why, how, who of the item.

The answers to these questions must be obtained from the best possible source, by avoiding generalities in order to find the best alternative by value analysis approach.

TECHNIQUES OF VALUE ANALYSIS

Brainstorming

The holistic brainstorming technique combines the best of both brain hemispheres, the left brain and the right brain.

In the first four phases of value analysis, namely, information, creation, development and presentation, the left and the right brains are alternatively needed, and during the implementation phase, a combination of the two is used.

Blast-create-refine is the focus of brainstorming where existing ideas are blasted to create and refine a new system. Supplier seminars and customer conferences can serve as a spring board for new ideas.

DARSIRI — Data collection, analysis, recording ideas, speculation on alternatives, review, and implementation — a popular method used in value management EXCHANGE (Eliminate, Keep and Change) and MISS (Modify Improve, Subdivide, Substitute) are the vehicles for introduction of value analysis.

Challenging specifications or putting a rupee value on tolerances is found to be an effective method.

A TO Z PROBLEM AREAS

(a) Like any other scientific method there is always resistance to change in the people responsible for implementation,

(b) While sound common sense is the basis for value management, it is not so common and execution of this method becomes difficult.

(c) Due to normal preiudices of people creative thinking gets prohibited.

(d) People tend to rush into judgments before considering all the alternatives.

(e) Creative ideas are often nipped in the bud by continuous interruptions.

(f) Members of the taskforces need infinite patience, imagination, flexibility, and tenacity.

Supplier Value Analysis

(g) Cost reduction can be achieved only when suppliers of components also apply value engineering concepts.

(h) Suppliers are apprehensive of mentioning their efforts to the consumers, lest they demand a share of the cost reduction.

(i) The top management's support may not be fully forthcoming for the efforts of value engineering.

(j) Every one praises the technique with a proviso — "but it will not work in our place."

(k) Lack of ideas for viable alternatives is an usual excuse.

(l) Classifying their industry as high tech, too big or too small are the usual excuses for not giving a trial to innovations.

(m) Overselling value analysis as a panacea for all ills of the company sometimes produces a negative effect.

(n) Individual lack of motivation, particularly in procedure oriented organisations, disallows value analysis.

(o) Some managements feel that the workers will go on strike if the benefits of value analysis are not shared with them.

(p) Managements firmly believe that conformance is best and are afraid to change established systems.

(q) Some companies abdicate their responsibility and feel that the competitor should give a lead in cost reduction.

(r) FEAR (Failure to understand, Experience, Anticipate difficulty and Resistance to change) is a major problem.

(s) Many profit-making companies feel that it is not the time to think about cost reduction when the going is good.

(t) It is easy to talk about team work, but difficult to execute it in practice. Educational facilities for value engineering are non-existent in India.

(u) Communication and collaboration between different departments within the company, as well as with external suppliers and customers, need to be improved for creating the climate for value analysis.

(v) The tendency is to anticipate more difficulties and kill the idea at innovation stage, before giving a trial.

(w) Questioning the specifications which is the primary step for value analysis is opposed by the technical people on grounds of safety and guarantee.

(x) Often it is difficult to measure the tangible benefits of value analysis in financial terms.

(y) The external government and beauracratic controls on administered price of inputs and outputs act as a major deterrent in many cases.

(z) In spite of these drawbacks, there has been growing awareness of value analysis in India during the last three decades due to the efforts of the Society of Value Engineering and the Society of Value Management. The concept of discovering lower cost substitutes to improve the performance and function of the original products is badly needed in a developing economy like India and hence value management is bound to succeed in the years to come.

TEN POINTS TO PONDER

1. Discuss the scope of value engineering, value analysis and value management.
2. Identify the benefits of value management to an industry.
3. Examine the meanings of price, cost and value.
4. Illustrate primary, secondary and tertiary values by means of an example.
5. State the objectives of value management.
6. Discuss the organisation for value engineering.
7. Describe the concept of brain storming.
8. Identify the techniques of value engineering.
9. Analyse critically the problem areas in the application of value analysis.
10. Elaborate on the items that are to be taken for value management in a company.

20. Right Incoming Quality

SIGNIFICANCE OF QUALITY

Today, most industrialists are following the motto: 'Manufacturing and marketing of goods/services of the highest quality at reduced cost.' Quality control principles, aided by statistical techniques, aim at improving the product quality and decreasing the production cost by concentrating on high standards of quality and controls which ensure that the standards are always maintained.

Quality control techniques are based on the principle that quality should be built into the product and it cannot be inspected into the item. It aims at detecting defects at inception and not at the inspection stage. The word *'quality'* denotes the quality characteristics involving the measured and attribute types. Examples of measured characteristics include height, weight, thickness, diameter, volume, fuel consumption, stores consumption, yield per cent, etc. Examples of *attribute characteristics* include number of defects per item, defective pieces, number of blow holes in a casting, number of mistakes in a typed sheet etc. wherein we classify the item as good or bad.

RESPONSIBILITY FOR QUALITY

In most organisations, an independent quality control department, reporting to the top management, is incharge of inspecting the materials at the incoming stage, in process stage and evaluating the outgoing quality. The warehouse manager is incharge of quantity accounting while the purchase manager looks after contractual obligations of the invoice. However, a major complaint of the user against the materials department is about the substandard incoming material quality. The user expects best quality and buying by brand name; the buyer is interested in right quality. The problem arises because procurement is done by the materials department. The warehouse incharge, however, must be acquainted with the procedures of incoming quality control.

The buyer's job ends only after he delivers the right quality material to the user. It should be noted that the rejected materials accumulate in the warehouse as the supplier thinks that they will be used in times of real emergency by the organisation. The warehouse manager has to ensure the early removal of the rejected items.

OBJECTIVES OF INSPECTION

Quality control inspection is undertaken in industry in order to ensure that products conform to the specified or pre-established standards. Such inspection is carried out in the purchase of raw materials, in the inter-departmental flow of materials, and in the final acceptances of finished goods for distribution. Such inspection helps to prevent substandard items reaching the next user department, and the final inspection prevents defective items reaching the consumer. Through this process, it becomes possible to detect the various sources which cause defective production and take appropriate remedial measures. Inspection of incoming materials is usually done by acceptance sampling procedures and the same principles can be applied for the outgoing quality evaluation, giving quality assurance to the consumer.

ACCEPTANCE SAMPLING

All kinds of inspection, even counting the items, cost money. Regarding incoming materials, the following three alternatives are available for the buying organisation:

(a) Accept the lot without any inspection whatsoever. This is done when preshipping inspection has been done at the suppliers' end or in the case of standard items from reputed manufacturers.

(b) Inspect each and every item of the lot. But this 100 per cent inspection may not be very efficient due to fatigue. Further, this is not applicable in the case of destructive items. The time and cost will also be prohibitively high. Hence this alternative is ruled out.

(c) The third alternative is to choose a random sample or a portion of the lot and inspect only the items included in the sample. The lot is accepted or rejected on the basis of sample inspection. The lots which are rejected are sent to the supplier. Sometimes the defective items are weeded out from the rejected lot, by subjecting them to cent per cent inspection, as per the agreement with the vendors. This decision alternative of accepting, rejecting and rectifying the lot, by means of sampling schemes, is known as *acceptance sampling method.* The object is also to cumulate information about the manufacturing process of vendors, which will be fed back to them later.

QUALITY SPECIFICATIONS

The right quality is usually indicated by specified tolerances. Often indents do not contain all the needed specifications, leading to unnecessary delays in procurement decisions. Specification by brand names must be avoided as it forces many items as proprietary single source items. Specifications must also be built while developing internal codification in the organisation. Internal codes can replace detailed specifications in a purchase order.

Adequate care should be taken to ensure that the specification is appropriate to the functional requirements. Overspecification in terms of tighter tolerances can cost the organisation dearly. Underspecification leads to problems of wastage and increased consumption. The physical and chemical properties of specified raw materials must take into consideration the environmental factors, where the end product is going to be used. For items of repetitive usage, the specifications can be standardised. In the case of spare parts, it is desirable to attach drawings of the part in order to clearly understand the product. When the items are urgently needed, the rectifiable items can be accepted after clearance from the user and the charges debited to the vendor. Specifications can also be expressed in terms of accepted and understood standards like the Indian Standards or Standards of the American Society for Testing Materials or the Indian Pharmaceutical Standards etc.

SAMPLING SIZE

The entire lot should be judged, based on the evidence provided by the sampling inspection results. Hence, a common problem in the context of acceptance sampling is the number of items to be inspected. Many industries check a fixed per cent, say 1, 5, or 10 per cent of the lot, for arriving at a decision of accepting or rejecting the lot of items. Such *fixed per cent sample* involves heavy inspection cost for lots of large sizes, without commensurate benefits as increasing the sample size beyond a particular size does not yield proportionate additional information. The effort is not commensurate with the gain in terms of protection to the buyer or the consumer.

For small lot size, it is not uneconomical to inspect all the items rather than depend upon the results of inadequate inspection based on fixed per cent sampling. Let us consider a pack of 52 playing cards in this context as the incoming lot of goods and picture cards as defective ones. We know that there are four sets of king, queen, jack, or twelve picture cards, or the lot is 23 per cent defective. But experience indicates that a 10 per cent sampling or a hand of five or six cards on 35 per cent occasions — does not contain any of the picture cards. This forces us to conclude on the basis of 10 per cent sampling that the pack is free from any picture cards. Thus, whenever the lot size is too small or too large, fixed per cent sampling is not helpful.

Further, the sample size depends upon which characteristic is measured. In the case of measured characteristic, we incur more cost and get more quantitative information of the deviations from the specifications. Hence, we need samples of smaller sizes compared to the attribute types of inspection, where the items quality is assessed merely as good or bad. Statistical methods are available to assess the reliability of tests based on sampling schemes and are useful in fixing the sample size, depending on the desired accuracy of the estimate.

CONSUMER AND PRODUCER RISKS

The decision of acceptance or rejection of the entire lot is taken on the basis of a sample. As a result of sampling, it is likely that some good lots are rejected and some bad lots are accepted. The risk that a lot of items of acceptable quality will be rejected by sampling as a result of a 'pessimistic looking sample', is known as the *producer's risk* as the producer is the loser. Similarly, the consumer's risk is defined as the risk that a lot of items will be accepted by the sampling scheme on the basis of an 'optimistic looking sample'. The consumer requires to be protected against the acceptance of an unreasonable number of inferior lots just as the producer needs to be protected against rejection of an unreasonable number of good lots. Therefore, appropriate precautionary measures need to be taken. We shall now discuss some of these precautionary measures.

OPERATING CHARACTERISTIC CURVE

In judging various acceptance sampling plans, their performance is compared over a range of possible quality levels of submitted product. A picture of this performance is given by the *operating characteristic curve*, called the *OC curve.* This curve shows the long run percentage of submitted lots that would be accepted by the given sampling plan, if many lots of any stated quality were submitted for inspection. The protection afforded both to the consumer and the producer under any plan can be judged by the performance over a range of possible quality levels of submitted lots and the operating curve of a particular plan gives a picture of its operation. On the horizontal axis, the lot quality per cent defective is plotted. The vertical axis indicates the proportion of acceptance of the lot, calculated with the help of Poisson distribution. The steepness of the curve gives a measure of discrimination between good and bad lots. The nearer the curve is to the vertical line, the higher the discriminatory power. The larger the sample size for a given acceptance number or the smaller the acceptance number for a given sample size, the steeper the operating characteristic curve. Thus the operating characteristics curve enables a comparison of performances of different sampling schemes so that an appropriate one could be selected.

ACCEPTANCE/REJECTION/RECTIFICATION

In the case of attribute characteristics, the quality of the lot is usually expressed as a percentage defective or average number of defects per item. For measured characteristics, the quality of the lot is expressed in terms of the mean and standard deviation of the measured characteristics.

The usual practice is for the buyer and the vendor to specify the good/acceptable lot, as well as the bad/rejectable lot, in terms of percentage defectives, known as Acceptance Quality Level (AQL) and Lot Tolerance Per cent Defective (LTPD). This implies that the buyer/consumer is willing to accept the lots whose quality is superior to a level of quality, called the *acceptance quality level*. Similarly, the buyer would like to reject lots inferior to a certain level known as the *lot tolerance per cent defective*.

If a lot rejected by the sampling scheme is merely sent back to the producer, the sampling scheme has little effect on the material and sometimes the supplier will even send the same lot to try his luck. Further rejection of the entire lot is not practicable, when sampling is in operation between different departments within the same organisation. The usual practice in such a situation is to screen the rejected lots, item by item, and get the defective items replaced by good ones. This rectification process is possible when inspection of the entire lot is economical and the test involved is not destructive.

The average per cent defective or number of defective items in the accepted lots after rectification of the rejected lot is known as the *Average Outgoing Quality* (AOQ). Whenever the incoming quality is good, the outgoing quality will also be good. But if the incoming quality is bad, a large number of lots will be rejected and replaced by good items, and hence the average outgoing quality is again good. In between these two extreme situations, a limiting value or peak value of the average outgoing quality is obtained. This limiting value is known as the Average Outgoing Quality Limit (AOQL).

MILITARY STANDARD 105 D PLANS

Based on one or more concepts of AQL, LTPD, AOQL, consumer risk and producer risk, which are stipulated by the buyer and vendor on the basis of market conditions, several sampling plans have been published. Military standard 105 D plans, developed by the American Department of Defense, is based on the AQL value. The defects are classified as the *critical major* and *minor* so that different AQL values are chosen for each category. Tables are available for AQL values ranging from 0.15 per cent to 1000, the former to indicate the per cent defective and the latter to include defects per 1000 items. Provisions for altering the sampling scheme to become tighter in the case of poor performance by the supplier and to reduce the amount of inspection, when the performance or process average is better than the AQL, are inbuilt in this scheme. IS-2500 part I has been similarly developed by the Indian Standards Institution (now called Bureau of Indian Standards).

DODGE AND ROMING PLAN

Dodge and Roming Plans apply strictly to schemes that submit the rejected lots to 100 per cent inspection. A knowledge about the average fraction defective, estimated by the average per cent defective in the incoming accepted lots during recent past, defined as the process average, must be had for using these tables. The plans are based on AOQL and minimise the average amount of inspection for a given AOQL value. Plans are available for AOQL values ranging from

0.1% to 10%. The corresponding LTPD values are also indicated. Similar plans, based on LTPD values ranging from 0.5–10 per cent have been worked out to minimise the amount of inspection for a given LTPD value and a specified process average.

BOWKER AND GOODE PLANS

These plans are intended for measured characteristics. The items are measured for comparing as to whether they meet the specifications or not. Compared to attribute plans, the number of items to be inspected is much smaller, as the variable inspection yields more information on the lot quality. The measured item quality has either an upper specification or a lower specification or both, while the lot quality is expressed in terms of per cent defective. The two cases, when standard deviation is known and not known, are distinguished. The plans are specified in terms of two numbers — n and k — n corresponding to sample size and k a factor to be used in judging the acceptability of the lots. IS 2500 part II has been developed by the Indian Standards Institution for variables.

In all the above cases, the concept of acceptance — rejection, or acceptance — rectification schemes can be extended to multiple sampling plans, where the decision-making process is postponed to one more stage.

QUALITY CIRCLES

In the context of materials management and right quality, quality circles could play a useful role. The quality circle is a small group which voluntarily tries to improve the performance in a regular basis, within the organisation. Quality circle aims at self development, motivation, improvement and control. It is usually a small group about 10 employees, voluntarily searching for solutions of the organisation's problems, particularly for those encountered in their own sphere of work. It is a regular voluntary meeting or a small group of people interested in the overall performance of the materials department. Quality circles enable the people to improve the significance of their jobs and build pride and a sense of belonging to the organisation by making suggestions for improving the overall materials management activity.

TEN POINTS TO PONDER

1. Discuss the significance of quality of materials in the Indian context.
2. Identify the responsibility for quality in company.
3. Describe the objectives of inspection.
4. Explain the concept of acceptance sampling.
5. Evolve a procedure for obtaining the sample size.

6. Define consumer risk and producer risk in the context of using acceptance sampling.
7. Elaborate the significance of OC curves.
8. Examine the meaning of AQL, LTPD and AOQL.
9. Discuss the relevance of published sampling plans.
10. Clarify the concept of quality circle and its applicability to the materials management function.

21. Waste and Scrap Control

CONCEPT OF WASTE

In every Indian city and town, we find persons gainfully employed in collecting and processing scrap and wastes of every kind. This scrap industry is a big business indeed. "Waste not want not" goes an old saying. We observe wastages in every walk of life. Some of these are as follows:

(a) Waste of water
(b) Waste of electricity
(c) Use of fire wood (contributing to deforestation)
(d) Use of cowdung as fuel instead of generating energy from it
(e) Conspicuous left-over food in dinner parties
(f) Waste of time in offices
(g) Under-utilisation of skilled personnel by delegating them for unskilled jobs
(h) Waste of space.

It has been estimated that in India rats, birds and insects eat up to 10 per cent grains which can convert the deficit into surplus. According to the Food Processing Ministry, about Rs. 5000 crores worth vegetables, flowers and fruits perish annually due to the absence of linkages from growing fields to the market, improper warehousing, inadequate cold storage and lack of appropriate timely transport facilities to the consuming centres.

In the industrial sector, wastages arise because of overspecification or underspecification. The process of overspecification forces us to purchase costlier items, while underspecification leads to a higher percentage of defectives causing frequent replacements. This necessitates a value analysis approach so that the right quantity can be purchased. Wastages occur because of overbuying, wrong identification of requirements, unnecessary buying, inaccurate planning etc.

LIMITED RESOURCES

It is well known that the total resources in this world, like those of a company, are very limited necessitating the optimum use of available resources. The present guesstimates of the availability of major metallic raw materials are given in Table 21.1. These are likely to change if more discoveries are made or the consumption pattern alters. They are the known global reserves that are non-renewable. The number of years these resources will serve mankind on the basis of present consumption levels is also indicated in the Table. Obviously, the realistic future consumption rate will be much higher as the world population and industrialisation grow, thereby reducing the time these known resources can serve us.

Table 21.1 Metallic Global Reserves

Type of Resource	Known Global Reserves	Expected Years of Consumption
Aluminium	1170 million tons	100
Chromium	775 million tons	420
Coal	5 million tons	2300
Cobalt	4800 million tons	110
Copper	308 million tons	36
Gold	353 million tri ounces	50
Iron	100,000 million tons	240
Lead	95 million tons	40
Manganese	800 million tons	98
Mercury	535,000 flashs	30
Molybdenum	10,800 million pounds	85
Natural gas	1,150 cub. ft.	50
Nickel	15,000 million lb	150
Petroleum	1 million million barrels	50
Platinum group	450 million tri ounces	150
Silver	10,500 million tri ounces	50
Tin	550,000 lg tons	30
Tungsten	3000 million lb	50
Zinc	300 million tons	30

From Table 21.1 we can see that we do not have unlimited resources and hence must not waste the same. We owe it to posterity to preserve the meagre natural resources. We have two options before us: (a) either we can discover new deposits by extending our mining efforts to virgin soil and ocean depths of this planet as well as other possible areas of the solar system; (b) or alternatively recycle the wastes. The former proposal is highly capital-intensive and can turn out to be technologically problematic. Further the risk of not striking a new source is always there. The latter course of action is well within the scope of present day researchers, scientists and technologists. Organisations should ensure that the materials are used in an optimum fashion without unnecessary waste.

A TO Z ASPECTS OF SORTING OF SCRAP

In Industry, scrap usually originates as a result of waste or as by-product of the operations in the production shops or foundries. Certain kinds of scrap are also procured to be used in foundry, blast furnace, open hearth furnace, electrical furnace etc. These categories of scrap include the following items: (a) wrought iron steel, (b) steel scrap, (c) black steel sheet scrap, (d) galvanised steel scrap, (e) wrought iron turnings, (f) iron borings, (g) machine shop turnings, (h) shoveling turnings, (i) plate scrap, (j) punching scrap, (k) cut structurals scrap, (l) hard steel, (m) unstripped motor blocks, (n) copper bearings scrap, (o) brass scrap, (p) copper turnings, (q) brass rod turnings, (r) brass clippings, foil, (s) old scrap zinc, (t) tin scrap, (u) scrap lead, (v) battery plates, and (w) nickel-chrome stainless stray, (x) used empties, old packing material, used paper and other waste material have also to be disposed of, (y) scrap has to be sorted according to the type of alloy, grade, size, greasiness, cleanliness, weight, etc. (It is necessary to avoid mix-up, whether the scrap is to be used in the plant or disposed of outside.) (z) If necessary, sorting can be done by magnetic testing, spark testing, chemical spot testing and spectrographic testing.

WASTES IN WAREHOUSING

One of the prime sources of wastage is spillage of materials which are transported in bulk. In the case of volatile elements like petrol, wastage is in the form of evaporation losses. Improper handling can lead to breakages, thereby resulting in waste. Pilferage of materials in transit and during storage is a common feature leading to wastage. During storage, wastages and deterioration in quality occur, due to improper storage facilities, corrosion, improper use of preservatives, humidity, dust, excessive heat, cold, rain and other environmental factors. This is particularly true in the case of electronic items and delicate instruments. While we may not be able to eliminate the losses completely, attempts should be made to control the waste through the application of scientific warehousing methods. Inventory carrying charges of waste and scrap — including cost of capital and storage charges, is about 30 per cent per year.

CONTROL OF OFF-CUTS

It is essential to develop good control procedures on production planning and material utilisation which can reduce the losses from waste and scrap. The problem in material utilisation is in specifying the minimum material for the specific final item. For example, since the weight of a round bar varies as the square of the diameter, the use of stock larger than necessary is more wasteful than apparent. This implies that if a part requires 1 cm diameter and 11/10 cm is used, then the material converted to scrap is roughly 10 per cent higher, although the diameter is only 6 per cent greater.

In the engineering industries, we often observe that due to non-availability of the correct size of sheet plates and rods, bigger sizes are used, leaving the remaining off-cuts as waste. These off-cuts can be utilised for smaller components. Linear programming approach can help us to reduce the incidence of off-cuts even at the planning stage.

For instance, let $X_1, X_2, X_3, \ldots, X_n$ be the number of cuttings according to plans $p_1, p_2, p_3, \ldots, p_n$ respectively. Let $L_1, L_2, L_3, \ldots, L_n$ be the loss for each plan. Hence the total loss is given by $L_1X_1 + \ldots + L_nX_n$, which should be minimised. Let C_{ij} be the number of pieces if width W_1 is obtained from plan 1. This should equal our requirements, which are $R_1, R_2, \ldots, R_n$. Hence we have to minimise the loss function ΣL_1X_1, subject to $\Sigma C_{ij} X_i \geq R_i$. This is the traditional linear programming approach to be solved by the simplex method.

TECHNICAL AND MANAGERIAL RESEARCH

Industrial wastes form only a fraction of the material purchased by the company, and do not have any end-use. Wastes may be gaseous, liquid or solid, but whatever their state, they cost money for disposal. Whenever research can turn a waste into profit-making stable product, the gain is double, since the expense relating to waste disposal is eliminated. Frequently, through research a waste is turned into a by-product that becomes a profitable co-product of the operation. In this context, ethylene glycol is an example as it was considered as waste some time back, but, presently it is a premium antifreeze motor coolant. Many highly integrated chemical and metal processing operations have been made possible through the finding of commercial values for what were originally waste products. Hence research activities at the micro level and national laboratories at the macro level must consider utilising the waste by-products into input raw material, particularly in our developing, capital scarce economy. For this purpose, it is desirable to have a reclamation and refurbishing programme at the unit level. The records have to be periodically examined with regard to rejections, movement of materials, scrap disposal and waste accumulation.

Thus, the solutions are partly technical and partly managerial. Technical solutions such as adequate utilisation of material, better design, proper handling and improved warehousing enable one to reduce the wastages. These require a longer time for proper implementation. Managerial solutions concentrate on the proper analysis of requirement, innovative organisation, better control of consumption and a well-designed information system with inbuilt checks on the bottlenecks. Both the solutions must be implemented to a time schedule on a war footing.

Suggestion schemes have been implemented in a few organisations to reduce the incidence of scrap and waste. It is also necessary to introduce norms for each category of scrap/waste at every generating point, with a view to reducing the scrap. "A problem well identified is half solved", goes an old saying; hence identification of causes can reduce scrap and waste.

DISPOSAL OF SCRAP AND WASTE

In many organisations, waste and scrap disposal is attached to warehouse manager as he controls the scrap yard. Production control generating the scrap should be associated with the disposal action. Two basic questions have to be answered in this context before disposal action is contemplated: (i) Does the scrap generated have any value? (ii) Is the value significant enough to warrant process classification? For example, copper scrap has a high market value, sometimes reaching 80 per cent of the selling price of the base product. Hence it is economically sound to segregate this scrap and prevent its contamination. It is also essential to keep a proper account of scrap generated by the different sources.

Disposal follows when scrap and waste cannot be utilised within the plant. Before disposal is initiated, the feasibility of use by sister organisations, returning to the vendors, selling to outside companies or perhaps donating to a charitable organisation may be explored. When the scrap is mixed, the price will be lower than the lowest element in the mixture. When costly scrap such as copper, aluminium, tungsten are involved, it is absolutely essential to segregate the scrap and waste, according to various grades. Since scrap is generated processwise, it comes out in a segregated condition and there should be no difficulty in sorting.

Disposal of scrap is indeed a tricky business. On the one hand, the warehouse in charge has to deal with a set of hard bargainers, and at the same time, he has to face audit queries with regard to pricing. It must be mentioned that scrap market is very erratic. Hence one has to do adequate preparation and planning, before initiating disposal action.

Many methods are used for disposal. Two methods frequently used are open auction and tender, periodically, once in three months or so. Clearance depends upon the size of scrap yard and accumulation of waste. Parties in both cases are normally required to inspect the scrap in the yard and deposit earnest money. Very often the company, based on its experience, insists on a basic price, depending upon the category of scrap. In some cases, the disposal section enters into a long-term contract with industrial end users like foundries or steel plants.

The contract procedure and legal formalities adopted in purchasing are applicable in the sales of scrap/waste also.

Many companies have found to their dismay, components scrapped by them appearing in the market and competing with their products. Automobile spare parts and bearings are prone to such dangers. This is the price which organisations pay for not dismantling and disfiguring the scrap before disposal. For this purpose, a few organisations go to the extent of requesting vehicle users to demolish filters and spark plugs before scrapping them.

Thus disposing, salvaging and utilisation of scrap amount to an art. No amount of theoretical inputs can substitute man's intelligence which has the knack of converting waste into money.

TEN POINTS TO PONDER

1. Identify the concept of waste by giving examples.
2. Examine the necessity of optimum resource utilisation.
3. Classify the different varieties of scrap.
4. Enumerate the wastes in warehousing.
5. Explain the procedure of off-cut utilisation.
6. Describe the technical solutions for scrap/waste reduction.
7. Discuss the managerial approach to reduce the incidence of wastes.
8. Elaborate on the concept of recycling.
9. Examine the problems of disposal of waste and scrap.
10. Name the departments to be associated in waste/scrap disposal.

22. Right Lead Time

CONCEPT OF LEAD TIME

"Justice delayed is justice denied" is a common complaint of most customers/users. They invariably complain to the warehousing, purchasing and materials department that the items which must have been delivered are yet to arrive. In this context the users must appreciate the concept of lead time of an indent to receipt of material. There is bound to be some delay between the identification of the need and the satisfaction of the need. Thus lead time is defined as the duration that lapses between the recognition of the need for an item and its fulfilment. This lead time is obviously more in the case of a procedure-oriented organisation, specially manufactured items, imported material, and when hand-to-mouth buying is practised. Inventory is more, if the lead time is more. Hence, it is necessary for the materials manager to reduce the lead time. The time at which purchases are made determines the price level for an item. If the material is not delivered at the right time, stockouts, delays and loss of usage result.

RIGHT TIME

The right time for procurement of an item is said to be the point of minimum stock, the point at which materials are expected to arrive in the store house. The responsibility for procurement of materials lies with both the materials control and purchase departments. The materials control department sends the requisition to the purchase department, as soon as the inventory touches the reorder level. Then the purchase department takes immediate steps to replenish the stock by calling quotations, approving one of them, choosing the supplier, pricing the orders, following up so as to procure the material by the time the inventory reaches the minimum level.

LEAD TIME CONTROL

Lead time is defined as the time elapsed between the receipt of a requisition and the receipt of goods required. It includes the time to understand the need, time to obtain quotations, time to place an order, time for the vendor to manufacture, time for transportation from vendor's plant, and time for inspection to approve the item. Since there is a direct relationship between inventories and the lead time, it is necessary to reduce the lead time with the help of an organisation and methods study. If the powers of delegation have not been increased to cope with high inflation, then approval of higher authorities may be necessary to trigger off purchase orders, thereby increasing the lead time. It is necessary to study the four components of the lead time, namely internal administration, manufacturing, transportation, and inspection, with a view to reduce the total lead time.

ADMINISTRATIVE LEAD TIME

The *administrative lead time* covers the duration from recognition of need till the purchase order is triggered off. Identification is easier for repetitive items for which the stock level could form the basis. The increase in this component will be due to wrong specification, correspondence with the users, discussions, low powers of delegation, advertised tenders, meetings, retendering, import licensing procedures, approvals, financial concurrence, etc. Conversion of an indent into a purchase order, in some procedure-oriented, central purchase organisations takes up to 24 months. This can be drastically reduced by critically reviewing the pre-purchase systems. Other steps are as follows:

(a) Specifying the need of the items unequivocally.
(b) Choosing the right procedures.
(c) Defining the negotiation process clearly.
(d) Buyer using his authority without referring to his bosses.
(e) Allocating adequate powers of delegation.
(f) Ensuring less file movement.
(g) Holding weekly meetings with user and finance.
(h) Standardising contract formalities.

In these days of zero inventory and just-in-time inventory, organisations will do well to reduce this internal administration lead time by increasing the delegation of powers and improving the systems.

MANUFACTURING AND TRANSIT LEAD TIME

These two components of the lead time are outside the control of the buying organisation. If it is an off-the-shelf item, then the *manufacturing lead time* is zero. But if it is a huge casting or forging or a jobbing industry, then it takes a long time. There are only a few good fabricators in India. Some suppliers like the Heavy Engineering Corporation take a long manufacturing lead time to supply. This lead time component also depends upon the production cycle time for manufactured items. Information at the supplier's works on the progress of the job is necessary to reduce this component.

Transportation lead time or getting the item from the supplier to the buyer's plant is another uncontrollable lead time component. Selecting from the various modes of transport like air, rail, ship, inland waterways, trucks, pipeline, etc. depends upon the item and the urgency of the requirement. Formalities such as sales tax, octroi, customs, entry tax, other levies, retiring of documents, etc. can be done simultaneously to reduce the transit time. Care should be taken to ensure that the items are not pilfered during transportation; some buyers cynically define RPF as railway pilfering force! It may be necessary to send a messenger to accompany costly intransit material. In all the above cases adequate care has to be taken on the follow-up measures.

A TO Z TIPS FOR FOLLOW-UP

(a) In this era of shortages, it is not sufficient for a buyer to issue a purchase order and hope that the vendor would make deliveries on time.

(b) Hence, it is necessary for the buyer to have a system to follow up the purchase orders, to ensure promised delivery obligations.

(c) Many large organisations have a separate follow-up section.

(d) Buying in India is known as purchasing.

(e) The objective of follow-up is to secure delivery in accordance with the delivery promise. Ensure purchase order acknowledgement and follow up if it is delayed.

(f) Obtain delivery status information from suppliers.

(g) Provide such information to the buyer, warehouse and the user.

(h) Keep close liaison with the buyer on changes or modifications in purchase orders and changes in delivery schedules.

(i) Initiate expediting actions on purchase orders with the suppliers as and when needed.

(j) The buyer often sends a copy of the purchase order after a fortnight of placing the order as a routine follow-up.

(k) A printed double postal reply card requesting specific delivery information of the order is usually the first step.

(l) Personal letter, telegram, telephone and personal interview are adopted by the expeditor, as the need becomes more acute.

(m) In the case of machinery manufacturing, it is necessary to follow up at each stage and compare with the promised dates.

(n) One copy of the purchase order is sent to the follow-up section; if supplier's acknowledgement is not received within a fortnight, the follow-up section sends periodic reminders.

(o) When a delivery has been effected, the material-receiving section advises the follow-up section.

(p) Since there are a large number of orders/suppliers/items in any organisation, various follow-up systems are adopted.

(q) The pigeon-hole system stores the copies of purchase orders, according to the date of follow-up.

(r) The filing system, with dates marked on the corner, files the purchase order according to the date for follow-up.

(s) The diary system enables follow-up of orders according to the dates entered in the diary.

(t) The follow-up function may be centralised and located at the headquarters. Fewer number of personnel can do the follow-up activity if the communication is effective.

(u) The decentralised follow-up system exists when an assistant is attached exclusively to an individual buyer for follow-up activity. This system enables closest coordination.

(v) Integrating the buying function with the follow-up function makes the buyer responsible for the total function, even though this may involve more high-priced buyers.

(w) The field follow-up organisation is utilised in times of shortages as an emergency measure. This may help to apprise the vendors on the spot.

(x) Follow-up activities also include aspects such as postponement of deliveries, liaison with transport, liaison with clearing agents, inspection at suppliers' plant, and close understanding of the supplier's problems.

(y) Follow-up personnel should be familiar with corporate long-range purchase objectives, inventory policies, effect of lead time on inventory and coordination with the user department.

(z) A successful follow-up system ensures reduction in lead time and inventories in the organisation, by asking questions like "Can we follow up the costly items from the moment they are despatched till they reach the destination?"

Often hugepiles of material are quarantined at the central receiving section. This may be due to (i) insufficient number of inspection staff, (ii) lack of testing facilities at the warehouse, and (iii) poor communication between the receipt section and inspection teams. The problem is more in large organisations, where the distance between buyer and inspection staff is large. The transaction of satisfying the user is not complete until inspection is over. Hence, organisations will do well to eliminate the *inspection lead time,* by identifying the bottlenecks. Lead time increases enormously, if items are rejected at the inspection stage.

FLUCTUATIONS IN LEAD TIME

The buyer should ensure the delivery of material at the right time by (a) reducing the internal administrative lead time, (b) coordinating with inspection and eliminating inspection lead time, and (c) following up with vendors for minimising the manufacturing lead time and transport time.

Lead time is bound to fluctuate, not only from item to item, but also for the same item over time dimensions, depending upon market conditions. While guesstimates of lead time are enough for low consumption value items, the buyer should have data on minimum average, and maximum lead time for high consumption value, long lead time, critical items. Steps should be taken to reduce all the four components of lead time of these items. The user should also be informed of the lead time of these items only.

RIGHT PLACE OF DELIVERY

In multiplant organisations, it is not uncommon that the supplier delivers the item to the wrong sites or the headquarters which has issued the orders. This may arise due to wrong communication or misunderstanding but transporting the material to the required site not only increases the cost, but increases the lead time as well. Hence, it is desirable to specify unequivocally the right place of delivery in the purchase order itself.

Similarly, the geographical location of the supplier affects not only the cost of transportation, but also increases the delays in the lead time. It is more difficult to communicate and follow up with vendor located at a far off place than local supplier. Accidents, strikes, floods, terrorism, transport bottlenecks, supplier's negligence, clerical errors of buying organisation, communication bottlenecks, etc. are not uncommon. The above factors result in increased lead time, more inventory carrying charges and higher freight. Hence the vendor located in farflung areas should be considered only if he is competitive in all aspects. Thus geographical location of unit as well as the vendors affects the lead time considerably.

TEN POINTS TO PONDER

1. Define the term *lead time.*
2. Examine the necessity of right time of buying.
3. Examine the necessity of controlling the lead time.
4. Describe the term *administrative lead time.*
5. Enumerate the steps to reduce administrative lead time.
6. Discuss the concept of *manufacturing lead time.*
7. Identify the procedure to reduce *transportation time.*
8. Clarify the concept of *inspection lead time.*
9. Elucidate the follow-up procedures in buying.
10. Examine the importance of *right place for buying.*

23. Right Packaging

BUYER'S ROLE

Packaging plays a vital role in the field of materials management, from the view of application of cost reduction techniques. The cost of a tooth-paste tube is more than the cost of the paste. Same is the case with several fashion accessories, cosmetics, toiletry items and some pharmaceutical items. The average packaging cost in the industry may be as high as 26 per cent of the product cost. The buyer is interested in reducing the cost of packing, as he not only buys packing items for selling his company's products, but also pays the vendor for packing the bought out items. Packaging is an integral part of production, marketing and physical distribution.

IMPORTANT ASPECTS OF PACKAGING

A packaging committee, with representatives from different departments may help to reduce the costs.

1. The Indian Institute of Packaging gives guidance for the right method of packing and for reducing the packing cost.

2. At the unit level it is desirable to have a complete packaging list of the following aspects:

(a) Planning and coordination of packaging development activities in the company.
(b) Conducting research with a view to reducing costs.
(c) Developing and engineering associated with packing.
(d) Testing.
(e) Writing specifications.
(f) Selecting and controlling the colour, and controlling the quality of packing components.
(g) Assembling and storing information, samples, etc.
(h) Reviewing of packaging costs by value analysis.

To sum up, the functions of packaging are to contain, preserve, protect, present and offer the case for handling.

Packing Information

The following information is necessary to identify the packing:

(a) Detailed description of the use of the product.

(b) The conditions in which product is used.

(c) Possible packing sizes.

(d) Minimum order size for packaging.

(e) Existing manufacturing and packing system.

(f) Requirement of storage.

(g) Promotional requirements to attract consumers with pleasing features.

(h) Legal aspects such as defective packing, or drug controllers approval etc.

(i) Pharmacological requirements like sterility, dosage, infection and shelf-life.

(j) Labelling on direction of usage.

(k) Physical state, weight, density, fragility, rigidity and surface finish.

(l) Physico-chemical effect of moisture, oxygen, light, flame, bacteria and chemical action.

(m) Odour, flavour, colour, ease to open/close/identify and pilferage proof are other factors.

(n) Corrosion proof.

(o) Minimising damage due to breakage and fire.

TYPE OF PACKAGING

Good packaging is essential for protection of goods against pilferage, damage and deterioration. The degree and type of packing needed depend upon the nature of goods being moved. One of the basic requirements is availability of sturdy and balanced containers. The materials used for containers include cardboard, laminated paper board, plastics, regenerated cellulose films, rayon, nylon, aluminium, corrugated paper board, wood, wood and wire, tin plate, combination of metal-plastic, tetra pack, kraft paper and styrofoam. Internal paddings in the form of rubber, wood shavings, saw-dust, straw, jute, crumpled newspaper, cotton, foam, cellulose, etc. are used to provide anti-shock protection for delicate and high value materials.

Bundling is often applied in the case of sawn timber metal rods and other loose/bulky commodities that can endure exposure. Sacks, bags, crates, cases, cartons, drums, etc. are used, where the risk of damage or loss through exposure is great. Tensional strappings made of wire steel or synthetic fibres are also used to bind the containers for extra strength and stability. The transport packing should withstand journey hazards and be easy to handle.

The weight of each package must be suited to the capacity of the available tools and equipments. Bulky and heavy packages must generally be avoided as they are expensive and difficult to handle. The risk of dropping and damaging the contents of bulky packages is generally high, especially when they are handled manually.

Even though lighter packages are easier to handle, their processing is not necessarily cheaper. They create a need for more movement and loading. The larger the number of packages to be handled, the greater the need for more clerical efforts. Small packages induce pilferage. A midway approach, regarding the size of the package is advisable.

RAILWAY PACKING CONDITIONS

The Indian Railway Administration under section 77(0)(4) prescribes the conditions for both internal and external packing. These packing conditions and special conditions have been notified in the Goods Tariff. It is in the interest of the trade to ensure that the goods are adequately protected from loss or damage in transit and the prescribed packing conditions are carefully observed. All packages must be marked both with individual and railway marks. Labelling of packages as prescribed in the rules is essential. In the case of wagon loads, 10 per cent of the bagged consignments are to be marked. The labelling and marking help the carriage of goods to the correct destination, without any mistake in despatch and render identification of goods easy at destination. If packed in gunnies, sufficient portion must be *left vacant at* the ends, so that handling becomes easy. The articles have to be declared to the railway authority while despatching, as it facilitates easy claims settlement in the event of damages or loss.

CONTAINERISATION

Containerisation has facilitated packing, handling and transporting. A container is a framed metal box with a strong door, floor, panelled sides and roof. It has an exceedingly good water-proof arrangement for the cargo with locking arrangement. The strong steel frames can support stacked fully laden containers, kept one over another in ships and container parking bays. The bottom frame has hollow sections to permit forklifts to move them for loading and unloading operations. They are usually made out of stainless steel, mild-steel or aluminium. The standard size used by the shipping companies in India is 6 m × 3 m. The gross weight, net weight, size, owner's identification marks, country of origin and other details are marked on the container for identification purposes. Insulated containers with fibreglass lining to reduce heat losses and gains are available for transport of temperature-sensitive cargo. Refrigerated containers with an integral refrigeration plant or connected to the ship's refrigeration system for transport of perishable items cargo are also available.

The advantages of containerisation are as follows: General cargo which would be shipped as separate, packed cargo for a destination port can now be grouped and consolidated into unit-loads of containers, which can be handled easily and faster, thereby reducing the loading, unloading time and also avoiding transit losses and pilferage. Packing expenses are considerably reduced as packing cases and crates are dispensed with and light packing material like corrugated board boxes, cartons, thermocole moulds, polyurethene packs, coir moulds, plywood chests are accepted by Shipping companies for stuffing into the containers. Shipping charges are also reduced and only minimum documentation is needed.

LABELLING AND PACKING

Marking packages for easy identification is a necessary requirement to guide the packing to its destination. Marking can be done by branding, embossing, etching, printing, painting, stenciling and other methods which ensure that marking does not rub off easily.

The marking must include the following information:

(a) Name and address of the consignee

(b) Brief description of contents (this may induce pilferage; hence this can be kept inside the container)

(c) Package number and quantity of packages making up the consignment

(d) Date of despatch

(e) Special instructions on shelf-life, handling of glassware, method of lifting explosives and possible damage

(f) Name and address of the issuing warehouse

(g) Labelling of each article for speedy delivery

(h) Location of the address, and not just post box number to ensure safe delivery

(i) Markings of suitable size for easy reading

(j) Dark, bold and distinguishable marking

(k) Black colour markings in general; red colour for caution/danger

(l) Ends to provide caution markings and sides for general markings

(m) Directional arrows to be indicated for handling

(n) Caution marking indicating "fragile-handle with care"

(o) No use of hooks

(p) Keep the consignment in dry and sling position

(q) Cargo be classified as:

(i) inflammable material, (ii) radioactive material, (iii) corrosive material, (iv) poisonous material, (v) explosive material, and (vi) fragile glass material.

A TO Z TIPS FOR PROTECTION

(a) It is necessary that the buyer instruct the supplier to pack the materials in such a way that they are protected from corrosion.

(b) Before packing materials which are liable to deterioration during transit should be treated with protectives, and

(c) Cleaning and drying the surfaces which are to be protected.

(d) Application by brush or spray of anti-corrosive compounds, films, wax or grease.

(e) Use of air-tight containers and dehumidifying compounds such as silica gel.

(f) Wrapping of material in water-proof paper or wax paper.

(g) Packing in water-proof bags.

(h) Lining the cases with water-proof paper or cloth.

Prevention of Damages

Methods to prevent damages are described below:

(i) Use sufficient cushioning material such as straw, paper-wool, wood shavings, corrugated cardboards.

(j) Use sufficient battens and blocks to hold the components in position.

(k) Pack the parts in the same manner as they are fitted on the machine.

(l) Pack delicate parts in individual cardboard boxes or metal containers.

(m) Bolt heavy machinery, motors etc. to take the base of the box.

(n) Suspend fragile items in boxes to absorb in transit shocks.

(o) Use wood and nails of correct sizes.

(p) Glass should be packed between corrugated cardboards.

(q) Threaded portion should be protected with tapes or a film of hot wax or gum.

(r) Instruments susceptible to damage should be packed in cardboard boxes with soft cushioning all around.

(s) When packing small packages in a large case, any vacant space should be filled with paper or soft, filling materials.

(t) Medium size rubber washers and gaskets may be put on a cylinder or corrugated cardboard.

(u) Glass or china jar must be packed in wooden cases with plenty of saw-dust all around.

(v) Thin or narrow parts like gaskets of special type may be fixed to plywood or plank before packing.

(w) Strong packing is to be used when trans-shipment is involved.

(x) Heavy boxes should be reinforced with hoop iron at two ends of the boxes of angles at various corners.

(y) Cases must be marked "This side up"/"Handle with care", etc., so that they are handled properly during loading and in-transit unloading.

(z) The cost and benefit of the packing efforts must be evaluated by a committee before deciding on high cost packing methods.

TEN POINTS TO PONDER

1. Examine the role of buyer in packaging efforts.
2. Discuss the steps to protect items from corrosion.

3. Describe the precautionary measures for protection against damages.
4. Explain the necessity of labelling.
5. Enumerate the factors influencing packaging.
6. Elucidate the concept of containerisation in packaging.
7. Compare the advantages of small-sized packages against large-sized ones.
8. Define the packaging conditions for railway transport.
9. Articulate on the scope of reducing packaging cost in Indian Industry.
10. Identify the documents needed in connection with transporting the packages from a foreign source to India.

24. Materials Handling

HANDLING COSTS

Contrary to the stress on manual labour in the construction of temples and pyramids in the past, today, materials are being handled mechanically everywhere. Warehousing is often considered as the centre of movement and fountainhead of materials in any industrial enterprise. Hence the warehouse incharge can ill-afford to overlook the technological development in this field. The total movement has to be planned and minimised, at the same time maintaining the service level to the satisfaction of the user/consumer.

Materials handling represents about 50 per cent of production — cycle time, an equal amount of the conversion cost, and about 25 per cent of product cost. It is one of the few areas where costs can be substantially reduced. In heavy engineering industries, the handling could go up to 60 tonnes for each tonne produced. Manager must try to integrate handling activities with those of production, packaging and finished goods movement.

A TO Z STEPS OF HANDLING

Any industrial process in materials handling can be considered as a series of steps such as the following:

(a) Packaging at vendor's plant
(b) Transportation from vendor
(c) Unloading
(d) Receiving
(e) Inspection
(f) Warehousing
(g) Issues to operation
(h) In-process handling
(i) WIP inventory storage
(j) Work place handling
(k) Handling within the department
(l) Handling from one department to another
(m) Handling from one plant to another
(n) Handling related to auxiliary functions
(o) Primary consumer packaging
(p) Protective packaging

(q) Loading
(r) Unloading
(s) Transportation to finished goods warehouse
(t) Unloading
(u) Shipping
(v) Record keeping
(w) Selling to customer
(x) Loading by customer
(y) Transportation
(z) Unloading at customer's end

CONCEPT OF MATERIALS HANDLING

Materials handling is concerned with moving the right quantity of materials — raw material, machinery spares, work-in-progress and finished goods — at the right time in a given space. It deals with the science and art involving the movement, packing and sorting of materials mechanically or manually. The movement of materials can be (a) vertical, (b) horizontal, (c) in batches or as a single unit. Materials handling function includes every operation except the actual processing. Handling materials increases only the cost of the product and not the value of the item. Hence materials handling substantially contributes to the hidden cost in industry and is one of the most lucrative areas of cost reduction.

MANUAL HANDLING

There is always a limitation with respect to handling by human effort. Extensive medical research has shown that the average man aged 20–35 years should not be subjected to a spinal disc stress of more than 30 kg/cm^2, i.e., 600–700 kg total load. The maximum permissible loads according to scientific research based on the spinal disc stress factors for different age groups are given in Table 24.1.

Table 24.1 Maximum Permissible Load

Age	Men	Women
16–18	44 lbs	26 lbs
18–20	51 lbs	30 lbs
20–35	55 lbs	33 lbs
35–50	46 lbs	28 lbs
over 50	35 lbs	22 lbs

Besides the limitation in terms of weight, several substances cannot be easily handled manually, necessitating handling by mechanical means. The Indian

Standards Institution has formulated several standards for materials handling process and handling equipments.

A TO Z PROBLEM AREAS

Let us discuss some of the problems encountered in material handling:

(a) Handling in any industry/administration is unavoidable.

(b) Handling increases the cost of product but the value of the product is not enhanced due to handling. This fact is not appreciated by the operating personnel.

(c) The handling equipment in warehouses is usually too old.

(d) The shop-floor handling equipments, after depreciation, are given to the warehouses for handling purposes in some organisations.

(e) A variety of brands is used, leading to excessive storage of spare parts.

(f) There is no proper coordination between handling equipments of shop-floor and warehouse. The equipments are bought without consideration of the movement in the warehouse.

(g) Preventive maintenance schedule for handling equipments not implemented.

(h) The distance to maintenance booth is too far for getting help during breakdowns.

(i) Storage space is either wasted or inadequate for free manoeuvrability of handling equipments.

(j) Bottlenecks exist in the operating systems due to lack of flow.

(k) Breakages occur during handling.

(l) There is a mistaken impression that contract labour is sufficient for material handling.

(m) Waiting time for availability of handling equipments is excessive in some cases.

(n) Handling equipment is not properly utilised for the intended job.

(o) Improper technology is adopted in handling.

(p) The manuals on equipment handling are not readily available.

(q) Ideas to avoid handling operations are not easily available.

(r) The first-in-first-out method of issue is not possible as heavy plate materials are stored in bottom and this results in unnecessary handling.

(s) Poor material handling results in delays leading to idling of equipment.

(t) The warehouse incharge is never consulted before procuring handling equipment, with the result unusable equipments are bought.

(u) Maintenance and after-sales-service are not up to the mark.

(v) It is difficult to procure genuine spare parts for equipments from the original manufacturers.

(w) The original manufacturer upgrades the technology often and the user is starved of spare parts.

(x) Warehouse is integrated with shop-floor for the purpose of handling but the former gets low priority in availability of equipments.

(y) The safety devices and manuals are not usually adhered to.

(z) Obsolete handling equipments of the shop-floor are passed on to the warehouse.

A TO Z TIPS FOR REDUCTION

Let us now discuss how to reduce the handling cost:

(a) Since handling the materials increases the cost and not the value of a product, the ideal goal should be towards minimum handling.

(b) Sequence the operations in the warehouse and the process in a logical manner to ensure smooth handling.

(c) Use gravity whenever possible as it results in conservation of power, fuel and energy.

(d) Standardise handling equipments to the extent possible, as it means interchangeable usage, better utilisation of handling equipment and less spares holding.

(e) Install a regular preventive maintenance programme for material handling equipments so that downtime is at the minimum.

Application of Scientific Methods

(f) In selection of handling equipments, criteria of versatility and adaptability must be an important factor. This will minimise investments in special-purpose handling equipments.

(g) Weight of unit load must be the maximum so that each 'handling trip' is productive.

(h) Workstudy aspects such as elimination of unnecessary movements and combination of processes should be considered while installing a material handling system.

(i) Non-productive operations in handling such as slinging, loading etc. should be kept at a minimum through appropriate handling system.

(j) Magnetic cranes for scrap, movement and loading in furnace, combination of excavators and tippers in mines for ore loading and unloading are examples in this respect.

Need for Coordination

(k) Location of warehouses should be close to major consuming centres and stores near the plants.

(l) Safety is the most important aspect in the design of material handling system. The system designed should be simple and easy to operate.

(m) Wasteful movements, transfers and repeated handling must be avoided by conducting method-study.

(n) Ensure proper coordination through a proper selection of equipments and training of workers.

(o) Synchronise related production operations. This must be planned even at the layout stages.

(p) Use mechanical aids like containers and pallets wherever possible.

(q) Flexibility is the key to any materials handling system and one has to consider future expansion of activities.

(r) Application of scientific management techniques like queuing can be effective in optimal utilisation of material handling equipments.

(s) There should be a constant review of the existing handling system in order to improve the overall performance.

(t) Equipment specially designed for motion should be kept in motion. Continuous material movement is most economical. Ratio of dead weight to payload must be kept minimum.

Organisational Aspects

(u) Proper organisation of materials handling is necessary to ensure adequate coordination and control of handling equipments, so that the utilisation is optimal both in factory and warehouse.

(v) It is interesting to note that the materials handling activity has been organised in many organisations as a line or staff function. In a few cases, a separate department like industrial engineering or design department looks after materials handling activities.

(w) In some multi-product undertakings, a central staff cell serves in a consulting capacity, with day-to-day materials handling controlled directly under the line functions.

(x) There is a pressing need for training professionals, practising executives, fresh personnel and skilled workers in material handling.

(y) The emphasis should be on cutting the operating as well as maintenance costs and at the same time to reduce downtime of men and machine by inviting suggestions for improvement.

(z) Cost effectiveness should be measured by expenses incurred per unit and weight of material handled and improvements can then be effected in the overall material handling.

MANAGEMENT ROLE

The plant management in collaboration with the warehouse incharge, design, finance, and top management may bring out a material handling manual giving general guidelines for easy handling. The following is a list of recommendations submitted by a team sent by the National Production Council, some time ago, which are even now valid.

HANDLING RULES

(a) Materials should be kept as far as possible at a level at which they can be worked upon. Floor level is generally not the most convenient level.

(b) Material should be placed on pallets, trays or platforms.

(c) Distances over which materials are to be handled should be kept as short as possible.

(d) Where distances are long, attempt should be made to handle as large a quantity as possible so as to cut down the frequency of handling.

(e) Layout should be arranged as far as possible for straight flow.

(f) Use of gravity should be made to maximum possible extent in material handling.

(g) Gangways and aisles must be kept entirely clear.

Systems Design

(h) Greater emphasis should be laid on the design of industrial architecture and proper plant layout for reducing material handling to the minimum and for facilitating the use of proper handling equipment, whenever necessary.

(i) The use of belt conveyors where load factor is high, should be encouraged. It is interesting to note that accidents take place during the stationary position of conveyors.

(j) The use of rail cars for bulk material transport by heavy industries must be speeded up to avoid using the rail cars as storage points.

(k) Unitisation, including palletisation of loads of materials such as bricks, aluminium ingots, small packages etc. facilitates handling by forklifts within an industry and the same practice could be extended further if transport carriers as well as railways are equipped with these.

(l) Indigenous manufacture of low-cost, battery-type and engine-driven fork trucks and low trucks should be encouraged for indoor and outdoor short distance haulage.

(m) Low loading trucks, wherein the truck loading platform could be lowered down to floor level and raised again by a hydraulic system can find several applications in all warehouses.

Storage Aspects

(n) Bins or wire baskets or other suitable containers should be designed so as to facilitate easy movement by fork trucks or fork lifts in the stores and the rest of the plant to which material has to be supplied by systems other than conveyors.

(o) Side-collapsible containers are particularly useful, if the empties have to be sent outside to obtain supplies from long distances. Rotating, table-type containers can be used for storage of very small components.

(p) Incoming materials should be received in an area covered by overhead crane or truck lifting facilities to permit unloading of trucks or railroad cars and transfer to storage in a single operation.

(q) In-process handling and handling within the warehouse should be as far as possible in pallets or in suitable containers, wherever direct transfer is not possible.

(r) Pallets or other containers should be standardised to the maximum extent possible.

Integration with Plant Activities

(s) While buying handling equipments, the needs of warehouse, and all parts of the plant must be considered for optimum investment.

(t) Manufacturing operations should be integrated as far as possible, so that the last resting point of the components from one operation becomes the starting point of the next operation.

(u) Considerable economy in handling can be effected by the use of sample handling equipment like barrel, trucks, portable trucks on wheels and various types of simple movable units on wheels, which can be fabricated to suit individual requirements.

National Policies

(v) A committee must be formed with the assistance of competent organisations like National Productivity Council, Bureau of Indian Standards and Professional Associations, to establish standards for performance of materials handling equipment within the country.

(w) This agency should be entrusted with the task of compilation of data of equipment available at present including technical features, capacity limitations, operations cost, etc.

(x) Emphasis should be laid on the introduction of various types of technical courses in materials handling within the existing technical colleges, national, professional associations and industrial associations.

(y) A National Institute on Materials handling should be established with the cooperation of engineers, institutions, associations and government agencies in order to carry out research, consultancy, and training in materials handling.

(z) The institute, should be devoted to the task of better handling methods, must undertake to collect all technical data, knowhow and literature in order to disseminate knowledge on improved handling methods.

TEN POINTS TO PONDER

1. Identify the needs for materials handling in a warehouse.
2. Define the concept of materials handling in a warehouse.
3. Identify the limitations of manual handling.
4. Enumerate the problem areas in handling.
5. Examine the bottlenecks relating to space in handling.
6. Express handling cost in terms of total product cost.
7. Discuss the steps to reduce handling cost in warehouse.
8. Explain the role of different departments in materials handling.
9. Articulate on the process of integrating warehouse with the plants.
10. Illustrate the need of proper policies for material handling.

25. Mechanised Handling Systems

NEED FOR MECHANISATION

Mechanisation of handling may sound ridiculous in a country like India where there is an abundant supply of unemployed unskilled labour. In fact, most of the warehouses, transport organisations and other agencies involved in bulk material handling rely mainly on contract or casual labour. But the improvement in social awareness has made the idea of manual labour unpalatable. Besides, the damages to property and the materials handled warrant the introduction of mechanised storage and handling. One of the major reasons for the slow growth of the Indian economy is the improper handling of materials and the unnecessary costs incurred.

The origin of mechanisation can be traced back to the use of lever principle and implements, such as crow bar, rollers, tongs, hooks etc. The present technology can boast of equipments designed on the basis of movement requirements, types of materials, safety and economy of operations. A large number of equipments is available to move materials up and down, or along a path while the routes can be fixed or variable, straight or curvilinear. We shall now consider the various types of equipments as well as the advantages and disadvantages of using them in the warehouse.

A TO Z INFLUENCING FACTORS ON EQUIPMENTS

A large number of material handling equipment is available and many of them are being manufactured in India. India also exports handling equipments worth about Rs. 500 crores. The choice of the equipment is influenced by a variety of considerations. Let us now identify and list some of the influencing factor. The first activity is to do a survey of material handling needs. The other aspects include the following:

(a) Location of plant/warehouse, (b) product layout, (c) process layout, (d) integration of raw material stores, manufacturing plants and finished goods warehouse, (e) floor structure, (f) transportation logistics, (g) packaging, (h) unit load, (i) industrial relations in the organisation, (j) inspection methods and procedures, (k) location of vendors, (l) location of warehouses, (m) Capital and operating costs of the equipments, (n) desired movement of materials, and availability of process flow charts/string diagrams, (o) material parameters or size, density, viscosity, flow ability, hazardous nature, safety, security, ruggedness, fragility, temperature, angle of repose etc., (p) distance to be travelled, (q) rate per hour/frequency, (r) loading and unloading levels, (s) slope, inclination, and aisle width, (t) cross-traffic sequence, (u) technological obsolescence and standardisation, (v) after-sales service and spares availability, (w) plane of movement location, speed cost of floor space, space constraints, (x) congestion/free flow, (y) location of high tension points, (z) these aspects could be summarised into the questions move what, in what, where, how far, how often, how fast, as what, by whom, by what and why. These questions relate to material characteristics, nature of movement, existing state of handling equipment, cost of equipment, and characteristics relating to noise while operating, speed, life, flexibility, reliability and availability.

COST BENEFIT ANALYSIS: A–Z ASPECTS

The best handling method is to avoid handling altogether but, when handling is necessary, a detailed cost benefit analysis has to be carried out before deciding on the choice of handling equipments. It may sound surprising that Orissa Cement has found it advantageous to use elephants for shunting wagons of materials and material handling purposes because of the cost involved was about Rs. 15,000 per month to maintain an elephant. The use of elephants for handling logs in the Indian forests is well-known.

The costs can be divided broadly into two categories: (a) capital cost, and (b) operating and maintenance costs. Experience indicates that these two costs move in opposite directions. Increased capital costs are usually associated with lower operating costs and vice-versa in a given environment. Suitable remedial measures are necessary if the costs exceed the norms.

The cost depends upon the distance moved, load carried, equipment condition, etc.

In order to reduce the cost, one has to collate the cost data pertaining to the following: (a) demurrage charges (and compare the same with the standards), (b) utilisation of warehousing space, (c) breakage in materials due to mis-handling, (d) inefficient use of materials, (e) down time adequacy of handling equipments, (f) excess maintenance, (g) manufacturing machine down time, (h) accumulation, of raw material/WIP/and finished goods inventory, (i) safety of operating personnel, (j) depreciation of equipments, (k) rate of obsolescence, (l) maintenance cost, (m) spare parts consumption, (n) cost of power and energy, (o) cost of fuel and lubricants used in handling equipment, (p) direct labour cost, (q) indirect labour, (r) departmental supervision, (s) clerical record keeping, (t) training cost, (u) investment in equipment, (v) salvage value of the handling equipment, (w) anticipated after-sales-service charges, (x) cost of documentation and systems. One has to compare the above with the benefits to be obtained by the different equipments before deciding on the purchase of the equipments.

FIXED PATH EQUIPMENTS

Broadly, material handling equipments can be classified into two categories, namely, fixed path equipments and variable path equipments. Fixed path equipments are those which move in a fixed path. Various types of conveyors, mono-rail devices, chutes and pulley drive equipments belong to this category. A slight variation in this category is provided by overhead cranes, which can move materials in any manner within a restricted area by virtue of their design. Overhead cranes have good range in terms of hauling tonnage and are used for handling bulky raw materials, stacking and times palletising. Conveyors are used where loads are uniform, and materials can move continuously as cross-traffic can be by-passed by conveyors. Lifting equipments such as winches and hoists are used where the area of movement is fixed. Elevators as well as screw conveyors play a key role in uncontaminated movement of powder and granular materials in pharmaceutical industry. Mono-rail systems are a highly efficient means of handling loads on an intermediate basis wherever the routes are fixed.

VARIABLE PATH EQUIPMENTS

Variable path equipments have no restrictions in the direction of movement, although their sizes have to be given due consideration. Trucks, forklifts, mobile cranes, skid trucks, gantry cranes, industrial vehicles and tractors belong to this category. Forklifts are available in many ranges and they are manoeuvrable and various attachments are provided to increase their versatility. Industrial trucks are used where materials of varied sizes have to be moved over short distances in a given area. Others in this category include yard vehicles like cranes, straddle carriers, side loaders, power shovels, front end loaders, bulldozers, dump trucks, highway trucks and other outdoor vehicles.

ELEVATING DEVICES

The simplest lifting devices are hand operated pulleys which can be single on multiple or a block and tackle arrangement and can be either fixed or movable. The next stage in mechanisation is the use of electric hoists which are chain type, powered by electronic motors. A combination of these elements, mounted on columns or rails, is the basic constructional feature of jib cranes and hand operated mono-rail cranes. When required for loading or unloading in the open or into trucks, these operate in the form of open gantry cranes.

At times, especially in a production shop, mere elevation is not sufficient. It has to be accompanied by transverse motion. In these cases, the travelling cranes come in handy. They can be either hand-operated or electrically operated. They are mounted on rails and there is a trolley car that can travel in between, which carries the hook and chain arrangements. The loaders and crane operators have to be qualified. The capacity can go up to hundreds of tonnes and these are usually custom built equipments. When there are more than one crane in the same bay, they can be used in tandem to increase the capacity. While hooks are normally used, cranes have also been designed with magnets for ferrous materials, vacuum lifting devices for sheet-glass and grab buckets for bulk material like sand etc. The crane can also be made mobile by mounting on moving equipments. The design and adjustments of the boom with which the range of operations can be varied should be made according to the needs.

ROLE OF PALLETS

A prerequisite for handling materials efficiently is to get proper and firm hold of the optimum weight that can be carried by the handling equipment. But, materials have their own physical properties and they must be transformed into holdable packages without impairing their physical attributes. Palletisation and containerisation are the two common methods employed to achieve this purpose in warehouses.

Pallets are basically metallic or wooden pieces holding the materials. They have a variable number of faces and entries. The simplest example is a wooden box. Pallets are normally used along with forklifts, trucks and the number of entries refers to the way in which the forms can enter. The design of pallet depends upon size or package, material to be handled, weight to be endured by the pallet, type (recycling or not), layout, available space and capacity of handling equipment. It is preferable to standardise the size of pallets. Storage patterns on the pallets are generally of three types — (a) block pattern, (b) brick pattern, (c) pin wheel pattern. In the block pattern of loading, similar blocks of materials are loaded uniformly in a line, but the structure is slightly unstable. Hence it has to be tied. In the brick pattern, the storage in alternate layers is like masonry brick work, while the pin wheel type of loading is for uneven packages.

The major advantages of using pallets are that, the items stored on these can be moved about with ease, incurring low costs. Pallets are useful for packages handled frequently. An additional advantage is that accounting and stock verification of items stored uniformly on pallets are very simple. Large organisations like the Central Warehousing Corporation have adopted pallet system. Bamboo mats are spread on the pallets before laying the rice and wheat bags. The pallets cost about Rs. 100 and can be depreciated in about six years.

CONTAINERISATION

The role of containers in transportation and handling is indeed a major one. Materials like liquids, delicate instruments and small components have to be placed in suitable containers, before they can be transported from one location to another. The purpose of the container is to hold the material, protect it from damages, external shocks, spillages as the case may be. At the same time it should be easy to handle and economical.

Containers are normally made of wood, metal, plastics, PVC, jute fibres, glass, ceramics and rubber. Depending on the characteristics of the material, mode of transportation, distance to be travelled and other environmental conditions, the materials of the container is chosen. For example, to transport naphtha-based fertilisers that absorb moisture, polythene-lined gunny bags are used. Containers should also take care of any hazards that may be intrinsic to the material. Explosives, inflammable liquids, poisonous chemicals, solids, gases, acids, alkalies, and corrosive materials need special protection so that they do not cause any harm in transit. A major purpose of containers is to prevent breakages. There have been many occasions, when due to some leaking valves, wagon loads of oils reached their destination empty. Bhopal gas tragedy due to leakage of MIC is well known.

Containers usually cost more than the pallets and can be recycled effectively. The empties also have a resale value. It has to be decided whether it is worthwhile to get back the empties and reuse them or buy fresh supplies. The factors to be considered are the collection costs, transportation costs and the purchase price. When using recycled containers, it should be ensured that they are of the required quality. A container of full truck trailer size, separate from rolling stock, opens up tremendous possibilities not only for lowering loading and unloading costs, but also for reducing breakage and spoilage. The Container Corporation of India is doing useful work in servicing the consumers.

TRANSPORTING EQUIPMENT

Use of the wheel barrow is the first step in mechanisation of transportation of material from one location to another. With the development of locomotive power, the wheel barrow has given way to battery-operated and power-driven trucks. If the source of power is a diesel or petrol engine, exhaust fumes will give rise to problems within the covered areas of factory and warehouse. It is here that electrically operated trolleys come in use. But, the disadvantage is the frequent recharging of batteries. Efficiency can be improved if rails are laid, but this introduces a restriction in the freedom of movement. Besides powered trucks we can also utilise tractors for transportation. The advantage is that the trailers can be attached and detached as needed and a better utilisation of the costlier prime mover can be achieved.

FORKLIFTS

Forklift trucks, whether manually operated or power operated, perform both the functions of lifting and transportation. These are useful in handling heavy materials, which are in the form of pallets, containers, sillages or unit loads. The lift height can be up to three metres or even more in specially designed models. As the name implies, these trucks have a pair of forks that carry the load. These forms can be moved up or down by either hydraulic, electric or manual lever operations. If necessary, the fork can be replaced by other carrying arrangements such as the ram for holding coils, hook for hanging, scoop and shovel for sand, etc. But these changes have to be done in the initial purchase stage itself. The driverless tractor train can be used to despatch individual loaded cars to remote points where they are all dropped and empties picked up. Vacuum and pressure systems are also treated as material handling equipments.

CONVEYORS

Belt Conveyors

Conveyors are generally used for handling materials which are required in a continuous flow in larger quantities, especially in a line production set-up. The major components are the belt, carrying and return idlers, head-end and tail-end pulleys, the supporting structures, drive unit, belt tension unit, feed and discharge equipment, etc. Tensioning is an important aspect of belt conveyors because it prevents undue sag between idlers and maintains sufficient tension for smooth functioning. The flat belt conveyor is normally used for component assemblies and similar functions which are carried in a line. If the material is in granular form, the troughed belt can be used. Belt bucket conveyors can be used as elevators. If the unit loads are heavier or if the material is of abrasive or corrosive nature, chain bucket conveyors are most suitable. Mono-rail systems are highly efficient for handling loads on an intermittent basis as rigid floor space is not needed.

Roller Conveyors

Rollers and wheels can be advantageously used for handling items like packages, boxes, assemblies, baggages, steel and wood wherever the movement is on a level or slightly inclined plane. Usually gravity is made use of but movement against gravity can be introduced either by manual efforts or by pioneering the rollers through chain or belt drives. If fragile materials like glass sheets are to be handled, the rollers can be rubberised.

Chain Conveyors

When the unit load becomes heavier and when the environment is corrosive, chain conveyors serve better. In this case, endless chains of various types, keep travelling the entire length of the conveyor system. The items to be handled can either be hung or carried in trolleys tied by the chain. The overhead chain conveyor is one of the most versatile types of conveyors. Since the load is hung from a chain, they can function effectively in much wider temperature range. In this the whole length of the chain is used, unlike others having a non-productive return run. In addition, installation, maintenance, and relocation do not create any technical problem. Sophistications of flexibility and automatic control can be introduced in all types.

AUTOMATIC WAREHOUSING

Mechanisation of handling has led to automatic warehousing. Electronic computers can be used effectively for automatic control of warehouses. The operations involved in the warehouse are: (a) sorting the goods received, (b) locating them in their proper positions, (c) picking the requirements according to the orders, (d) packing and (e) despatch. The handling mechanism can be controlled remotely by using electronic counting system and use of computer memory and centralised data input.

The entire operation of such a system is programmed. Inputs flow into the warehouse along a guided path and are counted electronically. The computer can activate the bin doors for storing materials. The empty containers are returned. The computer's memory has the data on what is available, where. Whenever requirements are fed to the computer, its central processing unit locates the position of the items required and activates the gate for releasing them. These are collected, packed and despatched. Such systems presently in vogue in developed countries can be applied most economically for multi-product finished goods warehouses in India where there is a very high turnover. In this process, adequate care should be exercised for forecasting, planning, provisioning, spare parts for handling equipments, controlling and customer order processing.

TEN POINTS TO PONDER

1. Identify the role of mechanised handling in warehouses.
2. Examine the factors influencing the choice of handling equipments.
3. Discuss the scope of cost-benefit analysis of handling equipments.
4. Elucidate the difference between fixed path and variable path equipments.
5. Explain the role of elevating devices in handling.
6. Describe the utility of pallets in warehouse operations.
7. Amplify the role of containers in material handling.
8. Articulate on the utility of transporting equipments and forklifts.
9. Discuss the relevance of different conveyors in warehouses.
10. Illustrate the utility of automatic warehousing in Indian context.

26. Transportation Systems

SCOPE FOR COST REDUCTION

The materials manager should be aware of the various modes of transport available for despatch of finished goods as well as bringing of raw materials. There is a wide scope for reduction of freight charges, particularly in items like ore/coal/ash in which the transportation cost is more than the materials cost. The average freight cost is about 20 per cent of the materials cost, and this figure is higher if suppliers are located in remote areas of the country. Transportation safety is also an important aspect, as more than one lakh Indians meet with fatal accidents every year.

In a large number of organisations, the transport manager is responsible for clearing the goods from the transport agencies. He also handles the despatch of scrap and rejected items. His duty includes transporting goods within the warehouse and in the plant. At times, he also handles the despatch of finished goods to the customers, in addition to arranging the transportation of items. Some large companies have a separate Tariff department to coordinate freight transportation. We shall examine in detail in this chapter how to reduce the cost of transportation.

MACRO LEVEL DECISIONS

Mass movement of men and materials has become an important prerequisite for the economic development in any country. Every Five Year Plan allocates more investment in various logistics systems such as railways, road transport, air transport, tankers, inland waterways shops, pipelines etc. These investments have ensured that we have about one million km service road, 65000 km of rail, 7000 km of pipelines, jumbo jets and airbuses for air cargo, state bus corporations, as well as growing tonnage of shipping. These constitute the backbone of our economic activities. Obviously the transportation has developed more towards major metropolitan cities rather than interior areas. One can well imagine the difficulties in movement of goods or personnel from Kanyakumari to Itanagar in India. The problem will be the same as travelling from Faulkland to Auckland, both lying in the same hemisphere, but one has to go to Rio de Janeiro, New York, London, Singapore, Sydney to reach Auckland.

A TO Z PARAMETERS ON CHOICE OF TRANSPORT

The important considerations in the choice of transport are as follows:

(a) Type of material
(b) Rates
(c) Discounts
(d) Distance
(e) Time available
(f) Volume of material
(g) Location of vendor
(h) Location of factory
(i) Route selection
(j) Shelf-life of material
(k) Urgency of requirement
(l) Terms of contract
(m) Full lorry load/or wagon load
(n) Pilferage proofing
(o) Corrosive nature of material
(p) Type of packing
(q) Safety of the transport system
(r) Handling breakage losses in movement
(s) Communication system
(t) Loading/unloading charge
(u) Octroi/Entry taxes
(v) Payment terms
(w) Value of the material
(x) Exposure to mob attack due to riots and risks
(y) Insurance premium
(z) Settlement of claims.

The cheapest method of transport is shipping/ inland waterways, while Railway charges are about twice the charges of shipping and lorry transport about twice that of railways. The costliest is the air transport which is about 100 times that of shipping. However, the Commercial Managers of Airlines argue that airline is the cheapest, as it delivers the material in the speediest manner, so that the same can be used to generate profit whereas the material brought by shipping may get corroded due to wrong positioning and delay and may not be useful at all.

Many organisations prefer lorry service for inland transport, because of credit facilities, avoidance of delays, minimum disputes, immediate settlement of claims etc. Pipelines are generally used for liquids like crude oil, and natural gases. In view of the enhanced discovery of oil and gas, on shore as well as off-shore, pipeline transportation system is bound to expand tremendously in future. One can very well appreciate the transportation of the statue of Budha made of brittle granite rock weighing 450 tons for a distance of 60 km by road to be installed in the middle of Hussain Sagar Lake in Hyderabad.

INDIAN RAILWAYS

Railways are the life-line moving the wheels of the Indian economy, offering freight and passenger services. The route of about 65,000 km spanned by the Indian Railways make it Asia's largest and the world's second biggest state-owned railways.

As much as Rs. 8000 crores have been invested in railways and nearly two million people are employed directly and in the form of contract labourers. Roughly 1.5 per cent of the Indian population travel every day by rail and about one million tons of goods are carried. The gross revenue of the railways is about Rs. 500 million a day. The gross amount paid as claims is about Rs. 100 crore per annum. The Railway budget increases the freight rate about 10 per cent every year.

Problems in Railway Transportation

The railways are ideally suited for long distance hauling of heavy unit loads and the choice of the units is normally governed not by strictly economic considerations alone, but by social, political, regional, linguistic, ethnic and other aspects also. As a result, capacity utilisation of different routes is not always optimal. In order to cater to the needs of multitude of destinations, heavy marshalling is resorted to, resulting in so much delays that the goods wagons spend more time in idling or marshalling. The social objectives of the railways are such that critical commodity movements — coal, foodgrains, oils, fertilisers, cement etc. accounting for 70 per cent of goods traffic handled by them — are at rates near or below the cost. Weak culverts, poor track maintenance, single-line traffic in most places, slow speed due to non-electrification of routes, high accident rates, decreased loading in the first six months, trans-shipment from metre gauge to broad gauge, difficult operating terrains, non-availability of adequate capital for investment, pilfering in some parts of India, etc. are some problems hampering the progress of railway transportation in India.

RECENT SCHEMES OF RAILWAYS

The Indian Railways have introduced a lot of innovative schemes to attract increased market share of the freight transport. We shall discuss some of these new schemes.

The guaranteed quick transport service scheme has been introduced to overcome the consumer's apprehension about undue delay in the transport of goods. A small surcharge paid initially is refundable in case the consignments do not reach the destination within the target time. This service has been introduced on the main trunk routes connecting most of the big commercial centres. Even small consignments are being carried by fast Express trains on payment of surcharges.

Freight Forwarder Scheme

To compete with the door delivery service of the road transporters, the railways have introduced a scheme of rail-cum-road transport in containers of five tonne capacity for smaller and high rated goods in a few commercial centres.

One of the biggest problems faced by the consumer desirous of transporting small items by rail, has been regarding wagons loaded with small items being repacked at the various marshalling yards and sheds. This was a necessity as all small items in a wagon were not for the same destination. The freight forwarder scheme involves a collection agency which ensures that a single wagon is loaded with small items of various consumers for a single destination thus eliminating *en route* repacking with possible loss and damage. Another scheme of collecting smalls from specified points through mobile booking service is also being practised. Introduction of increased number of Parcel Expresses for quicker transport of perishables and other consignments by coaches, is another step to attract more freight traffic.

ROAD TRANSPORT

Road transport is particularly useful for transporting within city limits and railway yards to warehouses, a freight of 1–15 tonnes, over short and medium distances of about 500 km. The road transport industry in India is playing an increasingly pivotal job in the logistics of our country by carrying a sizeable share of passenger as well as freight traffic. A portion of the freight carried includes supporting materials to core industries, such as cement, bricks, perishable items, medicines, fruits, etc. Road transport is particularly useful for carrying cargo to ports for export purposes where time is of the essence as ships will not wait.

The process of leasing lorries by Finance companies has given a tremendous fillip to road transport. The country plans to build about a million additional kilometres of national highways in the next ten years.

PROBLEM AREAS IN ROAD TRANSPORT

The following factors hinder the smooth flow of traffic through road. The innumerable octroi revenue and state check posts cause considerable delays and result in increased corruption and uneconomical operations. The condition of national highways, state highways, city roads and roads linking villages leaves much to be desired. Only ten per cent of the total roads can be operated under all weather conditions. The roads are intended for single lane traffic only and portions of city roads are dug up by several departments. Narrow, weak and old culverts form a part of the national highway system. Damages to culverts and bridges take decades for repair, resulting in the diversion stretch becoming the real road. This may be one of the reasons for the higher accident rates in road transport.

Many lorry owners possess an uneconomical small fleet of vehicles. Only breakdown maintenance is done in the case of most of the lorries. Drivers drive the vehicles under the influence of alcohol on national highways. Maintenance of roads is poor everywhere. In the congested city areas, often one comes across the cyclist overtaking the lorries both from right and left side. The network in the rural areas cannot be used by commercial vehicles. Farm products have to depend upon primitive means of transport such as a bullock cart, camel cart, head load, asses, etc. Unlike the railways, the road transport in many cases is not organised properly and is in the hands of private individuals. In view of private ownership, maintenance of lorry is poor resulting in frequent breakdowns and accidents.

The road network and railways have elements of overlapping and competition in the matter of freight traffic. This competition is more prevalent in traffic moving into city areas. Logistic support is woefully inadequate in hilly areas, tribal belts and rural segment. Linkage with the north eastern states is restricted through limitation of traffic on Faracca bridge. Only limited number of lorries with the national permit can pass from one state to another. Other lorries with state permit have to transship the goods from one lorry to another. One can imagine the difficulties in transshipment and handling of goods transported from Tinsukia to Trivandrum. Lorry traffic is more vulnerable to attacks by terrorists, Naxalites, linguistic riots and ethnic disturbances.

However, road transport continues to play its complementing role to the railways in the massive freight movement in Indian economy. In this context, in the unorganised sector — the supporting role is played by the 15 million animal drawn vehicles, and 25 million people are employed in this sector moving 10,000 million tons in a year over a radius of 20 km. Simple modifications such as modernising the wheels can result in significant improvements in efficiency and economy.

INLAND WATERWAYS

Inland water transport and international shipping are the cheapest forms of transport. The important navigable rivers and inland waterways are as follows: (a) the Ganga, (b) the Brahmaputra, (c) the Godavari, (d) the Krishna, (e) the Backwaters and lagoons of Kerala and (f) the Buckingham canal. The total length of navigable inland waterways is about 15,000 km. The proposed Telugu Ganga, Ganga-Cauvery links and other schemes can give the much needed fillip to the neglected inland waterways system providing an integrated framework for movement of commodities in the cheapest manner.

Coastal water transport system is used for the movement of materials from one port to another within the country. For example, coal is moved from Calcutta to Tuticorin through coastal shipping. Other essential commodities like salt, foodgrain, iron ore are moved by this means to supplement the rail-road networks. Unfortunately coastal shipping in India has remained static in the past two decades in terms of number of ships and gross tonnage. The induction of coast guard to patrol the economic zone not only curbs smuggling, but also helps in rescuing the ships along the coast lines.

INTERNATIONAL SHIPPING

International shipping plays a crucial role in transportation. The total exports and imports are worth nearly Rs. 60,000 crores and a major portion is transported by Indian ships. However, there are many problems in this area. For instance, the structure and management of ports need to be modernised to meet the technological challenges. Berthing facilities are not adequately equipped to handle bulk carriers and giant tankers. Infrastructural facilities in the form of dry docking, repair work and handling equipments — need to be improved in all the major ports. In view of congestion in all major ports of India, sometimes the liners off-load the material in Singapore.

At times the cargo has to be transferred in mid sea due to non-availability of berthing facilities. We have acquired ships from a multitude of countries resulting in spare part problems. The number of berths in major ports are inadequate in relation to the number of ships calling on them. Ports such as Calcutta require continuous dredging operation. Cargo handling for loading and unloading is highly labour-intensive, creating periodic unrest amongst dock workers resulting in heavy demurrages and deterioration of cargo.

Several committees have been appointed to improve port facilities, as a result of which repair facilities and dry docking are being gradually improved. Indian economy relies on imports of fuel, fertilisers, heavy machinery and defence equipments. Considering the location of supplies and distances involved, there will be a heavy dependence on shipping as a medium of transport in the years to come.

AIR TRANSPORT

In spite of the fact that the airport industry is about eight decades old in India, it obtains a very low share of freight traffic in domestic and international arena. Indian Airlines, which handles domestic air transport, has a fleet of various models from the Fokker friendship to the latest Airbus models. Even though, it is one of the largest domestic air transport systems in the field, the volume of freight is only about a million tonnes per year. The freight includes mail, flowers, fruits, medicines, leather goods, livestock, instruments etc. Air cargo consolidation, or bulk purchase of cargo space from Airlines and retail of the same at bulk rates to individual shippers at rates lower than the normal airlines rates, helps to increase the volume of freight.

International air traffic, which is in the hands of Air India, has exclusive cargo flights to Europe and USA, but the freight carried by this airlines is much less than that of many other international carriers.

By the dawn of the 21st century, there will be a five-fold increase in passenger traffic to 1200 billion km, through all modes of transport. This not only calls for a manifold increase in investment for transportation facilities but also requires a balanced and integrated development of different modes of transport and ancillary infrastructure. The warehouse manager should be aware of the national and international regulations for transporting hazardous and dangerous goods. He should persuade the suppliers to despatch items by quicker and safer transport systems though they may be more expensive. The warehouse manager should determine the mode of transport after a thorough analysis of all the influencing factors.

TEN POINTS TO PONDER

1. Identify the need for transportation of goods to warehouse.
2. Enumerate the methods of cost reduction in transportation.
3. Explain the criteria on choice of a specific transport system.
4. Articulate on the use of railway transportation of freights.
5. Discuss details of freight forwarder scheme.
6. Describe the peculiarities of lorry transportation and its problems.
7. Specify the uniqueness of animal drawn vehicles in transportation.
8. Illustrate the utility of inland waterways and coastal shipping.
9. Identify the problem areas of international shipping.
10. Examine the scope of freight traffic by domestic and international airlines.

27. Materials Intelligence System

RIGHT MANAGEMENT INFORMATION SYSTEM

Flow of materials' depends on information flow and success in industry on timely response to the needs of the customer. Timely access to information can make or break companies.

In a business organisation, every transaction generates some kind of data. A solitary instance of a specific transaction is data. When the data is collected and a meaningful pattern is derived from it, it is called information. One way of defining management is as a process of converting information into decisions. A system is an organisation of meaningful elements for a specific purpose. Thus a business organisation, represented by men, materials, money, machines, consists of a system of procurement, conversion, selling and bringing cash. This system aims at providing the right type of information to the decision-maker at the right time at optimum costs. The management information system is used for strategic planning, and for exercising managerial control on all functional activities. The determinants in management information system include the opportunities, risks, available technology, available information sources, corporate strategy, corporate structure, manpower and style of leadership.

Thus, management information system can be defined as an organised network of information flow aimed at supporting management by furnishing relevant, reliable and timely information. Use of personal computers and electronic data processing equipments in integrated data processing systems has provided materials executives with the tools to handle information flows with unbelievable speed, facility and accuracy. The increasing use of computers in many organisations has kindled new awareness of information needs and materials intelligence.

MATERIALS INTELLIGENCE CONCEPT

The terms, materials intelligence, purchase intelligence, creative purchasing, purchase information system, materials research, materials information system etc. signify one and the same system which helps the materials manager to know where the required item is. Materials research becomes important accounting for the bulk of cash outflow in the company. According to our research study, this is the most neglected area of materials management.

Materials research includes the steps involved in systematic investigation for locating alternative supply sources. In this context, research is defined as the concept of purposive analysis that tries to establish a basis for prediction and control.

The materials manager depends on an uptodate and accurate exchange of information from other departments in the company as well as from external sources. On the other hand, if the materials executive is 'submerged' by paper-work, he reduced his position to that of a clerk. Some means must therefore be found to handle vast quantities of information quickly.

A TO Z OBJECTIVES OF MATERIALS INTELLIGENCE

(a) The materials intelligence should be a subsystem of the total corporate information system.

(b) The selection of right quantity of information emphasises the fact that the life blood of any business systems concept of operation is a flow of information.

(c) Information is power. The objective of materials information system should coincide with the objective of the materials function — namely to ensure continuous supply of uniform quality of the item with lowest cost of ultimate product/service rendered.

(d) In this process, one has to weigh the costs and benefits of collection, compilation, analysis, and presentation of the information.

(e) It may be worthwhile to concentrate on high cost, critical, long lead items initially for the purpose of materials intelligence system.

(f) Materials intelligence is the process of applying different scientific principles for the purpose of developing a system in sufficient detail to realise materials management objectives. The objectives of materials intelligence should involve the following steps:

(g) Utilise the available manpower resources effectively.
(h) Attain higher efficiency in all activities of materials management.
(i) Anticipate future problems in materials area.
(j) Identify probable alternative solutions in materials management.
(k) Find alternative sources of supply.
(l) Improve quality of the ultimate product/service.
(m) Reduce the cost of final product/services.
(n) Reduce the price paid for the initial product.
(o) Standardise, simplify and indigenously substitute the item.
(p) Avoid delay and stockouts.
(q) Reduce paper-work in purchase, stores and inventory control.
(r) Optimise inventory levels.
(s) Reduce material handling costs.
(t) Reduce transportation/freight cost.
(u) Reduce packaging charges.

(v) Organise library on technical literature, commercial references, trade association, handbooks, manuals, catalogues, etc.

(w) Work towards better employee motivation, personal development, job enrichment, and job rotation.

(x) Develop better communication within the department, organisation and with suppliers, transporters, bankers and relevant government departments.

(y) Create better understanding of the impact of social, political, economical, technological, natural, regulatory, financial, ethnic, geographical, religious, ecological, ethical, internal, external factors on materials management activities.

(z) Foster appreciation of the strategy, tactics, managerial control, operational control, corporate goals, opportunities, risks, and available technology with their influences on the managerial function.

A TO Z ASPECTS OF INTERNAL INFORMATION

Materials intelligence can be collected from sources within the organisation, as well as from outside sources. The internal sources, includes the data generated by the materials department. Given below is an illustrative list of the data that may be useful for materials intelligence. These pertain to information on specific items relevant to the individual organisation.

(a) Materials planning — short-term and long-term

(b) MRP, BOM and MRP

(c) Sales forecast — quantity and price, if possible for all major items

(d) Materials/budget and financial constraints, if possible for each user

(e) Accounting information costs, stock valuation, pricing of issues

(f) Policy criterion on make-buy-lease, replacement and project buying

(g) Contracts, legal procedures, systems and manuals

(h) Quality specifications and challenging engineering designs

(i) Production scheduling and purchase planning

(j) Inventory control, reserve, safety, buffer, minimum, maximum, stockout for raw materials, machinery spares, semi-finished and finished goods

(k) Receiving items, quality control rejects

(l) New product development, new material requirement

(m) Lead time and its components, fluctuations and follow-up

(n) Consumption norms and fluctuations in consumption

(o) Cost-criticality — Availability — MUSIC-3D view

(p) Credit, discounts, bill payment terms

(q) Freight, packing, handling and transportation costs

(r) Audit, financial concurrence, vigilance and negotiation aspects

(s) Shelf-life, preservatives, storage space, issue methods

(t) Source development, buyer-seller relations, vendor rating and ethics

(u) Obsolete items, surplus, slow-moving items insurance, spares, scrap disposal

(v) Stock verification missing documents, write-off and charge-off
(w) Cost reduction, codification, standardization and value analysis
(x) Customs, excise, insurance, sales tax procedures and claims
(y) Training, reports, use of computer and scientific techniques
(z) Managerial styles, structure, super-ordinate goal, secrecy of information

A TO Z ASPECTS OF EXTERNAL INFORMATION

The productivity of materials department depends on processing items at optimum cost. For this, knowledge of external environment is essential. Typical external information that contributes to the materials intelligence is given below:

The materials manager may not get his exact requirements and he may have to convert the information to suit his needs.

(a) Information on general market conditions can be obtained from suppliers' salesmen, trade publications, Associations, bulletins, professional seminars and economic dailies.

(b) Total demand and supply at macro level can be had from the planning Commission documents and market gossip.

(c) A constant updating of the information on prices, supply, demand, competitions, actions taken, is necessary.

(d) Information on possible sources of supply can be had from suppliers' salesmen, advertising media, special promotions, exhibitions, trade shows, conventions, financial reports of vendors and trade journals.

(e) Suppliers' capacity, suppliers' production rates, labour conditions in suppliers' plants, technological skill of suppliers etc. constitute the most important component in the materials intelligence system.

(f) Information on prices, discounts, sales tax, inflation and other aspects affecting the general price level — apart from getting this information from a large number of salesmen a trend analysis of the economy will be helpful.

(g) Economic indices like cost of living index, wholesale price index, Dalal street stock market index, Dow Jones American stock market index, Japanese stock market KKFI index and other useful indices on national and global scene are useful.

(h) The types, availability and cost of transportation have a direct bearing on the input cost.

(i) New products, new processes, substitute items, design improvements, technological innovation, research efforts at the national and global level can be had from publications.

(j) Agricultural production, monsoon failure, elections and volume of money in circulation have a tendency to influence price level of some items.

(k) Government's monetary policy and policies on taxation, dual pricing, subsidies, defence, investments and imports have a direct bearing on price and availability.

(l) Non-availability of power in summer months force organisation to use generator sets for manufacturing items resulting in increased price.

(m) Riots, terrorist activities, strikes, agitations, etc. cause law and order problem affecting the smooth movement of goods.

(n) It is better to anticipate and not to react to disruption of transport due to law and order problems.

(o) The balance of payment situation, volume of imports and exports generally govern the import trade control policy.

(p) The monthly status in foreign trade published by the Commerce Ministry is useful in this case.

(q) Information furnished by National Sample Survey reveals the economic status of the country in advance.

(r) Agricultural information can be had from the publications of Ministry of Agriculture on estimates of areas and production of principal crops as well as the agricultural scene in India.

(s) Reserve Bank of India's bulletin and weekly statistical supplement give a picture of currency and finance in a concise manner.

(t) Publications by DGS & D; DGTD, Bureau of Indian Standards — are useful.

(u) Publications by the Central Statistical organisation give statistics on various aspects useful to the materials manager.

(v) Economic scene published by Tata consultancy gives the picture of the economic scene of the country.

(w) Publications of various trade associations, Chambers of commerce and professional agencies may be helpful.

(x) Reports of different committees, Bureau of Public enterprise, Committee on public Undertakings, Public Accounts Committee, etc. serve as useful guides.

(y) Various directories published by national and international organisations giving address, affiliations, functions, products and other details prove useful.

(z) The materials manager has to concentrate on the topics of his interest and collate the necessary data before converting them to materials intelligence.

INTELLIGENCE TECHNIQUES

The following methods are available to convert the data into materials intelligence:

(a) General/market/demand/supply analysis

(b) SWOT — Strength, Weakness, Opportunity and Threat analysis (Refer Chapter 15)

(c) Cost-criticality-availability analysis

(d) Vendor rating and supplier analysis

(e) Transportation/freight charges analysis

(f) Price/cost analysis
(g) Purchase procedures/systems/follow-up analysis
(h) Value management/engineering analysis
(i) Cost reduction techniques

The materials manager has to choose the method that is suitable for his needs.

TEN POINTS TO PONDER

1. Define the concept of management information system.
2. Explain the relevance of materials intelligence system.
3. Enumerate the objectives of purchase intelligence system.
4. Discuss the internal sources of information.
5. Describe the external sources of information.
6. Elucidate the methods of data collection and analysis for creating a purchase information system.
7. Name some of the important publications relevant to the materials manager.
8. Elucidate the type of information available from suppliers.
9. Examine the relevance of fiscal and economic information in the formation of materials intelligence.
10. Select a person who would be efficient in collecting, collating, analysing and disseminating materials intelligence.

28. Computers and Materials Management

THE ROLE OF COMPUTER

The computer either in the form of a terminal or a personal computer — has invaded not only the corporate world in India but also schools and colleges. The most lucrative area of application of computers is in the warehousing and inventory management. Computer helps trigger off new ideas and increase the manager's capacity to learn, remember and make new decisions. In view of the low cost of less than Rs. 10,000, the personal computer has even invaded some middle class homes. PC based data base and spreadsheet type managerial aids are being introduced in many forms as per our research survey.

The hardware of first generation computers used vacuum tubes and the major elements of the system operated one at a time. The second generation machines, using transistors, adopted decentralised control. The third generation machines use silicon integrated assemblies thereby drastically reducing its size, but increasing the speed. There are many third generation computers operating in India. With the advent of large scale integrated chips, thousands of electronic gates, constituting entire submits, can be built into a single chip resulting in fourth generation computers. These fourth generation supercomputers have started their operations in India. The fifth generation computers, based on artificial intelligence, are expected to provide natural language communications, speech understandings, handwriting recognitions and computer-aided learning. Because of the speed at which computer does mathematical operations, sorting and printing, it should be used for work which would be prohibitively expensive if done manually.

LIMITATIONS

The computer is only a fast calculating machine, and not a substitute for the expertise of the manager. If incorrect data are fed into computers, wrong answers will result. Hence the saying GIGO — garbage in garbage out, and TITO — trash in trash out. Since a lot of corporate secret information is stored in the computer memory, or in floppy disks, data security has become an important issue in the western world. Computer virus or the results disappearing from the screen is one of the recent problems encountered by many organisations. All the same the computer has stood the test of time and has proved to be an effective communication media in many undertakings, if adequate planning is done before installing the computer.

SYSTEMS DESIGN

Each system which is proposed requires an in-depth study to ascertain its feasibility from the point of view of existing equipment and facilities, and the requirements. Once these parameters are identified, attempts should be made to improve and modify the procedures, cut down tasks and eliminate duplication. The inputs to the computer system and the outputs to be generated from the system, should then be identified.

Now, the various tasks, that have to be performed on the computer to achieve the desired outputs are conceptualised, through flow charts. Codification, input and output formats are decided upon. Detailed program specifications to carry out the necessary jobs are also developed at this stage.

PROGRAMMING AND TESTING

Programming or instructing the computer in a language it understands is the next task. The specifications developed form the basis on which programs are written. These programs are then tested with synthetic data to check the logic and performance of the entire system. Testing is done with the entire system and then it is introduced to run in parallel with the manual system. This enables us to check the accuracy and stabilise the functioning of the system.

SYSTEMS DEVELOPMENT

Thus, the systems development is usually planned over the following series of time-phased steps:

(a) Initial investigation and discussion
(b) Understanding the existing infrastructure, resources and requirements
(c) Systems feasibility study
(d) Systems specifications evaluation
(e) Systems detailed specification
(f) Presentation of proposals to management personnel
(g) File creation
(h) Detailed systems design
(i) Programming
(j) Systems testing
(k) Parallel running
(l) Live operations.

COMPUTER PROCESSING DATA

For computerisation, the coding structure required for different attributes in the system such as item code, supplier code, limit of issue, limit of order, unit of sales, job order numbers, cost centre codes, purchase order numbers, etc. must be finalised. This is the best method of unique identification in a computer.

Computer master file is a directory of permanent and semi-permanent information on a set of related items. For example, a parts master file or an item master file would mean that this file contains data pertaining to each part/item stocked in the store. The first step in computer processing is the creation of master files.

An item master and an outstanding purchase order file can be used as the starting point. Creation of the master file is a time consuming process and this has to be done in parallel with programming, so that by the time all programs are ready, the files are also ready.

In the computer system, the processing involved can be stated as follows: The monthly/weekly/daily transactions are fed to the computer. Based on these and the data available in the master files, the computer produces the desired output reports and simultaneously updates the contents of the master file. The different output reports are then produced at different stages in the system.

A TO Z AREAS OF APPLICATION

The computer is useful in generating a large number of reports in all the areas of management. We shall discuss some of the areas pertaining to warehousing and inventory management.

The application areas fall into two categories — periodic or one time and daily or continuous. The one-time application for each item to be done annually or monthly includes the following areas:

(a) ABC analysis
(b) MUSIC-3D
(c) Movement Analysis
(d) Obsolete items
(e) Material requirement planning
(f) Forecasting price
(g) Lead time analysis
(h) Economic order quantities
(i) Safety stock
(j) Reserve stock
(k) Buffer stock
(l) Consumption norms
(m) Quantity forecasting

(n) Wendor rating
(o) Sales analysis
(p) Economic lot size in manufacturing
(q) Product — customerwise — Itemwise sales
(r) Stock statements
(s) Inventory valuation
(t) Material budget
(u) Distribution systems and market intelligence
(v) Transportation systems
(w) Cash flow forecasts of money committed
(x) Exception reporting and cost reduction efforts
(y) Service level/stockouts
(z) Customer complaints.

The repetitive applications include: (a) stock status existing at a point in time, (b) outstanding bills, (c) status of a purchase order, (d) follow-up mechanism, (e) outstanding supplies, (f) rejection, (g) rework, (h) accounts receivable, (i) number of sales invoice handled, (j) number of issues, (k) value of sales, (l) value of issues, (m) value of receipts, and other information needed by the warehouse management on a day to day basis.

STOCK ACCOUNTING

We will discuss some selective application areas relevant to warehousing in detail. Stock accounting implies recording of all transactions which affect the stock status of different items. The stock status is constantly updated and the computer master file reflects the latest position. The output required is the materials abstract. The opening balance of each item is listed and all transactions pertaining to that period are logged and the final balance is indicated. The stock-ledger is printed by the line printer and is available as a record.

COST ACCOUNTING

In this phase, cost centrewise sales or consumption reporting as well as job order-wise reporting are done. In each period, under a particular job order, or cost centre, the various issues of materials are summarised and both the quantity issued or sold and the value of issues/sales are printed out. Simultaneously we can have the total sales/issues till that month for the cumulative issues/sales made under a particular job order/cost centre/customers.

ORDER PROCESSING

In this subsystem, all the purchase orders that are raised are recorded and whenever vouchers against these orders are received the outstanding order file is suitably changed to reflect the latest position. A print-out of these orders which have been executed is furnished.

Progress-chasing is a vital area of application, which acts as a feedback report to the management on the position of orders. All periodic orders within the next month — or a period fixed by the company, are listed for action.

In the order-raising subsystem, stock levels are reviewed and based on the inventory policy adopted, indents for raising orders are printed out. The ordering policy can be fixed by the user depending on the class of materials. The computer can be programed to analyse the sales/consumption pattern and give out exception reports regarding excess or below normal sales/consumption. Based on this, either orders can be expedited or cancelled.

TEN POINTS TO PONDER

1. Discuss the advantages of computer in materials function.
2. Explain the limitations of the computer.
3. Discuss the steps of computerisation.
4. Examine the concepts of systems design in warehouse.
5. Describe the master file relevant to an item.
6. Identify the necessity of programme-testing.
7. Enumerate one-time application of computer.
8. Name the accounting applications of computer.
9. Amplify the role of computer in order processing.
10. Articulate on the concept of exception reporting in warehouse.

29. MUSIC-3D — 20/80 Rule

20/80 RULE EXAMPLES

The 20/80 rule is based on the principle that the five fingers of the hand cannot be equal. Though an individual spends about seventy years or 70 × 365 × 24 × 60 × 60 seconds on this planet, only very few activities are important. Similarly, in any class room situation only very few pupils monopolise the discussions.

At the national level, most of the wealth and/or power is concentrated in a few hands. In the domain of sports, even though 190 countries compete, a few countries account for the bulk of the medals. Thus about 20 per cent of people account for about 80 per cent of activities and the remaining 80 per cent of the people are engaged in the balance 20 per cent of the activities. This is the 20/80 rule. This 20/80 rule is also known as *Pareto's Law* and is based on the principle — "few are vital" and "many are trivial" and aims to direct efforts where results are important. In the field of warehousing management, this is known as *selective control*. Just as equality of all individuals/or ideal socialistic pattern of society is an Utopian goal, the warehouse manager cannot give equal importance to all items.

ABC ANALYSIS

The 20/80 rule in materials management is known as ABC (Always Better Control) analysis. This analysis is usually based on the monetary value of annual consumption. At the end of the year, the total consumption figure is multiplied by the accepted price or average price per item. The annual consumption figures are arranged in a decreasing order. The top 10 per cent number of the items accounts for about 70 per cent of the total annual consumption value and is known as A category. The exact percentage varies from unit to unit according to the observed consumption value patterns.

The middle 20 per cent of the number of items, amounting to the middle 20 per cent of annual consumption value is known as B category. The remaining 70 per cent account for the items which are trivial in importance, accounting for only 10 per cent value. If the percentages are plotted graphically — with the number on the *x*-axis and the cumulative annual value on the *y*-axis — it will be noticed that the curve appreciably changes at two places giving the A, B, C categories. The cut-off points can be varied to suit the individual requirements of the organisations, like a category being more than one lakh annual consumption value, C being less than 10,000 and B lying in between. Computer programs are available to categorise the items if annual consumption and price per item are given.

LIMITATIONS OF ABC ANALYSIS

Our research study has shown that, in spite of its limitations, ABC analysis is popular with most of the materials managers. Some organisations categorise raw material, crude, major inputs as super A items. Some organisations also classify A_1, A_2, C_1, C_2, and create more classifications. This is confusing as ABC analysis has been cynically expanded as "avoid, bluff, and confuse". This is because the idea that ABC analysis is based only on monetary value of consumption or on control on monetary value alone, is misleading. Hence MUSIC-3D or Multi-Unit Selective Inventory Control — Three Dimensional approach is brought in. The three dimensions represent the functions of finance, user/maintenance and warehouse/materials. Let us now develop MUSIC-3D after examining the other selective control principles.

XYZ ANALYSIS

While ABC analysis is based on the annual consumption on value, XYZ analysis is based on the value of inventory undertaken during the closing of annual accounts. *X* items are those having high value, while *Z* items are of low value and *Y* items are of medium value. The percentages are similar to ABC analysis. This analysis helps find items with heavy stock.

HML ANALYSIS

This method is similar to the ABC classification. But, in this case, only the unit value of item is considered, instead of taking the annual consumption value. Items are classified according to the unit value as high value, medium value and low value and this is used to control the purchase value of items. Here also the cut-off points vary from unit to unit. Some organisations use both HML and ABC classifications for control of unit value of items and the overall consumption pattern.

Both ABC and HML analysis are used for controlling the average inventory as well as purchase value. Let us consider three items for which the annual consumptions are Rs. 200,000, Rs 20,000 and Rs. 2,000 respectively. If the policy is to place equal number of orders — say, two per year — the value per order will be Rs. 100,000, Rs. 10,000 and Rs. 1000. The average inventory will be half the quantity purchased at one point of time leading to Rs. 50,000, Rs. 5000, Rs. 500 with a total inventory value of Rs. 555,000. On the other hand, if ABC/HML analysis is done, then the number of orders will be more for A items, as it is proportional to the square root value. Thus we get the number of orders as 8.3, 1 for the three items. The value per order will be Rs. 25,000, 6666 and 1000 with the inventory value being 12,500, 3300 and 500. The total inventory value is Rs. 16,333 which is one-third of the original inventory value.

MOVEMENT ANALYSIS

The FSN — Fast-moving, Slow-moving and Non-moving analysis — checks stock rotation and identifies the obsolescence of items. This analysis is particularly useful for spare parts. The combination of the movement analysis and unit price/ HML analysis results in the conventional ABC analysis. The cut-off periods for fast-moving, slow-moving, and non-moving may be varied according to the class of items and organisations' culture. The pattern is different for selling organisations. For instance, raw material and components move faster than spare parts. Many organisations consider items with at least one withdrawal in a quarter as fast-moving, one issue in a year as slow-moving, and no issue in a year as non-moving. The duration can be changed to suit the individual organisation's needs. For instance, in a sales organisation, many may consider at least one sale in a week as fast moving and no sale in a year as non-moving.

CRITICALITY CRITERIA

Technical considerations form the basis for the above criteria. VEIN is the accepted classification for equipments as Vital, Essential, Important and Normal. A vital equipment is one which feeds a battery of equipments downstream. VED is the art of classifying parts into Vital part, Essential part and Desirable part. The classification is based on performance, guarantee, warranty, reliability, safety, maintainability, functional utility, criticality, etc., taking the operational point of view into account. This is particularly useful for spare parts control and has to be done by the maintenance and design staff. The tendency to classify every item as vital should be discouraged. For instance, in summer months, when there is power cut the diesel generator providing electricity for high tension industrial consumers is the most vital equipment and major parts of this are critical parts. Non-availability of the critical spare parts can cause havoc in the industry.

The VED and VEIN analysis can be combined to give the highly critical and non-critical items.

AVAILABILITY ANALYSIS

Material availability analysis is based on the following criteria:

GOLF — Government-controlled, Ordinarily available in the open market, Locally available and Foreign imported purchase.

SDE — Scarce item or Single source item, Difficult to obtain or Easy to obtain as it is an off-the-shelf item.

SOS — Seasonal and Off-seasonal items. Here the raw material or the finished product or the transportation may be seasonal affecting demand and supply.

Thus the items may be classified as easily available from any of the many sources with minimum lead time and not easily available due to single source/imported/canalised agencies/government-controlled and hence long lead time items. Here we are interested in the total lead time i.e., the time for converting indent to order, manufacturing, transporting and inspecting the item. Individual organisations can classify items with more than one year as long lead time items and less than one year as short lead time items.

A TO Z BENEFITS OF MUSIC-3D

So far we have discussed items after grouping them on the basis of finance, performance and lead time. It is possible to combine any two dimensions and interpret the results.

(a) MUSIC-3D stands for Multi-Unit Selective Inventory Control — Three Dimensional approach, the three dimensions being finance, operations and materials.

(b) Based on ABC, FSN, and HML analysis, we have HCV/LCV or high annual consumption value or low consumption value. Critical and non-critical classification is based on VED/VEIN and LLT/SLT — long lead time and short lead time classification is based on GOLF, SED and SOS. Thus we have all the items grouped into 8 categories. Table 29.1 describes the MUSIC-3D eight categories.

Table 29.1 MUSIC-3D

	HCV		LCV	
Critical	LLT 1	SLT 2	LLT 5	SLT 6
Non-critical	3	4	7	8

Cost Reduction by MUSIC-3D

(c) It may be noted that items in cells 1, 2, 3 and 4 number around 20 per cent with an annual consumption (or sales value) of 80 per cent — while the remaining items in other cells numbering 80 per cent account for an annual consumption or sales of only 20 per cent.

(d) There should not be any application of cost reduction techniques for items in cells in 5, 6, 7, 8 as the cost of cost reduction methods is likely to be greater than the cost of the item itself.

(e) It may be dangerous to apply cost reduction techniques for the highly critical items falling in cells 1 and 2. Cost reduction can be applied on items grouped in cells 3 and 4.

Managerial Control

(f) Managerial control has to be very strict on items in 1, 2, 3 and 4 and lose or delegated down the line on items in 5, 6, 7 and 8.

(g) Powers of delegation on procurement should be very liberal on 5, 6, 7, 8 and most liberal with the lowest level procuring on items in 5.

(h) For High Value non-critical items in cell 4, with the Board of Directors or super steering committees approving the purchase/sale, thereby causing maximum possible delays.

(i) The inventory levels should be very liberal for low value items in 5, 6, 7 and 8. One can even think of three years stock for critical items of cell 5, as the working capital commitment is very low. Two-bin system of inventory control is ideal for 5, 6, 7 and 8.

(j) The stock level may be on a zero inventory for items in 3 and 4 as they are higher consumption value, non-critical items.

(k) Application of all mathematical/statistical techniques to reduce stock level must be done for items in cells 1 and 2. The safety stock level may be low for these items.

(l) Storage must be nearest to the user for items in 5 and possibly for items in 6, 7, 8.

(m) Central storage is recommended for items in cells 1 and 2. A spare parts Bank or common storage outside the factory's warehouse is recommended for items in cells 3 and 4.

(n) The issues or service level should be maximum at 100 per cent without any stockout for items in 5 and 6. Even though, non-critical, no questions may be asked initially for issues on items 7 and 8.

(o) The issues should be on required basis for items in 1 and 2 after verifying and scrutinising.

(p) The issues for noncritical high value items falling in cells 3 and 4 can be delayed. Hence replies in the form of excuses and alibis, are given as below: "We are seized of the matter", "We will look into it", "The files have started moving", "The Managing Director is out of the country", "The paper is under serious consideration", etc.

Documentation

(q) The postings have to be accurate and up to date for items in 1 and 2, whereas group posting and travelling requisitions may be adopted for items in 5, 6, 7 and 8.

(r) Documents and clerical efforts have to be maximum in the case of items 1 and 2 whereas cost of document may be more than cost of items in 5, 6, 7 and 8.

(s) Materials planning, consumption norms, scientific forecasting, management information system and materials intelligence system should be maximum for items 1 and 2 only.

(t) The number of sources should be increased for items in cells 1 and 2 and source development efforts must be carried for these items.

(u) Vendor rating, buyer-seller relations, ancillary development, etc. must be directed towards items in 1 and 2.

(v) Movement analysis must be periodically carried out for items in 1–4 to locate the slow-moving, surplus and obsolete items.

(w) The feasibility of charging off the material without maintaining any record for all items in 5–8 should be examined.

(x) In the case of selling organisations, the per cent margin of profit could be low for items in 1 and 2 and perhaps in 3 and 4, whereas the margin of profit has to be higher for items in 5–8.

(y) Stock rotation and verification have to be checked more often in the case of items in 1 and 2 and perhaps in 3 and 4.

Writing off procedures and discrepancy determination should be more thorough at higher levels for items in 1, 2, 3 and 4.

(z) For percolation of information and control at all levels, it is essential to identify the eight categories by eight colour cards and explain the relevance to the lowest level personnel for successful operation of the system.

TEN POINTS TO PONDER

1. Explain the concept of 20/80 rule.
2. Give examples of 20/80 rule in warehousing.
3. Classify the concept and limitation of ABC analysis.
4. Deliberate on the utility of XYZ analysis.
5. Identify the meaning of HML classification.
6. Elaborate the principle of movement analysis.
7. Examine the criticality criteria in depth.
8. Discuss the scope of availability analysis in inventory control.
9. Examine the necessity and scope of MUSIC-3D.
10. Enumerate the benefits of MUSIC-3D in warehousing and inventory management.

PART III

INVENTORY CONTROL

30. Inventory Basics

MEANING OF INVENTORY

The term *inventory* is used in day to day life as manpower inventory, equipment inventory, inventory of documents, inventory of raw materials, inventory of spare parts, inventory of semi-finished items, inventory of finished goods, inventory of vehicles, etc. as the dictionary defines the word as "stock of goods".

Inventory, in management parlance, is an idle resource of any kind having an economic value. So long as the resource is kept in the warehouse, it is idle. But, the material has economic value in the sense that it is capable of being converted to semi-finished and finished goods and sold to the customers to buy more items. This cycle goes on. Even though in practice the term covers only industrial items, this concept can be extended to other things, such as water, forests, space, manpower, etc. For example, too much water causes flood and too little of it results in drought.

OBJECTIVES OF INVENTORY CONTROL

Inventory control signifies the conversion of idle resource into an ideal resource. The objective is to ensure continuity of supply of uniform quality of goods at lowest cost. We have to devise systems and procedures to reach the optimum level without stockout and minimise the funds locked up in inventory. A company's working capital is locked up in raw material, machinery spares, semi-finished and finished items. But inventory control in many companies signifies only the input raw materials and machinery spares. The same principles must be applied to semi-finished shop floor, work-in-progress, as well as, finished goods in warehouses and the in transit pipeline inventory. Inventory control answers two basic questions when to order and how much to order. Fluctuations can be measured and stock levels can be calculated statistically.

TRANSACTION MOTIVE

A logical question arises as to why organisations lock up the working capital in an idle inventory. The motives for this may be classified as follows:

(a) Transaction
(b) Precaution
(c) Speculation
(d) Decoupling

Transaction

Transaction implies the total time of converting an indent into the material; after finalising the order or processing the customer's order in a marketing situation or set-up time in manufacturing situation, producing the item, transporting the item and inspecting the item before taking into the warehouse. This transaction time is called *lead time*. In actual practice in procedure-oriented organisations, the maximum lead time may go even up to four to five years. During the period of transactions, the demands from the user/customer have to be satisfied and this necessitates idle stock. Commonsense tells us that the more the lead time, the higher will be the stock level, suggesting thereby that we should examine the feasibility of reducing the lead time. The inventory maintained to cater to normal consumption in the average lead time is known as *buffer stock*.

Precaution Factor

Precaution means safety or minimum stock as commonly understood in the industry. While transaction caters to the needs of average lead time, it is likely that the total lead time or any one of its components may shoot up. Consumption may be more than the anticipated level. These fluctuations can be measured statistically. Safety stocks provide the cushion for average consumption during any increase in lead time from the normal lead time. Reserve stock provides for an increase in the rate of consumption/sales during the average lead time. The total of reserve stock and safety stock is the same as the conventionally understood minimum stock in the industry and is intended for the precautionary factors of fluctuations in lead time and consumption.

Speculative Motive

The speculative element is more prominent in the case of sensitive commodities. Speculation could be due to non-availability of item as demand exceeds supply or fear of increase price levels in future, forcing the inventory controller to hoard the stock. Obviously hoarding is possible only for non-perishable items with availability of funds and warehousing space. Another factor to be considered in increased stock levels is the question of trade discounts or quantity discounts. The inventory controller should avoid obsolescence of items ordered in bulk. He has also to compute whether increased carrying charges are amply compensated by the decrease in ordering cost and decreased prices. The seasonal factor also forces higher inventories. In the case of agricultural products the lower price during the harvest and the increased carrying costs should be compared against the higher offer prices and lower carrying costs.

Decoupling Concept

A smoker can smoke a cigarette in his house or office and he is decoupled or segregated from the cigarette making machine. This has been facilitated by means of idle stocks at the smoker's end, with the retail merchant, area stockist, and wholesalers and in pan shops and factory godowns. Similarly, finished goods are decoupled from raw materials in engineering industries by means of the idle stock of assemblies, subassemblies, work in progress and semi-finished goods. Otherwise, every time an item is made, there will be changes of jigs, fixtures, punches and dyes, resulting in increased set-up costs. Hence a continuous run of one process is made and kept aside as semi-finished goods before the next process is started.

ORGANISATION OF INVENTORY

Internally, almost everyone in the organisation is interested in the inventory, while suppliers, bankers, carriers, insurers and customers are the external agencies interested in the company's inventory. The finance manager wants the locking of funds or working capital commitment to be minimum, while the user customer complains about the non-availability of the item. The banker is interested, in the inventory as the company's inventory has been mortgaged to the commercial bank. The supplier is interested in supplying more items, while the transporter desires a full wagon load. Government departments like imports, sales tax, excise and customs are interested in the accuracy of figures to collect taxes.

Under the integrated materials management, raw materials inventory is under the control of materials manager. There are organisations where this inventory is combined under the function of production planning and control function. The machinery spare parts stored in the warehouse are usually controlled by maintenance executives and semi-finished work-in-progress is controlled by production executives. Inventory of finished goods in warehouses and in retail distribution channels is controlled by the marketing executive. Since one component of inventory is related to another, it is essential to coordinate these functions for the smooth functioning of the total inventory system. Evaluation can be done, initially by the executives controlling the inventory and later by the top management team.

CONCEPT OF OPERATING CYCLE

Working capital can be defined as a gross working capital, which includes funds in all current assets, comprising mostly inventories. The most important factor in determining the working capital needs of a company is its operating cycle. The period from the stage of investing cash in raw materials and ultimately the realisation of cash from sales of finished goods is called an operating cycle. When purchases are made, cash goes out and raw material comes in; value is added in the raw materials to convert into semi-finished works-in-progress and then into finished goods. These finished products are sold and finally cash is brought to purchase further raw materials; thus the operating cycle is completed.

The level of inventories needed in a company depends upon the length of the operation cycle. The longer the operation cycle, the higher will be a company's requirements for lock-up funds for various stages of the operating cycle and vice versa.

The length of operating cycle depends upon the nature of business, production policies, manufacturing process, terms of purchase and conditions of sales. It is desirable to reduce the individual components and the total time of operating cycle in order to reduce the level of inventories and working capital commitment.

LEVELS OF INVENTORY

Levels of inventory in any organisation depend upon several factors including social, political, economic, ethnic, fiscal, governmental, monetary policies at the global and national levels, which determine the demand and supply parameters of an item. At the unit level, cost, criticality, availability, service level, stockout, lead time, powers of delegation, consumption pattern, etc. affect the levels. For instance, the stock level of a low cost — highly critical — long lead time spare part is much higher at around three years consumption level. But a non-critical-high cost-short lead time items have nil stock. In the finished goods sector, the marketing manager aims at a high turnover by keeping about two months stock. The ordering/order processing charges of an item on the average is around Rs. 600 per order. The carrying cost, including the cost of capital and storage charges, is about 30 per cent on the average value of inventory.

Stockout cost is always greater than inventory holding charges and hence, the tendency to stock more, particularly in the area of machinery spare parts, leading to increased incidence of obsolete spares. Organisations adopt a no-risk policy resulting in increased stock levels. Other reasons for higher stock level include fear of non-availability, overbuying, inaccurate forecasts, changes in production plan, lower consumption, delays in transportation, bottleneck in communication, accumulation of rejected material, inadequate intelligence system, technological changes and rationalisation. Scientific inventory control principles help to identify the reasons for high level of stocks and help the executive to arrive at optimum levels of inventories.

TEN POINTS TO PONDER

1. Define the terms *inventory* and *inventory control*.
2. State the objectives of inventory of raw materials and finished goods.
3. Identify the significance of scientific inventory management.
4. Explain the significance of transaction motive in inventory control.

5. Distinguish between precaution and speculation in the context of inventories.
6. Describe the relevance of decoupling in inventory control.
7. Detail the organisational aspects of inventory management.
8. Relate operation cycles with inventory control.
9. Examine the relevant factors affecting inventory levels.
10. Critically analyse the need for higher inventory levels in India.

31. Tandon Committee and Chore Group Norms

A TO Z OF BANK FINANCING OF INVENTORIES

Guidelines for working capital financing have been formulated by the Tandon Committee and implemented in the Indian Economy since 1976. These have been reviewed by the Chore group in 1980 and have been adopted by the industries borrowing money for inventory from commercial banks. Let us now examine the financial scenario in India during the early 1970s.

(a) The working capital for funding different forms of inventory — namely, raw material, machinery spares, semi-finished and finished goods — was arranged through short-term bank borrowings.

(b) Several committees such as Laxminarayan Committee, Deheja Committee etc. suggested that the bankers should take more interest in the finance of working capital, but these recommendations were not implemented.

(c) Prior to 1974, due to cheap credit policies, inventory carrying charges were about 20 per cent while the inflation rate was higher in India.

(d) This phenomenon of increased inflation necessitated hoarding of materials in a few companies.

(e) The hoarding of goods resulted in the shortage of materials for others.

(f) Bank finance for working capital is intended only for short-term inventory financing, but was used for other purposes, as there was no end-use control or inspection by bankers to ensure that the money was used only for the intended purposes of inventory financing.

(g) Short-term working capital was diverted to longer-term financing of projects.

(h) The credit flowing from commercial banks to the industry expanded significantly, but this was not reflected in the industrial scenario in India causing anxiety at the financial policy-making level.

(i) There were a number of bogus bills, by which process more than one party obtained bank finance by hypothecating the same item.

(j) Bulk of the bank money — which is public money — was cornered by the large and medium borrowers.

(k) As the total amount of bank finance for lending was limited, the weaker sections, artisans, marginal farmers, taxi drivers, self-employed personnel get very little financial help out of the banking system.

Tandon Committee Recommendations

(l) The Government of India appointed Mr Prakash Tandon, the Chairman of the National Council of Applied Economic Research, and fourteen other experts from different fields to suggest guidelines regarding follow-up and supervision of bank credit.

(m) The Tandon Committee interviewed a large number of professionals, industrialists and others, before formulating their recommendations. These recommendations have been implemented since 1976 for borrowers of 50 crores and above per year.

(n) The major implication to the banking sector has been that the style of credit should change from the conservative security approach to potential and growth. As a result of this approach, today a large number of taxi drivers, milk vendors, vegetable sellers, etc. can get bank finance.

(o) Margins of individual products vary from item to item. The average margin is about 30 per cent and the party gets only 70 per cent of the value of the item, so that the banker could use the security in the event of non-payment of the term loan.

(p) The loan was divided into anticipated fixed component and a fluctuating cash component. The cash component attracts more interest from borrowers.

(q) The Committee has mentioned that the bank should check, if the spare parts inventory exceeds 5 per cent of the total inventory value.

(r) The Committee has suggested norms for different categories of industries, but these have to be utilised with great flexibility without affecting production.

(s) Bunched receipt of materials from imports or from canalised agencies, bottlenecks in transportation, electricity, power breakdowns, delays in receipt of government receipts, and operations of *force majure* clauses have been noted by the Committee. But the report does not consider any of these to be in its purview.

(t) The Committee desires that the borrowing units should supply a management information system, on a quarterly role over basis of future manufacturing activities, cash inflows and outflows to the banker before getting the working capital loan.

(u) The Committee has prescribed broad norms for each class of inventory for 15 categories of industries accounting for the bulk of bank borrowings. The Committee has suggested that the purpose and spirit of these norms hold good for other types of organisations also.

Inventory Norms

(v) Raw material has been expressed in terms of monthly consumption, semi-finished WIP in terms of monthly production value and finished goods in terms of monthly sales. In some industries, finished goods have been separated from the accounts receivable. Fluctuations for items like cotton, bamboo, different classes of inputs for fertilisers, etc. have been included in the norms.

(w) These inventory norms are available in Table 31.1

Table 31.1 Inventory Norms

Industry	Raw Material	WIP	Finished Goods		Accounts Receivables
Textiles	2.50	.75	–	2.25	–
Man-made fibre	1.50	.50	–	1.75	–
Jute	2.50	.25	1.00	–	1.50
Rubber	2.00	.25	–	1.75	–
Fertilisers	2.00	.25	1.00	–	1.25
Vegetable oil	1.00	–	–	0.75	–
Dyestuff	2.25	.50	2.00	–	1.25
Chemicals	2.75	.25	1.00	–	1.25
Cement	1.50	.50	–	1.00	–
Pharmaceuticals	2.75	.50	2.00	–	1.25
Paper	2 to 6	–	1.00	–	1.00
Engineering auto	2.25	.75	–	2.50	–
Engineering durables	2.00	.75	–	2.50	–
Engineering ancillaries	2.00	.75	–	2.50	–
Machineries	2.75	1.25	–	3.50	–

(x) The Committee has also suggested three methods of permissible bank borrowings for working capital financing. The borrower has to contribute at least 25 per cent of the working capital gap — defined as current assets minus current liabilities other than bank borrowings according to the first method. The second method requires financing of 25 per cent of the total current assets. In the third method, he should further finance the core current assets. These methods have been suggested to strengthen the current ratio and to bring financial discipline in economy.

Chore Group Suggestions

(y) The Chore working group in 1980, in its proposed amendments to the lending system, has suggested the scrapping of loan and fluctuating components. But limits for separate peak levels and non-peak levels have to be fixed. The group has advised banks not to entertain ad hoc requests or temporary limits. Banks provide additional finance by separate non-operative cash credit accounts at a higher cost in a time-bound programme of repayment.

(z) Withdrawal of funds has been regulated through submission of simplified quarterly statements on projections of actual performance and half-yearly statements for operating and funds flow statements. The group has advocated the use of bill market system and throwing 50 per cent of the credit limits for raw materials through drawee bills, and toning up of assessment techniques by the banks. The Chore group has also suggested shifting to second method of reducing bank borrowings of industries, but this has not been implemented.

IMPLICATIONS TO USERS

The most important implication is that short-term working capital loan for financing the different categories of inventories has been drastically reduced, forcing the organisations to look into their flabby industries. Supplying management information system to bankers will imply passing on the secrets of manufacturing data. The spare parts-intensive organisation has to rely on the banker's flexibility as spare parts inventory constituting only 5 per cent of the total inventory, is woefully inadequate.

The policy of bank finance to be based on potential and not growth may land bankers in loan recovery problems from individual clients.

At present, the total deposits in the form of demand and time liabilities are over Rs. 300,000 crores and about one third is invested in the industrial sector, with the remaining going for the needs of priority sector, weaker section, agricultural sector, trade, government loans, etc. About Rs 3000 crores are locked in two lakh sick small scale industries. Hence the banker's role in guarding public funds has become very critical in the development of Indian economy. There is a permanent committee in the Reserve Bank of India to solve the difficulties of borrowing firms.

TEN POINTS TO PONDER

1. Identify the possible sources of working capital for financing the components of inventories.
2. Examine the necessity of credit regulations by bankers in inventory financing.
3. Describe the genesis of Tandon Committee in India.
4. Explain the Tandon Committee recommendations.
5. Discuss the Tandon Committee inventory norms.
6. Analyse the three methods of working capital financing and their implications.
7. Explain the Chore groups recommendations.
8. What are the implications of Tandon Committee and the Chore Committee to banks?
9. Examine the implications of Tandon and Chore committees to the borrowing organisations.
10. Identify the problems occurring in industries due to the financial discipline of bank borrowings.

32. Right Order Quantity

CLASSIFICATION OF INVENTORY

Classification of inventory is the first step to determine optimum inventory levels. As already seen one way is to classify them as raw material, machinery spares, semi-finished work-in-progress, and finished goods. Another way is to classify each of the above as ABC, VED, or MUSIC-3D categories. But, in order to solve the inventory problem, the usual method of classification is based on supply and demand.

The demand/consumption/usage can be either fully known or partially known expressed in terms of probability, or completely unknown resulting in an increasing amount of uncertainty. These three situations are known as *demand certain, demand risky* and *demand uncertain*. Similarly, on the supply front there are two distinct possibilities. *Supply* is known as *static* if only a single supply is possible during the consumption period. *Supply* is known as *dynamic* if more than one supply in repetitive fashion can be obtained during the consumption period. These states of nature in demand and supply result in the following six different situations:

(a) Supply static-demand certain
(b) Supply static-demand risk
(c) Supply static-demand uncertain
(d) Supply dynamic-demand certain
(e) Supply dynamic-demand risk
(f) Supply dynamic-demand uncertain.

In all these cases, the inventory controller has to answer the two basic questions of (a) how much to order, and (b) when to order, for determining the optimum levels of inventory.

SUPPLY STATIC SITUATIONS

Supply static situations represent a single shot, one-time supply in consumption period. Solution for supply static-demand certain is self evident. Certainty of demand implies complete information on requirement schedule, price and lead time. The inventory controller has to place order for the required quantity taking the shelf-life and lead time into consideration. In the ideal case, stock need not be held if planning is perfect. Typical examples include construction of a road bridge or inputs in process industries, which are fixed in practice.

As the level of knowledge decreases, the risk in decision-making increases. Printing of calendars, diaries in a year or a newspaper on a particular day and entering a service contract are typical single shot examples of risky situations. Availability of previous data or past experience can express risks in terms of probabilities, which help to measure the costs and benefits and hence the inventory pattern.

In demand uncertain situations, level of information will be the least. It is not possible to quantify information in terms of probabilities. The decision-maker has to use subjective probabilities on costs, benefits and reach the solution based on experience. Delphi techniques of pooling the expertise of many individuals by iterative process may improve the situation. Most of the research and development projects' machinery spares bought initially with the equipments, new projects executed for the first time with very little information are examples in this category.

SUPPLY DYNAMIC SITUATIONS

The dynamic conditions of supply enable us to place as many orders as possible. Supply repetitive or dynamic, demand certain condition results in the EOQ which is the topic of this chapter. Supply dynamic, demand risky condition will be discussed in the next chapter. Supply dynamic, demand uncertain situation does not exist in practice, as it is mutually inconsistent, since supply dynamic condition implies the occurrence of the same event earlier, from which relevant information can be collated.

Economic Order Quantity

In the demand-certain situation, the total ordering charges are a direct function of the number of orders. Demand from consuming centres is fairly well established. We can meet this demand by either having a few orders resulting in a large inventory, or by holding low inventories and correspondingly increase the orders. Since ordering and carrying charges move in opposite directions, we have to balance the two costs in order to get the optimum or economic order quantity.

ARRIVING OPTIMUM INVENTORY

There are several ways of arriving at the optimum order quantity which reflects the inventory of an item. The conventional method is by the economic order quantity of formula. Let us initially arrive at the optimum inventory by trial and error method with the help of an example.

Let the annual sales/consumption/forecast known demand be 40,000 units, priced at Rs 2 each. Let the ordering charges — with one order per delivery — be Rs 750/-. We note that this is the same as ordering processing cost in finished goods sales or the set up cost for the work-in-process inventory. Let the carrying charges be Rs 30 on the average value of inventory. The immediate availability of item and constant rate of consumption pattern are assumed. Let us try to find 'Q' — the optimum quantity to be ordered by trial and error method. If we order 40,000 units, then one order with ordering charges of Rs 750, for Rs 80,000 will be placed per year. This results in the average inventory of Rs 40,000 and carrying charges of Rs 12,000. These results are tabulated in Table 32.1 for various values.

Table 32.1 Costs and Consequences (in Rs)

Order Quantity	Orders/ Year	Ordering Charges	Average Inventory	Carrying Charges	Total Cost
80,000	1/2	375	80,000	24,000	24,375
40,000	1	750	40,000	12,000	12,750
20,000	2	1500	20,000	6000	7500
10,000	4	3000	10,000	3000	6000
5000	8	6000	5000	1500	1500
2000	20	15,000	2000	600	15,600

Thus the lowest total cost is Rs 6000 corresponding to an order quantity of 10,000 items. The stockout cost and understocking costs are irrelevant in this situation. This order quantity, which optimises economises the total systems cost, is known as EOQ — Economic Order Quantity. At this point, the inventory carrying cost and ordering charges are the same. We can plot the costs on the *Y*-axis, against the order quantities on the *X*-axis. We obtain EOQ (Economic Order Quantity) by the curve shown in Fig. 32.1. Computer programs are also

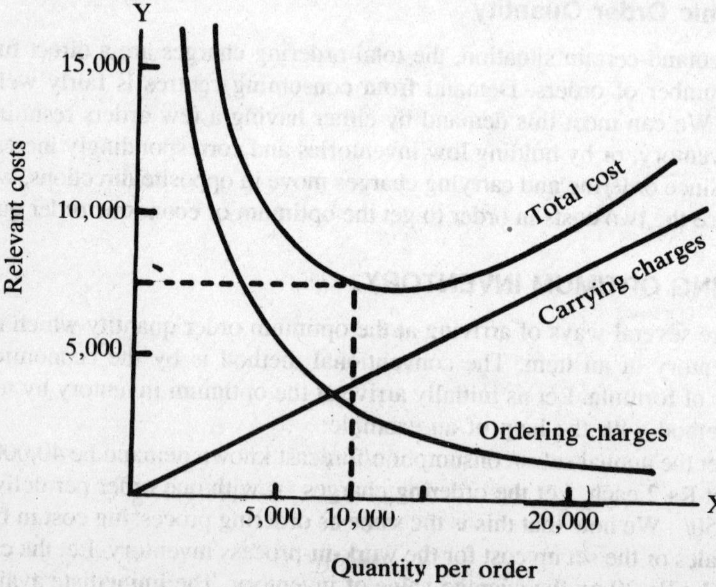

Fig. 32.1

available to decide how much to order. Our research experience shows that managers take a negative view of EOQ, focussing on its limitations, rather than using it as a guideline.

Derivation of EOQ

Let E be the estimated annual consumption of an item, priced at Rs P each. The value of total consumption is EP. Let C_o and C_c be the ordering and carrying charges. Q the unknown quantity, is derived as follows:

Number of orders per year = (E/Q)
Total ordering cost per year = $(E/Q)\ C_o$
Average value of inventory = $(E/2)\ P$
Total carrying charges, = $(E/2)\ PC_c$

These two costs are equal i.e.

$$E_2PC_c = (E/Q)\ C_o$$

The total cost of the system is

$$(E/Q)\ C_o + (E/2)\ PC_c$$

In order to arrive at the maximum, we derive the above total cost function with regard to the only unknown Q and equate to zero, i.e. $-(EC_o/Q_2) + P_2C_c = 0$. Thus the optimum quantity is obtained as

$$\text{EOQ} = \sqrt{\frac{2EC_o}{PC_c}}$$

$$= \sqrt{\frac{2\ (\text{annual consumption})(\text{ordering charges per order})}{\text{unit price carrying charges}}}$$

Substituting the previous values, we get

$$\text{EOQ} = \sqrt{\frac{2(40{,}000)(750)(100)}{2(30)}} = (200)(5)(10)$$

or 10,000 units.

$$\text{EOQ (Rs)} = \text{EOQ}\ P = \text{EOQ}\ (P)$$

$$= \sqrt{\frac{2EC_o}{PC_c}}\ p \times p$$

Simplifying, the above equations can be reduced to

$$\text{Constant} = \sqrt{\text{Annual consumption value}}$$

The EOQ value on inventory value varies as the square root of annual consumption value. Since the order quantity represents materials, and sales =

(materials + labour + profit), the inventory value increases as the square root of increases in sales. This implies that the inventory should increase only about the square root of the number of times the sales increases. This also shows that the inventory will be less if stores are centralised.

If sales increase four times in the coming years, the corresponding inventory will increase only 4 or two-fold for the same conditions of manufacture.

SENSITIVITY AND STAGGERING

Calculation of EOQ assumes constant rate of consumption and immediate availability of item, without any fluctuations in demand and supply. EOQ gives only a broad guidance in estimating order quantities. For practical situations, one can definitely deviate from the economic order quantity, or place an annual order with stipulated delivery schedules. But, staggering involves additional efforts on the part of the warehouse manager, in follow-up, documentation and other aspects the costs of which have not been calculated by any organisation. Hence it is advantageous to stagger in terms of economic order quantity.

If we examine the EOQ figure, the curve is very flat near the optimum value, suggesting we can deviate from the optimum in practical situations. This is also seen from Table 32.1 where 50 per cent deviations in quantities result in 20 per cent increases in the cost. By means of detailed calculations, the effect of the total cost increases in ordering and carrying, for various order quantities as shown in Table 32.2.

Table 32.2 Sensitivity Analysis

If you produce or order only this per cent of order quantity:									
10	20	33	50	80	100	125	200	300	500
Then the increased costs from optimum costs are as a percentage of:									
405	160	67	25	3	0	3	25	67	160

This sensitivity analysis enables the inventory manager to deviate from the theoretical optimum, easily up to ± 50 per cent of the economic order quantity, with increased costs being only 25 per cent. In other words, staggering may be done less than 50 per cent and availing bulk discount would be similarly up to 50 per cent from the order quantity. Beyond the Range of ± 50 per cent deviations from the economic order quantity, one has to calculate the costs and consequences before arriving at a decision.

But before accepting trade discounts for large quantity, the warehouse executive has to compute, whether increased carrying charges are compensated by decreased prices and ordering costs. Obsolescence of items ordered in bulk must also be examined.

TEN POINTS TO PONDER

1. Deliberate on the inventory problems on the basis of supply and demand.
2. Explain the analysis of supply static aspects.
3. Give example of supply static, demand certain, risk and uncertain.
4. Examine the meaning of supply dynamic aspects.
5. Prove that increase in inventory is proportional to the square root of the increases of demand/sales.
6. Identify the assumptions behind EOQ.
7. Draw the economic order quantity curve.
8. Derive the economic order quantity formula.
9. Discuss staggering and EOQ.
10. Describe the sensitivity analysis of an EOQ.

33. Quantity and Periodic Inventory Systems

PRACTICAL SYSTEMS

Any inventory system must answer the two basic questions — how much to order, and when to order. There are several mathematical, statistical and operations research methods which can be used to answer these questions. The most popular systems commonly practised are the 'Q' system or the quantity system for high consumption value items and 'P' system or the periodic review system, or cyclic system for low consumption value items. These models fall under the category — supply dynamic, demand risky situations.

The starting point is the EOQ model, but the EOQ assumes constant rate of consumption; it also assumes that materials will be available as and when ordered. In real life, however, such ideal conditions do not exist and the stores management will have to provide for various such contingencies. The consumption rate can suddenly increase or the time required for procurement can increase, or both may happen simultaneously. The economic order quantity also does not consider shelf-life, seasonality and consumption pattern.

BUFFER, SAFETY AND RESERVE STOCKS

To protect against such adverse conditions causing fluctuations in demand and supply as well as to avoid stockouts, additional stocks known as *safety* and *reserve stocks*, have to be maintained. The level of these stocks varies from item to item and depends upon the lead time, review cycle and service level required. The safety stocks provide for average consumption during any increase in lead time from the normal lead time. Reserve stock provides for an increase in the rate of consumption during the average lead time. In general, the industry uses the term *minimum stock* by combining safety and reserve. *Buffer stock* is that quantity which provides for normal consumption during average lead time.

Statistics can be used advantageously to take care of uncertainties in consumption and lead time. Consumption of regularly used items can be approximated by an average consumption rate per period with a standard deviation during that period. Similarly, the lead time can be considered in terms of an average lead time. Based on past experience, a probability can be attached to a maximum increase in lead time from this average.

FIXED QUANTITY SYSTEM

As the name implies, the order quantity in the fixed quantity system or 'Q' system is *fixed*, but the order period is *varied*. In other words, an order for the fixed quantity is triggered off whenever the reorder point is reached. This reorder point in the Q system is defined as the total of buffer stock, reserve stock and safety stock.

Let us understand the Q system by working out an example. The demand for a regularly consumed item, from previous records, indicates a normal distribution with a weekly average sales consumption rate of 400 items and a standard deviation of 50 units. The unit price is Rs 750 per order. One order signifies one delivery. The inventory carrying charges are 30 per cent on the average value of inventory carried. Lead time fluctuates between 1–7 weeks with per cent frequencies as 4, 15, 18, 28, 16, 13 and 6. Develop the 'Q' inventory system for the above

EOQ Computation

In the *Q* system, the reorder point is the total of average lead time consumption, safety stock and reserve stock. Since the weekly demand is 400 units, the annual demand is (52 × 400) units. The economic order quantity is calculated as follows:

$$\text{EOQ} = \sqrt{\left[\frac{(2)(52 \times 400)}{26}\,\frac{100}{30} \times 750\right]} = 2000 \text{ units}.$$

The average lead time is obtained by weighted average:

$$[(1 \times 4) + (2 \times 15) + (3 \times 18) + (4 \times 28) + (5 \times 16) + (6 \times 13) + (7 \times 16)]/100$$

or

$$[(4 + 30 + 54 + 112 + 80 + 78 + 42)]/100$$

400/100 or 4 weeks.

The average lead time consumption is (4 × 400) or 1600 units. Stocking up to the average lead time satisfies the demand with (16 + 13 + 6) or 35 per cent stockouts.

Safety Stock

Safety stock to cater to lead time fluctuations is calculated by multiplying the maximum delay, probability of maximum delay and the average demand. The safety stock is obtained as

$$(7–4)(.35)(400) \quad \text{or} \quad (3)(.35)(400) = 420 \text{ units}$$

Reserve Stock

The reserve stock for catering to consumption fluctuations is obtained as follows: If 50 is the standard deviation of consumption for individual weeks, then, by property of convulsion, the standard deviation for the lead time of four weeks is ($\sqrt{4}$) (50 = 2150) units. The service level depends upon the cost, criticality availability of an item, estimated by ($K_v/K_u + K_o$). This is usually taken as 95 per cent for regular items with a corresponding ordinate in the normal distribution as 1.64. The reserve stock is thus obtained as (1.64) (100) or 164 units.

Reorder Point

The buffer stock is (400 × 4) = 1600 units. The order point is the sum of buffer, safety and reserve stocks.

ROP = (1600 + 164 + 420) = 2184 units.

When this order point of 2184 units is reached, then place an order for the economic order quantity of 2000 units. This process is continuously monitored.

PERIODIC REVIEW SYSTEM

In the periodic review P system, the economic order quantity is converted into time scale, and this period is known as *review time* or *cycle time*. The period between the placement of orders is fixed, while the quantity ordered is varied. In the system we consider only the variation in the lead time. But in the P system, we have to plan in addition for any increase in consumption during the period between orders.

Obviously, the average inventory held in the P system is greater than that held in the Q system. But the great advantage lies in the reduction of monitoring labour and launching of orders. In the periodic review system, the buffer stock represents the average consumption during lead time and review time. The safety stock is the same as in the Q system to denote increased consumption as a result of possible extensions in lead time. The reserve stock is calculated on the basis of excess consumption during lead in time and review time.

DESIRED INVENTORY LEVEL

The Desired Inventory Level — DIL — or the virtual stock is worked as the sum of the three components, namely, buffer, safety and reserve stock. The review time is calculated as EOQ/400 or 5 weeks. If this figure happens to be in decimals as 5.4, then we have to calculate the ordering and carrying charges for five and six weeks before rounding off, to optimise the total costs. The new buffer will be consumption in average lead time and review time.

Buffer = 400(4 + 5) or 3600 units

The standard deviation for reserve stock calculation is

$\sqrt{(3 + 4)}(50)$

or

$\sqrt{7}$ (50) or 132 units.

The reserve stock is (1.64) (132) or 215 units, for a corresponding service level of 95 per cent. So the DIL is (3600 + 420 + 215) = 4235 units.

Under the periodic review system, stocks are reviewed once in five weeks. If stocks are more than the desired inventory level, no action is taken till the next review. On the contrary, if stocks are less than the desired inventory level of 4235 units at the time of review, an order for the difference between the DIL and existing stock is triggered off.

COMPARISON OF Q AND P SYSTEMS

In the Q system, the order quantity is fixed, but the time of ordering depends upon when the reorder point is reached. In the P system, the time of ordering is fixed but the quantity varies from one review cycle to another. The Q system requires continuous monitoring as to when the reorder point is reached. This is particularly useful for high value items. The use of computers is a must for the Q system. Stock levels are lower in the Q system. On the contrary, the lead time is automatically increased by the review time, resulting in more inventory to be carried in the P system. But the advantage lies in the bunching of materials and reduction in monitoring time. Hence, we can apply the periodic review system for low consumption value items.

Large number of low value items can be grouped into four categories as annual review, half yearly review, quarterly review and monthly review in the P system relieving the warehouse manager from daily monitoring of the records of a large number of items. The inventory manager can delegate the low value items to the subordinates to be reviewed at the time of issue or once a month/ or quarterly/or half yearly. He can better utilise his time by concentrating on high consumption value items using the Q system.

TEN POINTS TO PONDER

1. Deliberate on the statement that EOQ is not a practical system.
2. Distinguish the meanings of buffer, safety, reserve and minimum stock levels in warehouse.
3. Describe the working of the Q system.
4. Identify the meaning of R.O.P. in the Q system.
5. Examine the category of items for which Q system is applicable.
6. Explain the term *periodic cyclic review system.*
7. How do you compute the review time in the P system?
8. Detail the procedure for computing the desired inventory level.
9. Mention the items for which the P system can be applied.
10. Explain the difference between the P and Q systems.

34. Inventory and MRP

MATERIAL REQUIREMENT PLANNING

Material requirement planning (MRP) is defined as the scientific method of planning for arriving at time-phased requirements of materials, components, parts, subcontracted items for meeting the demand of finished goods at the stipulated time. Sometimes this is also referred to in the literature as dependent-demand inventory system. The material requirement plan derives the gross quantity of components required by using the product structure records, and then the net quantities of materials are calculated by considering on-hand inventory. The total lead time for each material and the matching time period when material should be available are also obtained. This enables the system to plan order releases, purchase orders, work orders and reschedule notices. The Americans perfected the applications of material requirement planning in the late sixties.

INPUTS AND OUTPUTS

Usually material requirement planning is done with the help of computer programs on material requirement packages. For this purpose, the input system requires the corporate business plans, aggregate production plans, sales forecasts, master production schedules, inventory status file, and the parts explosion list or the bill of materials. The primary computer output includes plant order schedules, order releases and changes to plant orders. The secondary output from computer comprises planning reports, performance reports, exemption reports and inventory transactions.

LIMITATIONS

The problem areas in using the computer material requirement packages include the following:

(a) Poor inventory level accuracy
(b) Inaccurate forecasts of total lead time
(c) Scrap factors and allowances at different manufacturing stages
(d) Production capacity constraints
(e) Economic batch size for manufacturing
(f) Acceptance of any level of scrap, lead time and stocks without challenging

(g) Modifications in computer packages to suit the needs of the firm
(h) Impractical assumptions of master schedule
(i) Incorrect information system
(j) Centralised planning.

However, the materials requirement planning is effectively used in assembly-oriented industries.

MANUFACTURING RESOURCE PLANNING

MRP-II or Manufacturing Resource Planning II, pioneered by Oliver White, is an improvement over the materials requirement planning discussed so far. The system uses a set of computer-supported planning and scheduling tools that enable management to control production scheduling, cash flow, labour, capacity utilisation, inventory levels, physical distribution and materials purchases. With MRP-II system, marketing, finance, materials, and manufacturing subsystems perform their individual functions as one integrated system. This is a departure from the traditional approach in which each department has its own individual stakes in all areas. It is based on getting the right material at the right time. This manufacturing resource planning system enables to continuously monitor production and preplan production changes to give the desired response with the help of computer systems. It is more than a planning tool, as it produces schedules which prescribe how to carry out plans by emphasising feedback loop.

This manufacturing resource planning has six levels, namely:

(a) Business planning
(b) Production planning
(c) Master production scheduling
(d) Materials requirement planning
(e) Capacity requirement planning
(f) Ordering systems.

By deriving the gross quantity of components using product structure records, the net quantities of materials are calculated by the stock on hand. Using the knowledge of lead time for each material, the date of material availability is calculated. This system formalises the decisions that were left to the individual's discretion hitherto. This enables releasing procurement orders, work orders, reschedule notes and inventory levels by effecting linking of subsystems and centralised planning in a planned fashion. MRP results in the improvement of product quality by capturing reasons for failures and returns and provides a basis for controlling the production environment. It optimises inventory investment by improving inventory accuracy and measures vendor performance adequately.

Many organisations in the USA report significant improvement in inventory turnover, delivery time and meeting delivery promises, reduced frequency of split orders and less number of progress-chasers as a result of application of MRP.

PRODUCTION PLANNING

Production planning is the management function of establishing levels of manufacturing operations applicable to some future periods of time. Production plan is made after considering sales forecast and resource requirements. The resource requirement includes availability of men, machines, materials, stores space, financial resources, etc. In this context, long-term planning deals with projections as far into the future as is possible with some logic. Short term planning deals with operational details of a period up to one year. Necessity of careful production planning arises from the following factors:

(a) Increased complexity of production and distribution systems
(b) The need for careful timing of interrelated activities
(c) The necessity for anticipation of changes and orderly reaction to them
(d) The desire to achieve the most economical combination of resources.

In this context, production scheduling-defined as fitting of specific jobs into a general time-table so that orders may be manufactured in accordance with contractual liabilities so that each component may be produced in proper sequence and time — assumes great importance.

The differences in the process of planning and scheduling are given in Table 34.1 below:

Table 34.1 Comparisons Planning and Scheduling

Planning	Scheduling
Precedes scheduling.	Follows planning.
Identifies resources and their availability.	Determines who does what, with what and when.
Estimates and allocates labour.	Serves as the plan's spark-plug.
Devises procedures for procuring tools, materials and personnel.	Decides on the work to be done within a specific time.
Simplifies communications between maintenance and production.	Specifies times, places and events.
Establishes approval procedures.	Follows through on priorities.
Indicates priorities.	Stimulates planned actions.
Coordinates activities with other operations.	Specifies personnel and crafts needed.
Sets up specific course of action.	Deals in small time intervals.
Remains flexible to meet changes.	Tends to be routine, not innovative.

BILL OF MATERIALS (BOM)

The bill of materials approach is an important aspect in production planning.

When a design engineer completes the design of a part or an assembly, he makes a list of all machines required to manufacture the item. The bill of materials with a production schedule can be sent directly to purchasing department as a notification of production department's need for materials. The production control department can also send a copy of the engineering bill of materials and send a copy to stores as the document authorising release of materials. A daily report of withdrawal is sent to the inventory section for deduction from current inventory balance.

Explosion of bill of materials refers to the splitting of requirements for the product to be manufactured into its basic components. With the codes and specifications, the demand forecasts of the number of items enable us to get the total requirements. Allowances for rejection at different stages are then considered to get the complete requirements. An explosion chart is a series of bills of materials grouped together by combining the requirements of different components. For instance, a table can be split into its legs, top, and screws.

GOZINTO MATRIX

The process of parts explosion for the correct development of demand schedules for dependent item can be done by matrix methods. The 'Gozinto' philosophy gives the matrix method of describing the relationship between the end product explosion and their subassemblies as well as with the individual components that make up the final product. This philosophy describes the 'goes-into' relationship of components into end products to obtain the bill of materials. This process involves developing immediate coherent relationship of individual components, intermediate assemblies and final product, in the form of a total matrix. The use of matrix methods facilitates computerisation. If the 'Gozinto' matrix is fed into the computer, then with the help of lead time, the overall requirements and schedule of deliveries of components can be obtained from the computer.

ADVANTAGES OF BILL OF MATERIALS (BOM)

The bill of materials, known as BOM, is the simplest technique of materials requirement planning. BOM drawn with required lead time and necessary contingency events finally turns into indents for procurement. It helps in keeping a watch over the delivery of matching components and spares going into the production process. It enables the evaluation of progress of the project undertaken and ensures the flow of needed materials. It helps in avoiding the locking up of funds unnecessarily by properly scheduling the orders, delivery and arrival of materials.

The bill of materials technique is ideally suited to engineering industries, as a large number of components are needed for assembling an end product which is to be obtained from various sources in varied quantity and quality. Quarterly planning, broken into actual monthly requirements, is ideal in the preparation of the bill of material.

PAST CONSUMPTION DATA

Materials requirement planning can also be done by the use of past consumption data. As referred earlier the frequency distribution of any measured characteristic, including consumption figures, follow the normal distribution pattern. We have also seen that the Normal distribution is uniquely identified by its mean and standard deviation and an interval containing (mean ± 3 standard deviation) includes practically all future values. Hence (mean ± standard deviation), could be used as the starting point for materials planning purposes. If the standard deviation is too large, the reasons for the high level of fluctuations must be studied technically in order to reduce the same. This technique can be successfully used in process industries, where the materials are being used regularly on a continuous basis. The Normal distribution has to be replaced by the Poisson distribution if we are dealing with maintenance spares. Since the Poison distribution is identified uniquely by its average only, (average + *k* average) gives the requirement determination for materials planning. Here the level of *k* depends upon cost-criticality and availability analysis of a particular spare part. Besides the above methods, the planning process, must be supplemented by the knowledge of the executives and vendors', information on failure, usage replacement and consumption patterns.

TEN POINTS TO PONDER

1. Define thc materials requirement planning process.
2. Specify the inputs and outputs needed for MRP-I.
3. Identify the limitations of MRP-I.
4. Discuss the utility of manufacture resource planning.
5. Explain as to how MRP-II is better than MRP-I.
6. Describe the scope of production planning in inventory control.
7. Examine the utility of 'Gozinto' matrix in inventory function.
8. Elucidate the significance of bill of materials.
9. Amplify the differences between planning and scheduling.
10. Examine the relevance of past consumption data in material requirement planning.

35. Just In Time Inventory

THE JIT CONCEPT

Just In Time (JIT), or the Japanese method of integrated philosophy by team approach, is applicable not only to the various components of inventory but also in most of the day to day activities. Taichi Ohrio applied this concept at Toyota manufacturing company, and so it is also known as the Toyota production system. In this system, the customers pull their requirements from the shopfloor, as against the conventional method of pushing the product to the market. This is a part of workaholic culture of the Japanese to make their products internationally competitive, by integrating the just in time cost with the total quality control system at the shopfloor level. In the *just in time* system, the advanced stage of production, draws the right amount of inventory from the preceding stage to sustain the activity. In this process, the production activity is planned upon the actual demand, rather than a predetermined schedule, since the cycle time for production of various models is given only to the final assembly point of mixed production line. This process is repeated at each stage, right down to raw materials or up to subcontracted items. The production stages are interlinked, in the form of a tree and each production order generates pull throughout the system. This system enables the market place to pull the necessary items from the plant, as against the conventional system of pushing the product to the market. Hence, this system is likely to be operated by most companies in future, as competition develops.

ZERO INVENTORY ZIN SYSTEM

The JIT system is similar to the conventional job shop or make-to-order procedures, with the customers specifying their requirements. At the operational level of inventory management, this is also known as the Zero Inventory concept or ZIN system, practised widely in American organisations. Even in our own country, this system operates in cycle-manufacturing units in Punjab, where the components are brought in the morning by a cycle rickshaw, and the assembled cycle leaves the factory by the same cycle rickshaw. The term *zero* does not imply literally nil inventories, as nil implies losing customers to competitors. Obviously, small quantity of inventory, which is required to sustain the production activity between two operations in the organisation, is maintained. This small quantity is equal to the irreducible minimum under the process motive.

This philosophy is at the extreme end of what is known as JIC — Just In Case system, of hoarding inventory to meet contingencies and uncertainties, particularly in the fields of machinery spare parts and imported parts, which is widely practised in Indian companies. However, the system is more useful for assembly-oriented engineering industries and the automobile sector in India. Both the JIT and ZIN systems emphasise the fact that the production of a part component or subassembly should occur just in time and not earlier or later. This is accomplished by the customers through the market pull mechanism.

A TO Z ADVANTAGES OF JIT

The Japanese companies using just in time system claim the following benefits to the organisation: The system eliminates the waste arising from the following areas:

(a) Over production.
(b) Waiting time in the assembly process.
(c) Transportation bottlenecks.
(d) Excess inventories.
(e) Increased process timings.
(f) Low labour utilisation.
(g) Generation of scrap and rework.

Cost Reduction

The other significant benefits reported by the industries using JIT and ZIN systems are the following:

(h) 25% to 60% reduction in cost of quality
(i) 35% to 70% reduction in raw material stock
(j) 40% to 80% decrease in space requirements
(k) Up to 25% reduction in indirect and direct labour
(l) 10% reduction in process lead time
(m) Up to 80% reduction in finished goods inventories
(n) JIT integrates the thinking worker, listening supervisor, supportive manager and dynamic leadership.

Consumerism

(o) The basic philosophy behind JIT system is that the source of information is always the customer and not the manufacturing capacity of the plant.

(p) Maximum importance is given to the consumer by practising the motto "customer is the king". The salesmen, with catalogue in hand and armed with latest information, probe the customer for his requirements, and the information gathered is fed to the manufacturing system.

(q) The concept of JIT applies not only to materials, but also to manufacturing, sales, marketing, merchandising and distribution by producing exactly what is needed in the required quantities.

(r) The JIT system has enabled Toyota to sell what they bought before paying for the inputs without any need for working capital.

(s) JIT system aims at reducing inventory levels, and exposing the fundamental problems which are bottlenecks by devising an appropriate system.

(t) The application of JIT concept has minimised the inventories at the work spot and the stores to minimum.

(u) Another spin-off from the implementation of JIT concept is the flow of defect-free parts and subassemblies through the production system.

(v) Apart from ensuring smooth flow of materials through the production system, JIT envisages the participation of all employees from top to the lowest level, and subcontractors, to maintain harmonious work and smooth material flow.

(w) The quantity produced is identical with that sold. JIT is based on the concept that in the market place, there is no par, but only competition.

(x) The JIT concept allows for logical relationship in the other management techniques like leadership, creativity, zero defects, total quality control, modernisation, rationalisation, automation, management information systems, operations sequence, setup time, cycle time of manufacturing, smooth material flow, etc.

(y) The JIT philosophy enables managers to ask the question what is our business? — during prosperity and not during depression.

(z) The main theme in the JIT concept is neither past nor future but the current needs of the consumer.

A TO Z STEPS IN JIT

(a) The philosophy behind successful implementation of the JIT concept, is the small lot sizes for handling material flow, through the process of manufacturing.

(b) In the JIT system, the advanced stage of production just draws the right amount of inventory from the preceding stage to sustain the production activity.

(c) The production activity is planned upon the actual demand, rather than a predetermined schedule since the cycle time for production of various models is given only to the final assembly point of the mixed production line.

(d) Each respective stage repeats this process right down to raw materials or up to subcontracted items.

(e) The production stages are well connected in a tree form, and each production order generates a pull, throughout the system.

Redesigning and set up Time Reduction

(f) JIT challenges large batch sizes at manufacturing, long lead times, high rejection rates, flabby inventories and low quality. Hence it is imperative to alter the configuration of machinery, change the components and relay the work place.

(g) This redesigning of work centres eliminates the redundancy in the system, and makes the system flexible to accept the changes required to meet demand fluctuations.

(h) The redesigned tooling and auxiliaries should allow rapid changeover set-ups in order to reduce the batch size.

(i) If four or five stages are involved in manufacturing, then the die or tool changes have to be accomplished in a short span of time, usually in less than 30 min.

(j) The reduction in set-up costs and manufacturing lead times, gives flexibility to the management to forecast the market accurately over a shorter period of time.

Kanban Control

(k) *Kanban* is the Japanese word for a small card or tag, which is attached to a standard bin of inventory. The use of Kanban is essentially a production management system for the production employees in the plant.

(l) The production Kanban authorises the preceding stage to produce the required quantity of components or subassemblies that are listed.

(m) The conveyance Kanban is attached to the container, when it is transported from one stage to the next operation stage. Use of the two Kanbans will expose production bottlenecks and encourage finding out solutions.

(n) The production control processes begin with the insertion of the job card or Kanban in each container of materials. The finished goods inventory of one production department becomes the raw material inventory of the next downstream department.

(o) The final assembly rate determines the production quantum for both vendor-supplied parts and for immediate predecessor or production department, which in turn, determines the production rate for all other production sections.

Subcontracting

(p) Toyota has about 250 primary first level suppliers, 2500 second level suppliers and about 25,000 third level suppliers, and all of them are connected in a tree form. Thus the corporate philosophy aims of subcontracting as much as possible instead of manufacturing everything. These subcontracting companies are the backbone of the JIT system and hold the key to its successful implementation.

(q) The external Kanban is used between the plant and the vendors, by providing relevant information on withdrawal, production, overproduction, work order, occurrence of defects, and inventory control.

(r) The effect of scheduling upstream department production in response to the demand of downstream departments, creates a pull beginning with final assembly and moving upstream through other production departments and subcontractors of parts.

(s) The next step is the preparation of valid plans for the sake of execution and also improving the execution mechanism. Excellent real-time planning and flawless scheduling of execution are the fundamental requirements of JIT.

(t) Since the system is dependent on the supply of good quality subassemblies without delay, it is imperative that defective parts are not supplied by the subcontractors.

(u) Quality control techniques like introduction of automatic inspection devices, use of 'fish bone' charts, ABC analysis, autonomy of shopfloor inspection, and other simple scientific approaches, should be used in the subcontractor's plants to ensure the zero defect philosophy.

Limitations of JIT

(v) It should be appreciated that the time framework required for introducing the JIT system to a total factory, goes beyond the normal time requirement for any system implementation. In other words, JIT is not a quick-fix item, as it calls for sequential decision-making and review at every step of implementation.

(w) The most important step is the commitment of the top management and line function to the philosophy of the worker-oriented JIT approach. The top management should issue policy statement in support of JIT and review the results of the pilot study in a phased manner. Indian organisations, accustomed to flabby inventories find it difficult to switch over to JIT system for all the four inventory components.

(x) In order to aim at the laudable objectives of zero defects and irreducible inventory at all stages, perfect coordination and management information system should exist not only between the different departments of the organisation, but also amongst the vendors and customers.

(y) It should be appreciated that the JIT system is not a panacea for all the ills of the organisation and bad management practices. The attitude that anything sells in India, is embedded in the minds of most managements and hence, the customer is the king philosophy is only a theoretical concept in most cases.

(z) Thus, the steps crucial for implementation of JIT are as follows: Starting the system educating the concerned personnel, accomplishing process improvements, achieving control improvements and extending the vendor-customer links.

The organisations which have introduced JIT system in Japan also claim the following advantages:

(a) Increased awareness of problems and their costs
(b) Reducing lot sizes
(c) Less work in progress

(d) Less raw material inventory
(e) Less indirect cost
(f) Faster market response
(g) Better forecasting
(h) Less finished goods inventory
(i) Higher productivity
(j) Less administration
(k) Faster feedback of defects
(l) Reduced inventory
(m) Smoother output
(n) Less materials waste
(o) Less rework
(p) High quality of finished goods
(q) Improved scrap control
(r) Better controlling of defects
(s) Improved overall working.

TEN POINTS TO PONDER

1. Identify the philosophy of JIT system.
2. Explain the concept of ZIN system.
3. Enumerate the benefits of JIT system.
4. Elucidate as to how JIT enables cost reduction.
5. Discuss the steps involved in JIT system.
6. Describe the process of set-up time reduction.
7. Define the significance of Kanban control system.
8. Examine the relevance of subcontracting in the context of JIT system.
9. Identify the pitfalls of application of JIT system.
10. Illustrate the relevance of application of JIT in Indian industries.

36. WIP Inventory

CONCEPT OF WIP

Ideally the inventory quantity and the value of the WIP — work-in-progress, semi-finished goods — should be zero or the least, but this figure is sizeable in engineering and assembly-oriented organisations and is one of the areas needing immediate control. The semi-finished goods include all materials, parts, subassemblies, and assemblies, which are being processed into finished goods. The items are those actually undergoing productive operations or temporary storage between two successive manufacturing stages. Unlike the process industry, where the raw material is converted directly to finished goods, in the case of engineering goods, the final product is decoupled or segregated by the WIP buffer to facilitate smoother manufacturing operations.

Inventory is a buffer between two or three subsystems like raw materials to manufacturing, manufacturing to finished goods, and finished goods to the ultimate customer. In contrast, work-in-progress acts as a buffer within the manufacturing subsystem itself, which may consist of groups of machines and or assembly lines that can be termed as work centres. The raw materials before they take shape as a saleable finished product, have to be converted by adding value and hence have to go through a combination of processes.

PROBLEMS IN WIP

The increase in varieties of work-in-progress inventory is due to technical bottlenecks. The rate of production at each work centre depends upon the technology profile, manufacturing facilities, capacity utilisation, operator skills and availability of supporting facilities and tools. While the production executive tries his best to balance the lines, a perfect balance is almost impossible. To overcome these problems, work-in-progress is stored as inventory at the work centre interfaces.

Another reason for this inventory is the long finite time period of conversion from raw material to output known as *manufacturing cycle time*. The greater this period, as in the heavy engineering industries, the more will be the work-in-progress which piles up due to lack of certain bought-out components. The shop incharge, blames the materials/warehousing function for not providing matching components at the appropriate time, resulting in increased work-in-progress. The strategy for controlling the work-in-progress inventory is effective planning and coordination of the manufacturing activities. The work-in-progress inventory is the primary concern of the manufacturing department, since this comprises all materials being worked in the shopfloor. Sometimes, off-cuts of heavy material are in the shopfloor and these are also classified as work-in-progress.

PRODUCTION PLANNING AND CONTROL

The production control has a major role to play in reducing the WIP inventory control. Production planning involves routing, scheduling, despatching and coordination of all resources of men/machine/material and methods to achieve the objectives of manufacturing. This planning includes long-term planning, current planning and short-term planning of all resources. Production translates the sales forecasts into production operations by setting levels of manufacturing operations by preparing production programmes and master schedules of production. Production control aims at the orderly movement of material through the entire manufacturing cycle from requisitioning of raw materials to the finished goods, with maximum manufacturing efficiency. The loading and scheduling activities are concerned with the flow of work to the shopfloor, and the relationship between the time required by production orders and available shop capacity. The production planning and control should consider methods such as intermittent manufacturing, custom order manufacturing, production for stock, production in continuous manufacturing etc. in order to evolve a suitable strategy of reducing the work-in-progress inventory.

GANTT CHARTS

Henry L Gantt has developed the Gantt charts for scheduling, despatching and control of the facility under consideration and the critical element of time, against which all production is carried on. In this chart, horizontal bars are drawn for each activity in proportion to the time required for carrying it out. Rectangles are used for tasks and circles indicate the accomplishment of the tasks. The Gantt chart shows the relationship among the milestones contained in the same task and focusses on those situations needing attention. Care should be taken that the primary data is not obscured by excessive detail. Considerable clerical effort is needed to draw the Gantt charts. However, this is a simple and useful mechanism for machine loading and production control purposes and is a pace-setter for the network analysis methods.

LINE OF BALANCE

Line Of Balance (LOB) is a management control procedure for collecting, measuring and presenting facts of time, cost and accomplishment, against a specific target. This plan indicates the progress status, background, timing and phasing of intra-work centre activities. As the name indicates, the line of balance is a graphic device that enables a manager to visualise, at a single glance, which of the activities are in balance. The chart indicates the activities scheduled for completion, but are lagging behind. LOB was established in 1941 to monitor the production of Good Year tyre and rubber company.

A TO Z TIPS TO REDUCE WIP

(a) Production scheduling and shopfloor personnel should consider reduction of WIP as their major goal. The ideal level is zero, as prevalent in Japan. However, the Tandon Committee norms may be useful initially. It is also essential to get stagewise payments from the vendor, so that the company's working capital is not locked up in the WIP inventory. The survey indicates that this is the most neglected area of inventory control.

(b) The time lost in waiting for facilities and bought-out components should be drastically reduced by better planning and coordinating the manufacturing strategy.

(c) The raw material is known as the WIP as soon as it enters the shopfloor, as well as when it is lying in the shop floor awaiting despatch, after conversion and value addition. Obviously a valuewise analysis will indicate the difference between the two.

(d) Inventory is taken from the information obtained on the quantity of work-in-progress and the quantity reported for payroll. This is compared at various operational stages or when WIP is delivered to the final despatch.

(e) Periodic financial summarisation of the value of work-in-progress, will alert the shop floor personnel, when increased WIP inventory value violates the established norms.

(f) It is essential to identify each item of the work-in-progress inventory, the location in the workshop and the reasons for its existence.

(g) A pictorial representation of the value and quantity of the WIP for all the work centres, over the time dimension, indicates the trends.

(h) The shopfloor operating personnel should carry out a preliminary check of the actual needs and the availability of materials to start the manufacturing operation.

(i) Periodic analysis of WIP on the basis of inventory value, ABC/XYZ/MUSIC-3D, as well as classifying WIP items lying beyond a certain time dimension will help to locate the high value WIP items lying on the floor. Action should be taken to convert high value WIP items into finished goods.

(j) Suitably designed job chasing card, regularly informing the manager on the status of the job, helps to reduce the WIP inventory. Application of economic lot size formula in manufacturing, may also help in this direction.

(k) Scientific management principles like Gantt charts, line of balancing method, computerised production scheduling methods, just in time, work study, time study, motion study, PERT, CPM, etc. may help to reduce WIP inventory.

Manufacturing Cycle Time

(l) The operations/production/production planning/production scheduling/ manufacturing personnel should first be convinced that the conversion time of manufacturing cycle, could be reduced by better planning and improving the set-up time.

(m) The set-up time, is the time to get the equipment ready, to do a new job by changing jigs, fixtures, punches, dyes, tooling and cleaning.

Set-up Time

(n) It is necessary to reduce the set-up time on the machine in order to reduce the works in progress inventory and the shop floor personnel should be educated on this aspect.

(o) The external set-up actions should be standardised and the adjustable gauges be replaced by permanent ones. The die receptacles should be standardised. QCT or quick changing tools must be adopted and the JIT concept must be incorporated.

(p) Put a bench at the side of the press at the same level as the press opening, to hold the next die.

(q) Use a check-off sheet in order to ensure the presence of all components and workers, before the commencement of the set-up operation.

(r) Involve tool and die in set-up reduction programmes so that all new designs incorporate quick change-over concepts.

(s) Use better communication devices like two-way radio, between the set-up man and lift operator, who removes and delivers items.

(t) Review materials flow chart with reduced movements in mind.

(u) Make the majority of the set-up activities internal to the run time by doing them without shutting down the equipment.

(v) Codify the parts on despatch list for major or minor set-up to aid smoother set-up and scheduling.

(w) Standardise the size of all bolts and other possible categories.

(x) Standardise the use of common parts in the product so that the number of different parts required is reduced.

(y) Avoid breakdown in different work centres by proper coordination.

(z) Suggestion schemes to reduce the manufacturing time and set-up time, should be encouraged by means of suitable rewards.

TEN POINTS TO PONDER

1. Identify the need for WIP inventory.
2. Discuss the problems in reduction of WIP inventory.

3. Define production planning and production control.
4. Explain the role of production planning in WIP inventory control.
5. Elucidate the significance of Gantt chart in planning.
6. Describe the concept and utility of line of balance technique in reducing WIP inventory.
7. Elaborate on the information requirement to reduce the WIP inventory.
8. Mention the steps to reduce the set-up time.
9. Clarify the concepts associated with reduction in manufacturing cycle-time.
10. Examine the scope of WIP inventory control in an engineering manufacturing organisation.

37. Finished Goods Inventory

NEED FOR CONTROL

Control of finished goods is normally with the marketing department. In many organisations the marketing department and the top management give only a low priority to the control of finished goods in the warehouse and in the channels of distribution. The ideal level of the idle finished goods inventory may be zero, without affecting the consumer demands, but it is impracticable. However, in view of the tight working capital situation, the marketing executive responsible for the physical distribution management has to be aware of the costs and consequences, associated with the finished goods inventory. Many organisations marketing spare parts have complaints from their consumers about the non-availability, high price, poor quality and low service levels in the supply of spare parts.

OBJECTIVES OF FINISHED GOODS INVENTORY

Materials, manufacturing and marketing departments are the three main operating subsystems of any organisation. The materials subsystem procures the input, the manufacturing subsystem converts it to the product and the marketing subsystem sells the output. The output of one subsystem forms the input of the other and hence all the three subsystems are mutually dependent. Inventory, which is primarily a stock of goods, forms a buffer at these interfaces and enables the smooth transfer or materials from one subsystem to the other. The finished goods stocks at factory, warehouse and in the channels of distribution network, serve as a bank and facilitate the supply of the item.

Finished goods also act as a buffer between the production department and the marketing department within an organisation. The input rate is quite predictable because of the production programme, but the consumption of the inventory depends upon the market behaviour. Even the best market intelligence, market research and forecasts, are not cent per cent accurate. The purpose of the finished goods inventory is to assure a constant supply in the distribution channels.

JOBBING INDUSTRY PROBLEMS

The market conditions determine the stock level of finished goods in the consumer industry. But the problems faced with regard to the finished goods by a heavy engineering or jobbing industry are different in nature. These industries usually begin their manufacturing operations with a firm order and a schedule of payment for each stage of manufacturing and hence, they do not have the problem of finding a consumer to sell the product. But long manufacturing cycle time and the numerous items which constitute the product, cause problems. The product has many assemblies, which are complete in themselves and which have to be transported to the site for final erection and commissioning. These assemblies are the finished products of the department manufacturing them. But, unless all the assemblies are complete, the final erection and commissioning cannot take place, sometimes, leading to the storage of completed assemblies as finished goods. The inventory then, has to be controlled by better coordination and scheduling of production.

SPARE PARTS MARKETING

Problems associated with selling spare parts are multi-dimensional and complex, needing prompt attention from various stand-points. The irritants in this sphere include high cost, inferior quality, spurious items, long lead time, unwanted spares dumped initially appearing in the market causing disrepute, poor after-sales-service, obsolescence, change in import policies, inadequate information system, improper maintenance by the customers, lack of communication, wrong coding, bad warehousing, etc. Quick response, prompt repair, least failure, better communications, informing consumers about technological upgradations, may lead to the desired high level of satisfaction.

COST MINIMISATION

The costs discussed earlier in the chapter of warehousing costs, are applicable for finished goods as well. The inventory carrying cost including capital and warehousing charges, will be slightly more than 30 per cent as witnessed in the case of raw materials. This is because the finished goods warehouses, are located in costly downtown areas, unlike the factories. The order processing charge, or work involved in connection to meet the customers demand is also very relevant. Besides salaries, space, communication and documentations, this cost should include the substantial advertising or promotional expenses as well. The stockout cost is very relevant, particularly for units selling machinery spares and represents not only the loss of sales but also the image of the organisation. The overstocking cost, leading to surplus and obsolescence is rarely observed, as organisations holding finished goods inventory expect the sales to inventory turnover, ratio around six, implying that they hold about two months stock levels. This is about the same level as recommended by the Tandon Committee norms for finished goods.

Large organisations have more than one production unit and/or warehouses. There are two types of problems. The first one is the allocation or transportation problem of finding the quantum to be supplied to each warehouse from each production centre. Obviously, this problem does not arise if there is only one production centre. The allocation or transportation problem simply a linear programming problem. There is a unique cost in transporting items from each production centre to a particular warehouse. Each warehouse has to be supplied its requirements. Within these constraints, the quantum of supply from each production centre to each warehouse, is to be arranged in such a way that the total transportation cost is minimised. The second problem is to decide on the reserve stock to be maintained at each warehouse and optimising the number of orders. EOQ can form the basis as the order quantity with Q and P systems of inventory levels operating for the finished goods. The reader is advised to refer to the chapter on Right order quantity (Chapter 32), Q and P inventory system (Chapter 33), LP for warehousing (Chapter 10), Business logistics (Chapter 13), Transportation systems (Chapter 26), Material handling (Chapter 24) for solving the above problems.

MULTIPLE WAREHOUSES

A multi-unit organisation with many consumption centres, can organise a chain of regional warehouses so located, as to optimise the procurement and disbursement of the materials. While deciding on the location, the future requirements, growth rate, potential sources of supplies and centres of market should also be taken into account so that a short-term optimal solution does not turn into a long-term dead loss investment. The number of expected transactions in a location is the major factor influencing the number of retail field warehouses.

Organisations that cater to a multiplicity of needs and cover large areas with established distribution channels often have a chain of warehouses. There is a central warehouse at the apex level, which caters to regional warehouses, which in turn cater to subregional warehouses and so on, up to the ultimate depot that caters to the customer. Each warehouse has a minimum stock and there are orders from the lower warehouse to the upper one and in-transit stocks.

The simplest solution is to consider each warehouse as a separate entity and optimise the inventory decisions individually. But, the best solution is to have a centralised order processing system, centralised storing of minimum stocks and centralised control of stock movement. This necessitates a well-established information system, but the reduction in inventory costs amply justifies the increase in information costs. This problem is similar to the economics that can be achieved by bunching together orders for multiple items on a single source, rather than ordering them separately according to their respective economic order quantities. However, the respective transportation costs and in-transit inventories have to be considered.

In a multiple warehouse system that has n warehouses, the safety stock can be reduced by a factor equal to the subsidiary warehouses. It can also be proved that if central procurement or bunching of orders is done, the total inventory cost is also reduced by a factor equal to the square root of *n* as per the EOQ formula.

PHYSICAL DISTRIBUTION MANAGEMENT

The finished goods inventory plays a crucial role in the physical distribution management, as the executive has to carry adequate stocks from the production centres to the individual consumer through the distribution channels. The inventory carrying charges in developing economies like India — with cost of capital around 18 per cent works to about 30 per cent including storage charges. Hence, the distribution executive is constantly in search of policies for inventories in the channels with minimum investment as well as minimum consumer dissatisfaction. In this context of search for optimum policies, multi-echelon models come in handy to him.

MULTI-ECHELON SYSTEM

Several industries have tried to tackle the finished goods inventory with multi-echelon inventory models. A multi-echelon model can be visualised by an irrigation dam across a river supplying water to the fields through storage tanks located in different centres. The irrigation dam is at the first echelon. It receives supply from river/rain which is outside the inventory system. The storage tanks at different centres are the second echelon. The fields receive their supply only from the storage tanks. Restrictions may be placed regarding particular tanks servicing particular fields.

Thus a multi-echelon system can be defined as a series of two or more production or supply facilities, where any change in the policy parameters in one facility affects the other facilities, either directly or indirectly. A typical multi-echelon system is one involving factory, central warehouse, regional depots and retail outlets. The retail outlets are termed the first echelon, the depots the second echelon, the central warehouse the third echelon, and so on. As can be seen, there may be more than one facility at each echelon.

Most inventory systems of finished goods are multi-echelon in nature. Because of the operational difficulties and problems of coordination, when the number of facilities is large, quite often the facilities are run independently. But in a number of situations, especially in industrial products, the system is controlled by a single organisation. In such situations the inventory should be designed to take a holistic view of the system. The periodic review systems, optional replenishment policies and simulation are helpful to evolve inventory policies. The research study indicates that top management wants to control only the raw material inventory and does not exercise the same amount of control on finished goods.

TEN POINTS TO PONDER

1. Examine the relevance of finished goods inventory control.
2. Identify the objectives of finished goods inventory.
3. Categorise the problems regarding finished goods encountered in consumer industries.
4. Catalogue the peculiarities regarding finished goods in the jobbing industry.
5. The spare parts needs of company are never fully met by the sales organisation of the original equipment manufacturers.
6. Discuss the cost reduction techniques in finished goods, inventory control and logistics management.
7. Elaborate on the needs of multiple warehouses.
8. Describe the problems faced by multiple warehouses.
9. Define the term physical distribution management.
10. Explain the role of multi-echelon systems in the area of finished goods inventory management.

38. Seasonal Inventory

PROBLEMS OF SEASONALITY

The six seasons in a calendar year, which include autumn, spring, summer, monsoon, fall and winter, constitute the natural seasonality. The effect of natural seasonality is mostly felt in agriculture, though the monsoon season affects industrial activity too. The purchasing and sales activities are at their peak near the end of financial year i.e. 31 March. This is because the yearly sales targets have to be met and the purchase budget gets exhausted. Pronounced waves of buying are noticed for items such as fresh fruits, food stores and vegetables, where the supply may show considerable variation, ranging from many times the annual average in one part of the year, to zero in another. Sales of gift items shoot up in the festival season from Dussehara to Christmas. Fashions, eating habits, dressing pattern and work culture change according to the seasonality.

Tax collections, import policy changes as well as fiscal and monetary pronouncements involve a seasonal pattern, forcing the organisations to change set methodologies. Research indicates that pronounced seasonals of more than 10 per cent average deviations, are observed in textiles, agricultural machinery, fertilisers, canning, pharmaceuticals, rubber, paper, transport and utilities.

In free market conditions, the prices change as dictated by the disparity between demand and supply. Hence, it is necessary to examine the seasonal behaviour of demand and supply.

SUPPLY SEASONALITY

The supply seasonality is applicable to agriculture-based materials such as foodgrains, linseed oil, groundnut oil, vegetable oil, etc. The demand for these products, however, remains uniform throughout the year. The demand seasonality is exhorted by agricultural inputs, pesticides, fertilisers, paints, construction material, pharmaceutical items, etc. Paints, construction materials, and hardware stores show very low demand during monsoon and winter seasons.

Sometimes, in commodity markets, the buyer enters into two transactions—a purchase and a sale in two markets, whose prices move up and down together. Thus, the profit or loss, sustained in buying transactions, is compensated by the profit or loss in the selling transactions. This implies that the two markets behave similarly, and by a perfect hedge, one can reach the perfect solution of zero loss. But the usual practice is to trade in the spot market and to protect against future losses due to vagaries in seasonal market conditions.

The materials manager and inventory controller are worried, when the supply is seasonal and demand in terms of sales is low during a particular season. The following steps are taken in this context:

(a) Assure continuous supply of material to the company.

(b) Make purchases at a reasonable price since the prices fluctuate widely.

(c) Ensure that the purchase budget is adhered to.

(d) Tackle the working capital problem of proper cash outflow within the prescribed monthly monetary limits.

(e) Rent warehousing space if the management decision is to purchase larger quantities to take advantage of low prices.

(f) Adjust the workload of manpower in the materials department suitably, without having more staff during the peak period.

SEASONAL INDICES

Government departments ensure constant supply of essential foodgrains, vegetable oils, imported non-ferrous metals, etc. These include the Food Corporation of India for essential foodgrains, State Trading Corporation for vegetable oil, MMTC (Metals and Minerals Trading Corporation) and other canalised agencies. These organisations indulge in forward buying in order to effect continuous supply when prices are low. This process is restricted by the available warehousing facility and the amount of liquid capital at hand.

Thus, seasonal variations are those influences, reflected in the past pattern, that are the result of man-made or physical phenomena. The major characteristics of seasonal variation are that they are repetitive and periodic, the period is less than one year, usually around three months. Establishing the past changes enables us to compare two time intervals which would otherwise be too dissimilar. The seasonal fluctuations help us project past patterns into future, particularly for short-term decisions. Once the seasonal pattern has been established, it is possible to segregate its effects from past data in order to predict for the future. The data or time series is called, deseasonalised, when the seasonal fluctuations are eliminated.

In order to measure the seasonal variation, we use the ratio to moving average method, which gives an index of the seasonal pattern. The index is based on a mean of 100, with the degree of seasonality, measured by fluctuations away from the base.

In the *time series analysis*, initially an estimate of the trend cycle is obtained by the use of a moving average, which combines twelve successive monthly figures, thereby eliminating the seasonal element. Division of the original data by this moving average yields a series of the so-called seasonal-irregular rations. An estimate of the seasonal factors is then obtained, by averaging the ratios, quarter by quarter, and assuming that the irregular factor will be averaged out in this process. Finally, the original observations are seasonally adjusted by dividing them by seasonal factors.

It is also possible to use the *regression analysis* for seasonal adjustment. The procedure involves deriving a relationship for each quarter, between the original data and the twelve months' moving average. Computer programs are also available to identify the seasonal indices.

DYNAMIC PROGRAMMING

Dynamic programming is a quantitative approach for overall effectiveness with a combination of optimal decisions over a period of time. The principle of optimality, according to Professor Bellman who developed dynamic programming model, states:

> An optimal policy has the property that whatever the initial decisions are, the remaining decisions must constitute an optimal policy with regard to the state resulting from the first decision.

ITERATIVE PROCESS

Dynamic programming approach is very useful for arriving at optimal policies on purchase, warehousing, sales, production scheduling, and inventory control of seasonal commodities. The problem is to determine the best plan for producing a seasonal product considering the demand fluctuations during different seasons. Dynamic programming minimises the cost of production and inventory carrying charges.

The objective of dynamic programming is to obtain the best decision at each stage of the sequence so that the overall benefit is maximised. Linear programming cannot be used here as the relationship between costs and inventory quantities is not linear. The step-by-step approach of review or iterative process is used in dynamic programming. Let us solve this by means of a simple illustration. Complications may arise when the computer's help is sought.

ILLUSTRATION

A company requires a certain type of edible oil at the rate of 100 tons per week. The storage space available is for an absolute maximum of 1000 tons or 10 weeks' demand. It is predicted that the price of this particular oil will fluctuate between Rs 25,000 and Rs 26,000 per ton. The market price forecast for a period of 50 weeks is given in Table 38.1.

Table 38.1 Forecasted Prices

Weeks	Weekly Price Per Ton (in Rs)				
1 – 5	25570	25650	25200	25090	25800
6–10	25340	25460	25020	25050	25040
11–15	25150	25400	25060	25870	25880
16–20	25170	25190	25770	25660	25140
21–25	25680	25270	25850	25110	25190
26–30	25260	25950	25680	25970	25730
31–35	25750	25640	25270	25450	25010
36–40	25870	25200	25020	25190	25260
41–45	25450	25420	25960	25720	25980
46–50	25780	25810	25520	25310	25880

If the company were to adopt a conservative policy of buying 100 tons, regardless of the market prices, it would buy 5000 tons over a period of 50 weeks at an average price of Rs 25,480 per ton. If the forecast of the price was perfect — without any major fluctuation — the company would buy oil at the lowest possible price in the long run. This hindsight policy considers the prices in the next 10 weeks.

ASSUMPTIONS IN SEASONAL BUYING

For the sake of simplicity, the following assumptions are made in solving this problem, even though with the help of computer, it is possible to find the solution without any assumptions.

(a) Time has been divided into successive periods of weeks.

(b) A planning horizon of N periods is arbitrarily taken. In this example, a storage restriction of 10 weeks is imposed. A company cannot have infinite periods, as the knowledge of prices will be diffused after a certain period.

(c) Demand is known for all periods in the planning horizon. In this case, a uniform demand of 100 tons per period is assumed. It can vary in other circumstances.

(d) No shortages are allowed. Whatever be the price, a company has to buy at least one week's requirements.

(e) No storage cost has been assumed. However, if the planning horizon is longer, then the storage cost is brought into the picture.

(f) There is only one price offer in a period. If this opportunity of purchase is foregone, the company has to wait for the next period to purchase its requirements.

(g) There are no supply restrictions. Quantities up to the entire requirement of planning horizon are available in the market during any one of the periods.

(h) The prices fluctuate randomly in the planning horizon duration and are independent of each other.

(i) Delivery is assumed to be instantaneous.

(j) Stock once bought cannot be resold.

(k) It is essential that the company should hold off when the prices are low.

(l) The above policy, obviously, depends upon the stock on hand.

BUYING IN STAGES

Let P_0, P_1, P_2, ..., P_9 be the prices of indifference when stocks are for zero week, week 1, ..., week 9. These are price levels when the company will be indifferent between buying and holding off.

(a) If a company has no stock on hand, it has to buy at least one week's requirement regardless of the price since the highest price is Rs 26,000 per ton, P_0 is Rs 26,000.

(b) If the company has one week's stock on hand, it has to decide whether it should buy or hold off. We are essentially looking at what the price that would prevail next week. Prices vary randomly between Rs 25,000 and Rs 26,000. Therefore, P_1 = Rs 25,500, i.e., with one weeks stock on hand, buy if the price is below Rs 25,500 for one more week.

(c) When the company has two weeks' stock, the possible ways of buying the third week's requirement are as follows: (i) Immediate buying, (ii) buying in the second week if the price is below P_1, (iii) buying in the third week with a probability 0.50. The chance of the price being below Rs 25,500 is 0.5 or 50% and its expected value is Rs 25,250. There is again 50% chance of the price in the second week being above Rs 25,500, and in such a case the company will buy in the third week. The expected price in the third week is 25,500. Hence

$$P_2 = .50(25{,}250) + .50(25{,}500) \text{ or Rs } 25{,}375$$

Similarly,

$$P_3 = .375(25{,}187.50) + .625(.5)(25{,}250) + .625(.5)(25{,}500) = 25{,}305.$$

Likewise, we can compute $P_4, P_5, P_6, P_7, P_8, P_9$, as Rs 25,258, Rs 25,225, Rs 25,200, Rs 25,180, Rs 25,160, and Rs 25,150. These can be plotted in the form of a graph. The company can estimate the average price paid in the long run, after comparing these prices to the 50 weeks forecasted price; the results would be as in Table 38.2.

The average price paid for all the 50 weeks' requirements of 5000 tons is Rs 25,139.20, while the conservative policy price was Rs 25,480 and the hindsight policy price Rs 25,096.40. The difference between hindsight policy and the optimal policy of Rs 42.50 per ton, can be viewed as the *cost of uncertainty*. Thus dynamic programming is helpful in seasonal buying and developing optimum inventory policies.

Table 38.2 Seasonal Quantities

Week	Opening Stock	Quantity Bought	Price Forecast
1		100	25,770
2		100	25,650
3		600	25,250
4	500	500	25,090
8	600	400	25,020
9	900	100	25,050
10	900	100	25,040
13	700	300	25,060
16	700	100	25,170
20	400	600	25,140
24	600	400	25,110
33	100	300	25,270
35	200	800	25,210
38	700	300	25,020
48		100	25,520
49		200	25,310

TEN POINTS TO PONDER

1. Discuss the meaning of seasonality.
2. Give examples of natural seasonality.
3. Identify the problems of seasonality and the effects on materials.
4. Describe the process of computation of seasonal index.
5. Explain the use of regression analysis is seasonality.
6. Clarify the concept of iterative process in decision-making.
7. Define the concept of dynamic programming.
8. Explain the role of dynamic programming in seasonality.
9. Elucidate the concept of indifference prices.
10. Illustrate the assumptions behind dynamic programming.

39. Project Inventory Management

CONCEPT OF PROJECT INVENTORY

A *project* is an organisation unit, established for the attainment of a goal like the launching of a new factory or the development of a product on time, within the stipulated resource constraints. These constraints include (a) resource limitations of the budget, (b) specified time schedule, and (c) pre-determined performance specifications.

The material required for any project, usually consists of cement, steel, equipment, plant and machinery. The material cost or cash outflow of any project is over 80 per cent of the project cost. But, the project manager, who is incharge of the entire project does not attach great importance to the project material. The purpose of purchase, follow-up and storage of project material is usually delegated down the line. In order to ensure that the project is on stream by the scheduled date, the project material is ordered sufficiently in advance. Sometimes the material reaches the project site, when the stores officer has not been recruited and or the warehouse is not ready, and the project inventory is lying idle due to faulty planning. Lack of coordination and improper and inept monitoring of project activities result in cost overrun and time overrun up to five years in the infrastructure sectors of power, coal, petroleum, steel, mining and transport. When the project is properly scheduled and lead times are estimated sufficiently in advance, the project inventory should be zero, as we are dealing with a certainty-static situation in the project management area. No wonder, seasoned project managers cynically sum up the different stages as: (a) initial enthusiasm on detailed project report, (b) political interference, (c) disillusionment, (d) panic (e) searching for the guilty, (f) punishing the innocent, and (g) rewarding non-participants.

A TO Z SPECIAL FEATURES

Understanding the peculiarities of project material helps in better management of the project inventory.

(a) The decision with regard to the purchase of project material is usually taken by the top management based on techno-economic analysis.

(b) The total lead time from need identification to the need satisfaction stage or conversion of indent to the end product is invariably large, sometimes beyond five years.

(c) In view of the high cost of project materials, adequate planning, sourcing, scheduling compatibility with existing machinery, follow-up and negotiations have to be done.

(d) The advantages and disadvantages of a single bid, two bid system should be studied carefully in order to develop suitable bid evaluation criteria.

(e) Legal aspects like contract, performance, negotiation, guarantee, disputes, risk purchase, liquidated damages, *force majure* clauses, disputes, arbitration and adjudication should be thoroughly considered.

(f) Availability of funds, credit facilities, aids from other sources, etc. are usually studied by the finance department while choosing the supplier. Currency fluctuation is an important aspect.

(g) Make-buy-lease criteria should be adequately explored from techno-economic-financial view.

(h) Life cycle costing, tax life, physical life, recommended life, life extension programmes, future availability of spares, technological considerations are reviewed critically before placing an order.

(i) Payment terms, including stagewise payment, penalty, bonus, freight, insurance, etc. should be finalised at the time of contracting.

(j) Agreement on the inspection procedures at the stages of manufacturing, erection, commissioning and operation at site is another important aspect.

(k) Overestimation of requirements, which results in project surplus materials and obsolete items in stores, should be avoided by better planning.

(l) Estimates of initial supply of spares and procurement of the same have to be carefully done before ordering in order to avoid dumping of unwanted spares by the suppliers.

(m) The purchase-parameters of right quality, right quantity, right price, right sources, right time and right material should be followed through right bill of materials, and requirement planning.

(n) A buy-back clause which enables the return of unwanted spares to the supplier should be introduced in the contract.

(o) Transport requirements of the material from the manufacturer to the project site must be critically analysed.

(p) Wherever possible, drawings of critical parts of the equipment must be procured with operating/maintenance manual.

(q) The storage of materials, equipments and other assets must be carefully protected against theft, rain, sun, heat, cold, insects, rats and ants. Arrangements must be made to receive, store, issue and monitor the project inventory as well as dispose of surplus items and old equipments.

(r) Proper accounting of receipt, issue, and accounting of the material has to be done at site. Whenever subcontractors bring or borrow welding equipment or other small machinery, proper allowances of the same must be done by the project stores staff.

(s) Provisions for contractor's stores must be made in large projects at the sites.

(t) In multi-project organisations it is very common to receive materials intended for other projects. This should be avoided through better coordination.

(u) The project manager should be aware of the cost of stockout, ordering charges, emergency of follow-up, expediting cost, and inventory carrying charges, in order to schedule the arrival of items from different sources to the project site.

(v) Some of project staff in the departments of stores/purchase/inventory control/ materials are employed only on temporary basis. This policy of hire and fire may result in low motivation and commitment.

(w) Since projects are located in far flung areas without adequate communication facilities, efforts should be made to improve the communication from the site to the different suppliers.

(x) The aim of any project, is to commission the project in time, without any overruns of cost and time, so that the ultimate product cost of the project is the least. The material subsystem should strive towards the overall project goal.

(y) The ideal project inventory should be zero. In other words, the materials should be used in the project right from the time it is transported. This ideal goal can be achieved if adequate planning, coordination, follow-up and monitoring are done by the project inventory staff.

(z) Planning and monitoring techniques like the network analysis as well as PERT/CPM techniques should be applied to the project materials.

NETWORK ANALYSIS

Management of any project involves planning, coordination, control and monitoring of a number of activities with limited resources of material, machine, money, means and time. Network analysis provides a dynamic planning and scheduling system which is capable of meeting any subsequent changes. It provides a framework which does not following: (a) Provides a broad work break-up structure of the project activities. (b) Define the details of the job to be done. (c) Integrates them in a logical fashion (d) Affords a system of dynamic planning to continuously monitor the progress of the project. (e) Identifies the critical jobs and minimises the trouble spots in the project. (f) Attempts to complete the project within the constraints of time and cost.

PERT/CPM

The most important network analysis techniques are PERT — Programme Evaluation Review Techniques, CPM — Critical Path Method, and CPA — Critical Path Analysis. The special project office of the American Navy introduced PERT on its Polaris weapon system in 1958, after the technique had been developed by the management consultant Firm Booz, Allen and Hamilton. At about the same time, the Du Pont initiated a similar technique known as the Critical Path Method in order to reduce the time and cost of new products.

PERT relies on *probability theory* to get three time estimates for forecasting any activity of the project and is particularly suited to Indian conditions. The CPM on the other hand, relies only on one time estimate and is known as the *deterministic model*. In spite of the minor differences in the terminology used, both are useful in completing the project in the context of given resources. It is necessary to seek the assistance of computers for planning and monitoring large projects.

A TO Z ASPECTS OF PERT

PERT aims at answering the following questions:

(a) When will the project be completed?

(b) What are the main activities of the project?

(c) When is each part of the project scheduled to start and finish?

(d) Which of numerous jobs of the project are critical?

(e) Is it possible to transfer resources to critical jobs in order to finish the project in time, without affecting the overall performance?

(f) Where should management concentrate its efforts, out of the numerous items, at any specific point in time.

The steps involved in developing a PERT diagram include the following:

(g) The project is split into small independent jobs.

(h) The order of precedence — which job should be accomplished before undertaking the others — is determined.

(i) The activity sequence is represented diagrammatically.

Events and Activities

(j) In this context, an *event* is defined as an accomplishment occurring at an instantaneous point in time without consuming any resource.

(k) Events are usually represented by circles or squares in the network diagram.

(l) An *activity* is defined as a recognisable part of work that requires time and resources for its completion.

(m) An activity is represented by an arrow in the diagram and lies between two events.

(n) Sometimes dummy activity is introduced to preserve logical sequencing, but it does not consume time or resources.

(o) Thus the network consists of a diagram of activities indicated by arrows and events by circles. The three time estimates required for each job are denoted by t_o, the optimistic time estimate; t_m, the most likely time; and t_p, the pessimistic time estimate. These are obtained from people responsible for the accomplishments of the respective activity.

(p) This concept is based on the well-known laws: (i) If a thing can go wrong it will. (ii) When things are going fine, something has to go wrong:

(q) The three time estimates are compressed into the expected time estimate, t_o

$[t_e = (t_p + t_m + t_o)/6]$.

(r) This time estimate is entered on the respective arrows indicating the activities.

Critical Path

(s) Addition of the time estimates in a path gives the total early completion time T of the project under normal circumstances. This is also known as the latest completion time T of the project. At the junction of more than one path, the maximum time is taken to ensure the completion of all the jobs at the junction.

(t) From the earliest total completion time, T_e, we subtract the time of each activity in the reverse order to get the latest time of completing the specific activity, taking care to put the larger number at the junction of two paths to ensure logical sequencing.

(u) The path will zero slack or minimum slack is known as the *critical path*, as delay of any activity on this path will in turn mean delay of the entire project.

Crashing

(v) The top management usually wants an early commissioning of the project, than that indicated by the PERT network. In other words, it becomes necessary to crash or shorten the duration of the entire project. This process is known as *crashing*.

(w) Initially crashing of activities lying on critical path is undertaken by deploying additional resources.

(x) In the process of crashing, it is likely that more than one path may be critical. Hence, the concept of time-cost trade-off has to be considered by judiciously allocating time for the different activities lying on the critical path.

(y) Crashing beyond a particular level increases only the cost without reducing the time. If 100 workers can construct an institution in 100 days, 10,000 workers cannot build it in a day.

(z) Activities involving project purchase, project stores, project material and project inventory, can be developed into a separate PERT diagram, which can be interfaced with the master network of the total project.

Illustration/Exercise

Readers are requested to draw the network and identify the critical path for constructing a new plant, with the help of the data given in Table 39.1.

Table 39.1 Activity Timings

Activity Number	Activity Description	Duration Months t_e	Activity Dependence
0–1	Prepare detailed specification	3	nil
1–2	Develop plant site	5	0–1
1–3	Complete civil works drawing	2	0–1
1–4	Prepare fabrication drawings of plant	4	0–1
1–5	Finalise for imported plant	4	0–1
5–6	Order imported equipment	10	1–5
1–7	Recruit personnel	3	0–1
7–11	Train personnel	8	1–7
4–9	Fabricate plant equipment	12	1–4
4–10	Procure local material	5	1–4
9–11	Despatch plant to site	4	4–9
10–11	Despatch local material to site	3	4–11
6–11	Despatch imported plant to site	2	5–6
3–11	Complete civil works	7	1–3
11–12	Erect the plant	6	3–11 9–11 10–11 6–11 7–11
12–13	Conduct trial runs	1	11–12

TEN POINTS TO PONDER

1. Define project inventory.
2. Identify reasons for delays in project commissioning.
3. Examine the distinguishing features of project material.
4. Discuss the precautions to be taken in project contract management.
5. Deliberate on the issues of project stores.
6. Clarify the management of project inventory by better planning.
7. Explain the significance of network analysis in project material.
8. Distinguish between PERT and CPM.
9. Enumerate the steps in drawing PERT network.
10. Explore the significance of critical path and crashing in relation to project inventory.

40. Spare Parts Inventory

A TO Z DISTINGUISHING FEATURES

In a large number of factories a few machines remain idle for want of machinery spare parts. A common feature observed is the accumulation of dormant and slow-moving spare parts and at the same time machines being idle for want of the necessary spare parts. Let us now identify the problems of spare parts from maintenance, materials and suppliers' points of view.

Maintenance Problems

(a) The maintenance manager cynically describes spare parts as not available when needed, but available in plenty when not needed.

(b) Spare parts management is a challenge as the literature on it is scanty, secretive, and theoretical. Spare parts account for about 80 per cent of maintenance budget.

(c) A paradoxical situation exists — there is a huge inventory of machinery spare parts while machinery lies idle for want of spare parts!

(d) The capital-scarce Indian economy has to extend the life of the machinery to over 125 years, beyond the economic life as in the case of Digboi Oil Refinery and Indian Railway system as well as resort to buying second-hand plants like Scooters India Ltd., at Lucknow. The maintenance of these old machinery becomes highly complicated, with the operational life extending beyond the recommended life or defying scientific concepts such as LCC — Life Cycle Costing, economic life, etc. [refer Chapter 49]. The concept of make-buy-lease analysis remains only theoretical in many organisations.

(e) Even though the maintenance manager is familiar with modern concepts like condition monitoring, signature analysis, vibration analysis, failure analysis, preventive maintenance, productive maintenance, predictive maintenance, design-out maintenance, corrective maintenance, terotechnology, hands-off maintenance, tribology, etc. in practice, he resorts more often to breakdown maintenance. Cannibalisation — or removal of a part of one machine and putting into another — is not an uncommon experience in many organisations. This problem is more acute in maintenance oriented-spares-intensive industries such as mining, transport chemicals and processing.

(f) Infrastructure facilities for reconditioning, recycling and repair, need to be strengthened both at micro and macro levels.

(g) The inspection and testing facilities — in the form of skills, manpower and equipments — have to be improved. For instance, many organisations find it difficult to segregate original bearings from spurious ones.

(h) The maintenance manual — describing the maintenance systems, documentation, organisation and control has to be prepared and updated continuously. The role of the maintenance manager and warehouse purchase/inventory/materials controller in planning, forecasting and controlling the spare parts function have to be demarcated clearly to fix the responsibility for spares function.

(i) The maintenance manager incharge of total maintenance function in some firms, is expected to be a specialist in civil, mechanical, electrical, electronic, instrument and other categories of specialisation which is a herculean task by itself.

(j) The maintenance manager should be involved in the purchase of machinery if he has to be responsible for maintaining it.

(k) The spare part function is a service function to the maintenance function, which in turn is a service function to the production or customers. This service-to-service, function is usually neglected and delegated down the line, with high fliers choosing financial and marketing disciplines.

Materials Problems

(l) The total lead time, which comprises converting an indent to getting the end product into the warehouse, including manufacturing, transporting, inspecting, averages around one year and goes beyond that range for specially manufactured and imported spare parts.

(m) Standardisation and variety reduction in number of parts remains only at the conceptual level. A case in point is the Indian Petrochemical complex, which has about 25,000 varieties of only one category known as bearings. This is primarily due to non-standardistion of mother equipments, with each collaborator insisting on his own design. Russians rate electrical motors only by weight, and not by horse power as done by other countries. British and Metric systems of measurement continue in most organisations, creating complications.

Problems in Specifications

(n) In spite of internal codification, many maintenance engineers persist with the suppliers' part number — like Fenners' V belt, making many spare parts single source proprietary items. Assigning temporary codes is another common feature, leading to difficulty in identification.

(o) The indent and the documentation are incomplete, when the mechanics want immediate replacement for a broken part from the warehouse in order to carry urgent maintenance work.

(p) Development of drawings is essential, particularly when the original equipment manufacturers do not provide drawings even for critical items. We have to go a long way in this aspect.

(q) In view of the small requirements of uneconomic manufacturing lot sizes, the laudable objective of two suppliers for every item, becomes only theoretical. But, in procedure-oriented organisations, the finance and audit insist on three quotations perhaps from the same typewriter.

The powers of delegation are obsolete and should consider the current level of inflation in prices.

(r) The starting point for scientific spare parts management is to get the failure data, from which, concept like Poisson distribution, mean time between failures, mean time to failure, etc. can be derived. But in many organisations, even the equipment history card is not maintained properly. It may be noted that unlike raw materials, the spares issued, may not be consumed immediately.

(s) In view of the large varieties and number of spare parts, identification becomes a big problem leading to differences between physical balance and book balance for many items.

Marketing of Spares

(t) Many industrial users feel that the original equipment manufacturer, exploits the users' initial ignorance by dumping unwanted slow-moving and dormant spare parts which become obsolete later on. Use of buy-back clause may reduce the incidence of this problem and improve the suppliers' image. The After-Sales Service is cynically known as ASS amongst the users, as even during the guarantee period it is costly and not rendered at the appropriate time.

(u) According to the customer, the supplier gives a whole module when only a small part of it is needed, rendering the rest useless.

(v) The original equipment manufacturer is unable to supply parts for the machines supplied by him due to technological and model changes at his end; when the part numbers prove insufficient for identification, he even calls for the photograph of the machine. This forces the user organisation to adopt a conservative policy of stocking more items. But the marketing executive feels that consumers are not ever sure of their requirements.

Pricing of Spares

(w) Every machinery manufacturer makes huge profits, up to 300 per cent, depending upon several factors: the Maruti car is a classic example. The annual spare parts sales are about Rs 7000 crores, out of which imports account for Rs 1500 crores. In spite of this, the original manufacturer does not have separate manufacturing capacity earmarked for the replacement market causing the advent of spurious items.

(x) Many original manufacturers buy parts from several vendors and assemble them. Often, the original manufacturer is unable to satisfy the replacement demands and forbids the subvendors from directly satisfying the demands of the ultimate user.

(y) The marketing and after-sales personnel of the original equipment manufacturers accuse the maintenance engineers of the user organisation of not adhering to the maintenance/replacements/overhaul programmes recommended by them.

(z) Stockout cost of a spare part is always higher than (at times, even 100 fold) the cost of the spare part, which in turn is obviously higher than the 30 per cent carrying charges. This forces many organisations to follow a conservative no-risk policy and hoard an average of about five years' stock even though according to the Tandon Committee, spares inventory should constitute only 5 per cent of the total inventory.

The overstocking leads to the following situation:

(a) Spares are bought from the manufacturer.
(b) Spares are stored in the warehouse.
(c) Spares are called as slow moving after some time.
(d) Spares are known as dormant after some time.
(e) Maintenance classifies them as insurance emergency.
(f) Finance categorises them as obsolete.
(g) Spares are disposed of by auction as no one wants to use them, or sometimes, obsolete spare part is removed by paying money.

Definition of spares

Spare part is defined as a part, identical to the part of the machinery, which needs replacement due to wear and tear as the life of the part is less than the operating life of the equipment.

Difficulties in Spares

It is difficult to forecast failure but every failure need not result in a replacement. It is difficult to generate exact consumption data and procure the spare part immediately. Numberwise, spares constitute a major percentage compared to other items. Variety is large, and standardisation and identification of spare parts are not easy. Up to 80 per cent of maintenance budget is accounted by spare parts. Obsolescence is greater in machinery spare parts. The margin of profit is higher in spares sales. Indigenisation is difficult due to low volume of uncertain requirements. Spurious spares create problems. Up to 10 per cent of value of machinery is imported as initial spares while importing equipment.

Commissioning Spares

The spare parts needed for commissioning an equipment or machinery are called *commissioning spares* or *project spares.* Since the time of commissioning can be estimated by planning techniques like PERT, and since these spare parts will not be required later on, their inventory should ideally be zero. But, most organisations overestimate the commissioning spares and dump them in stores as project surplus.

Overhauling Spares

Equipments/machinery, after running for many shifts/hours/months, should undergo complete overhaul. The time of overhaul and which parts are to be replaced are determined based on maintenance experience aided by suppliers' recommendation. These parts are overhaul spares. Since the time and quantity are known in advance, the organisations should not carry any inventory of this category. But the no-risk policy of units carry the stock two or three months in advance of the overhauling date.

Rotable Spares

Engineers, rotors, stators, motors and big assemblies come under this category. For example, the workshop of transport organisation stocks engine assembly, transmission system and gear box assemblies. If a bus with a damaged engine comes to the workshop, it is immediately removed and replaced by an overhauled engine from the stock. This rotation process yields the rotable spare part.

The inventory of the number of finished engines is done by the mathematical queuing approach; and depends upon the average service time, pattern of the input and the discipline followed by the queue as first-in-first-out or emergency basis. A case in point of cannibalisation in this context is that of Indian Airlines reserving two Airbus 320 aircrafts only to meet the spares requirement of the remaining twelve planes.

Insurance Spares

An insurance spare part is defined in terms of its stockout. Even though the reliability of the part is very high, the non-availability of insurance spare part may prove catastrophic resulting in permanent damage with huge stockout cost. *Insurance spares* are also called *standby spares*, and the policy is to carry one standby. But the tendency in many firms is to carry more than one i.e. standby on a standby, and classify many spares as insurance spare parts to avoid risk. The insurance spares can be depreciated over a period of time and hence these are called capital spares. These parts can also be kept in a central spares bank if the equipments are similar.

Maintenance Spare Parts

The common category of spare part, used for repetitive routine maintenance purposes, is called the *maintenance spare*. It is possible to collect the failure data over a period of time.

The failure of maintenance spares is similar to the mortality rate of human beings: During infancy, the mortality rate is high, during adolescence and middle age, the rate is low, but during old age the mortality rate increases.

A similar phenomenon is also observed in the case of spare parts with increased failures in the initial adjustment phase, low steady failure rate for a long time followed by increased failure rate due to wear and tear. By better scientific management techniques like condition-monitoring, the maintenance engineer has to increase the life of the part in the third phase. The statistical Poisson distribution is applicable in the long, steady middle state. This distribution is uniquely determined by the average failure rate. Hence, the inventory under control may be, [Average + $K\sqrt{\text{Average}}$], where K is the service level constant depending upon cost-criticality of the item.

A generic formula of always stocking [1 + $KLTC$] items in the warehouse is optimal for internal consumption or external sales in many firms. Here LT is the total lead time; C is the average consumption; K is the service level factor, depending upon cost — criticality — availability.

If for example, the lead time for a spare is 10 months and average consumption is five per month and K is two, then $KLTC$ becomes $(2 \times 10 \times 5) = 100$ or 100 items will always be under the custody of the warehouse manager including in the pipeline (1 + $KLTC$) is 101. The additional one stock is brought to avoid zero stock when consumption is nil as $KLTC$ is zero if C is zero. Application of MUSIC 3-D also help to control the spares inventory. The maintenance/warehousing/purchasing inventory managers should categorise spare parts in the above manner for avoiding stockouts and obsolescence and solve the problems relating to spare parts.

TEN POINTS TO PONDER

1. Examine the statement that the spare parts inventory is different from other categories of items.
2. Explain the role of maintenance manager in the field of spare parts.
3. Locate the difficulties of maintenance management in relation to spares.
4. List the problems of spare parts in the area of materials management.
5. Discuss the problems of specifications in the field of spare parts.
6. Narrate the problems of spare parts marketing.
7. Identify the specification bottlenecks in the spare parts field.
8. Discuss the pricing and availability of spare parts.
9. Elaborate on the distinguishing features of spare parts.
10. Classify the various categories of spare parts used in any company.

41. Obsolete Inventory

EXAMPLES OF OBSOLESCENCE

The phenomenon of obsolescence is prevalent in all walks of life. Technological obsolescence is well known, particularly in the electronics field.

The obsolete item is disposed of and gets converted into another form. The presence of obsolete items cannot be eliminated but can be controlled by better planning.

No proper estimate of the value of obsolete items is available in the Indian scenario. The first estimate was provided by the then Finance Minister of Government of India Mr. C. Subramaniam, who mentioned on December 10, 1974 (refer the *Hindu*, December 11, 1974) that the amount of money blocked in obsolete and surplus spare parts was estimated around Rs 2,500 crores in India. Presently, with further industrialisation and inflation, this estimated figure is expected to be over Rs 10,000 crores. We have to define surplus and obsolete items separately for better control purposes.

Surplus Items

Surplus items are those materials which have no immediate use but have accumulated due to faulty planning, forecasting and purchasing. Surplus stocks at a specific time include items, which are in excess of requirements and may be consumed at a future date. Surplus stocks result from overordering and significantly contribute to the quantity of obsolescent and dormant stock, if stored over a period of time. Surplus is the state of an item when the stock is likely to last longer than a reasonable period or when it is no longer required for the job for which it was purchased, but the material is in good condition. Surplus materials are merely in excess of what is needed, while obsolete materials are those which are not required at all due to changes in design. Dormant or slow-moving stocks are these items whose usage is infrequent. The items usually become dormant when large quantities of spares are held as safety stock.

A TO Z OF OBSOLESCENCE

Definition of Obsolete Items

(a) The basic problem in obsolete items is the definition of obsolescence. The user classifies all the obsolete items as non-moving whereas the warehouse manager tries to classify them as obsolete.

(b) All obsolete items are non-moving in nature, while all non-moving items are not obsolete, as non-moving items includes insurance/emergency spare parts.

(c) The problem of obsolete items is more acute in spare parts-intensive industries like transport, chemical, mining and process sector.

(d) Obsolete items are those items which have gone out of date because of various reasons, and cannot be used economically or safely in the company. Keeping the items in the warehouse may entail a loss to the organisation. It may have its own economic worth, but no longer can be used for the organisations' operations as advantageously as it has hitherto been used.

Identification of Obsolete Items

(e) The items need proper enlisting, scrutiny and expert opinion before classifying them as obsolete.

(f) The warehouse manager has to periodically analyse the movement of items and classify them as fact-moving, slow-moving, and non-moving, depending on the frequency of issues/sales from the stores. The duration of non-moving may be fixed about five years in spares-intensive companies and one year for sales organisation. This duration can be varied to suit the individual company's culture, products and philosophy.

(g) The non-moving items are subjected to the XYZ inventory value analysis. It is noted that few *x* items, usually about 20 per cent, account for the bulk or about 80 per cent of the inventory value, and the large number of *z* items account for a trivial inventory value of about 20 per cent.

(h) The list containing the small number of items falling in the non-moving and *x* category, is circulated to all the departments, requesting them to indicate as to whether those items are likely to be consumed and if so by what date.

(i) In the above process, one set of the real insurance spare parts is omitted, after getting the user's views. Further, the purchase manager should cancel all pending orders for items falling under non-moving and *x* category.

Reasons for Obsolescence

(j) One of the most common reasons for obsolescence in many organisations is the sudden introduction of new technology, rationalisation and design changes, without adequate preparation, communication and planning.

(k) Adoption of standardisation, elimination of non-standard varieties and introduction of cost reduction methods, have led to obsolescence in a few companies.

(l) The initial procurement of spare parts is the most error-prone area, and is usually based on the suppliers' recommendations. The original equipment manufacturer exploits the ignorance on the part of buying organisation, by dumping slow-moving and non-moving spare parts, leading to obsolescence later on.

(m) Sometimes equipments become obsolete when they are cannibalised to feed the spares requirement of a battery of other equipments. Similar modular replacement renders the good parts of the replaced module, obsolete.

(n) The spares-intensive industries want to adopt a safe policy of no stockout and classify most of the spares as emergency insurance spares because the consequence of stockout is much more than the cost of the spare parts.

(o) In many organisations there is no specific department for controlling obsolescence.

(p) Faulty forecasts, overbuying, buying in bulk, faulty store-keeping, inadequate application of preservatives, inefficient materials handling, poor manufacturing method, improper codification, and use of supplier's part number are the other common reasons for obsolescence.

Remedial Measures

(q) Whenever an equipment is purchased, it is necessary to introduce buy-back clause, to enable exchanging the initially dumped wrong spare parts by the required spare parts.

(r) Stock records should be periodically combed for identifying surplus, slow-moving, non-moving, insurance and obsolete items.

(s) Non-moving items falling under *x* category should be circulated to the maintenance/operations department, for declaring the items obsolete.

(t) This list containing non-moving obsolete items should be circulated to sister organisations and vendors on a time-bound basis.

Disposal of Obsolete Items

(u) Delay in disposing of redundant stocks not only increases the inventory carrying cost, but also creates the risk of deterioration and loss of sale value of the stock. Stock verification also becomes laborious.

(v) Due to faulty planning, an item, which is bought for usage in the organisation, becomes non-moving and unserviceable and has to be disposed off.

(w) Before disposal is planned for, a minimum value, to be approved by the finance, should be fixed to avoid future audit queries.

(x) The obsolete items should be disfigured before disposal, so that the same item but in a new packaging, is not purchased by the company: bearings is a case in point in this context.

(y) Disposal is effected by (i) inviting offers from time to time from possible sources, (ii) entering into annual contracts and (iii) public auction. It is necessary to associate the finance and users in the process of realising the highest value from the obsolete items.

(z) Since complete eradicatiòn of obsolescence is not possible, it is necessary to allow for this in the annual budget to avoid accumulation of huge obsolete stock in the warehouse over the years.

TEN POINTS TO PONDER

1. Comment on the statement 'Obsolescence is unavoidable in any sphere for purposes of improvement.
2. Distinguish the identity of surplus items.
3. Define obsolete items in a warehouse.
4. Describe the identification of obsolete items in warehouses.
5. Explain the valid reasons for obsolete items.
6. Examine the role of FSN/XYZ analysis in obsolescence.
7. Discuss the scope of applying buy-back clause in Indian situation.
8. Illustrate the procedure of disposal of obsolete items.
9. Illustrate the roles of finance, users, purchase, warehousing, and design in obsolete items.
10. Explain the best methodology of controlling obsolete items.

42. Inventory Simulation

EXAMPLES OF SIMULATION

Simulation method which reflects actual working conditions, is used in many fields of activities. Flight simulators wherein pilots undergo aviation training, Monte Carlo simulation in gambling activities, flow-chart or set of rules used for computation and war games used for instruction in tactics are some illustrations of simulation method. The model sampling that statisticians use to generate bar charts or histograms for probability distributions, is essentially a simulation process. Scale models of dam, wing, and tunnel testing of aero-models, pilot plant experiment, laboratory tests, etc. are typical examples of the simulation process.

MANAGEMENT GAME

Use of computerised management business games, is an important pedagogical method in many executive training programmes, and represents the simulation process. In simulated environment, a number of teams or firms compete for a known consumer market. Each firm is represented by a team of players and has information on past sales, prices, market share, production costs, inventories, budget allocations, competitors' performance, etc. and constraints in the form of working capital, skilled manpower, overtime, strikes, supply delays, changes in consumer's forecasts, etc. In a specific duration of about one hour, the team should take decisions on manpower, materials, and manufacturing strategies to meet the demand. The status is fed into the computer to determine the cost per unit and the process goes on. Situations like strike, floods, transport bottlenecks, and forced exports are built in to simulate the reality. At the end of about 12 to 15 trials each team is given a chance to explain its mission strategies, failure causes, etc. to serve as a learning process.

CONCEPT OF SIMULATION

Simulation is a quantitative process, developed for studying alternative courses of action, by building a model of the system and then conducting a series of repeated trial and error experiments, in order to predict the behaviour of the system over a period of time. Simulation can thus be defined as any representation of a system or a process by means of a model. Simulation can be performed with analogue computers or even by hard computations.

NEED FOR SIMULATION

Simulation is the only method available for solving some problems as it is difficult to observe and quantify the actual environment. This method is helpful when it is difficult to develop a mathematical solution and when actual observation of a system may be too expensive or time-consuming. When it is not feasible to conduct experiments on real life situations as they may disrupt the entire process, simulation enables the decision-maker to evaluate the alternative strategies for determining the optimum policies.

Monte Carlo Simulation

This method derives its name from Monte Carlo, a place in Southern Italy, famous for its gambling dens, it is frequently used in the simulation process. This method is a computation process involving a statistical sample of random numbers. It is used to solve problems involving conditions of uncertainty and where mathematical evaluation is a difficult process. Monte Carlo simulation is a simulation process in which the random elements in the original process are represented. In this context, the random numbers are defined as those numbers selected in such a way that every number has an equal chance or probability of being included in the selection process. These numbers generated from the computer indicate the demand profile.

Two-Bin Inventory System

We will apply the simulation method to the (S, s) inventory system known as *optional replenishment system* or *minimax inventory system* or the *two-bin inventory method.* In this system, two parameters known as S and s are obtained, based on past data. When the stock on hand and the stock on order fall below s, an order is placed to bring the stock to a level s which is the maximum or the desired inventory level. The reorder point is s and the reorder quantity is $(S-s)$. This is a combination of the quantity and periodic systems of inventory control discussed in the earlier chapters. One bin of size S is chosen for each item with a marking under s which serves as reorder point. Thus such a procedure avails two bins of sizes S and s for each item. In practice, during receipt or issue, the warehouse incharge can observe as to whether the level has fallen below s in order to trigger off an order for quantity $S-s$.

APPLICATION OF SIMULATION

Let us consider a spare part costing Rs 50 per item. The past consumption/demand/sales as well as supply, total lead time have been analysed and the pattern given in Table 42.1 is obtained.

Table 42.1 Demand and Supply Patterns

Number of Units Demanded	0	1	2	3	4	5
Frequency %	47	36	11	3	2	1
Cumulative Frequency %	47	83	94	97	99	100
Lead time Days	5	6	7	8	9	10
Occurrence %	16	20	27	22	10	5
Cumulative Frequency %	6	36	63	85	95	100

The carrying charge is one rupee per item per day. The stockout cost is Rs 10,000 per unit per day. The ordering cost is Rs 1000 per order, with one delivery in an order. The future delivery cycles have been simulated as 7, 8, 5, 9, 6, 7, 8, 10, 9, 6, 5, 7, 8, 10, 7, 8, 6, 5, 10, 9, 8, 7, and 5. The future demand has been predicted for 100 trials and is given in Table 42.2. The reader is requested to determine *S* and *s* and simulate the same. For example if we take *S* as 30 and *s* as 40, then we continue meeting the demand with an initial stock of *s* or 30 units as computed in Table 42.2. The cycle indicates the ordering cycle and the lead time as given earlier. For instance, the first order will be placed as soon as the final stock reaches 10 units and it is called ordering cycle 1 with a lead time of 7 days. The last column is intended for entering the shortages. The final charges, including the totalling of shortage cost, ordering cost and carrying charges are added, for a policy of (30, 10). Similarly, we can compute these costs, for different sets of [*S*, *s*] in order to determine the optimum charges.

Table 42.2 Simulation Methods

Period	Cycle	Lead Time	Initial Stock	Demand	Final Stock	Shortages
1	—	—	30	1	29	—
2	—	—	29	2	27	—
3	—	—	27	0	27	—
4	—	—	27	3	24	—
5	—	—	24	0	24	—
6	—	—	24	1	23	—
7	—	—	23	0	23	—
8	—	—	23	2	21	—
9	—	—	21	0	21	—
10	—	—	21	0	20	—
11	—	—	—	1	—	—
12	—	—	—	1	—	—
13	—	—	—	1	—	—
14	—	—	—	0	—	—
15	—	—	—	2	—	—

Table 42.2 (*Contd.*)

Period	Cycle	Lead Time	Initial Stock	Demand	Final Stock	Shortages
16	—	—	—	2	—	—
17	—	—	—	1	—	—
18	—	—	—	3	—	—
19	—	—	—	0	—	—
20	—	—	—	0	—	—
21	—	—	—	5	—	—
22	—	—	—	0	—	—
23	—	—	—	1	—	—
24	—	—	—	0	—	—
25	—	—	—	4	—	—
26	—	—	—	0	—	—
27	—	—	—	1	—	—
28	—	—	—	2	—	—
29	—	—	—	0	—	—
30	—	—	—	1	—	—
31	—	—	—	0	—	—
32	—	—	—	4	—	—
33	—	—	—	0	—	—
34	—	—	—	0	—	—
35	—	—	—	0	—	—
36	—	—	—	0	—	—
37	—	—	—	1	—	—
38	—	—	—	0	—	—
39	—	—	—	1	—	—
40	—	—	—	1	—	—
41	—	—	—	0	—	—
42	—	—	—	2	—	—
43	—	—	—	0	—	—
44	—	—	—	1	—	—
45	—	—	—	0	—	—
46	—	—	—	1	—	—
47	—	—	—	0	—	—
48	—	—	—	1	—	—
49	—	—	—	0	—	—
50	—	—	—	2	—	—
51	—	—	—	3	—	—
52	—	—	—	0	—	—
53	—	—	—	1	—	—
54	—	—	—	0	—	—
55	—	—	—	0	—	—

Table 42.2 (*Contd.*)

Period	Cycle	Lead Time	Initial Stock	Demand	Final Stock	Shortages
56	—	—	—	4	—	—
57	—	—	—	0	—	—
58	—	—	—	0	—	—
59	—	—	—	1	—	—
60	—	—	—	0	—	—
61	—	—	—	0	—	—
62	—	—	—	1	—	—
63	—	—	—	0	—	—
64	—	—	—	2	—	—
65	—	—	—	0	—	—
66	—	—	—	2	—	—
67	—	—	—	1	—	—
68	—	—	—	0	—	—
69	—	—	—	1	—	—
70	—	—	—	0	—	—
71	—	—	—	0	—	—
72	—	—	—	0	—	—
73	—	—	—	0	—	—
74	—	—	—	1	—	—
75	—	—	—	0	—	—
76	—	—	—	1	—	—
77	—	—	—	1	—	—
78	—	—	—	2	—	—
79	—	—	—	0	—	—
80	—	—	—	0	—	—
81	—	—	—	0	—	—
82	—	—	—	0	—	—
83	—	—	—	1	—	—
84	—	—	—	2	—	—
85	—	—	—	0	—	—
86	—	—	—	3	—	—
87	—	—	—	0	—	—
88	—	—	—	1	—	—
89	—	—	—	0	—	—
90	—	—	—	0	—	—
91	—	—	—	0	—	—
92	—	—	—	2	—	—
93	—	—	—	1	—	—
94	—	—	—	1	—	—
95	—	—	—	0	—	—
96	—	—	—	1	—	—
97	—	—	—	2	—	—
98	—	—	—	0	—	—
99	—	—	—	1	—	—
100	—	—	—	1	—	—

Total number of orders — Total value of inventory — Ordering charges — Total inventory carrying costs — Total number of days stockout — Total stockout costs.

Mathematical Optimum

We can derive the optimum (*S, s*) in this example by mathematics and simulate around these values to check the same. For instance, in the given example, the average lead time can be calculated as

$$[(5\times16)+(6\times20)+(7\times27)+(8\times22)+(9\times10)+(10\times5)]\,/\,100$$

$$(80 + 120 + 189 + 176 + 90 + 50)/100 \text{ or } 7.05$$

We also observe that by keeping the average lead time of 7 days, the stockout will be for 37 days, i.e. [22 + 10 + 5] = 37.

Similarly, the average consumption can be obtained as follows:

$$[(0 \times 47) + (1 \times 36) + (2 \times 11) + 3 \times 3 + (4 \times 2) + (5 \times 17)]$$

or

$$(36 + 22 + 9 + 16 + 5)/100$$

or

$$88/100 \text{ or } 8.8$$

The average lead time consumption is (8.8) (7.05) = 6.2. This is rounded off to the higher number of 7 or the lead time consumption represents the *s* or reorder point.

The economic order quantity denotes the $(S - s)$:

$$\text{EOQ} = \sqrt{\frac{(2)\ (365)\ (.88) \times 1000}{1 \times 365}} = 42$$

$(S - s) = \text{EOQ} = 42$ $(S - s) = 7$. So we get '49 and 7'.

We can simulate around (50, 7), (48, 7), (49, 8) and (49, 6) in order to find the real optimum solutions from the total cost minimisation point of view.

TEN POINTS TO PONDER

1. Give examples of application of simulation in actual practice.
2. Define the concept of simulation.
3. Identify the need for simulation.
4. Examine the scope of management game in training exercise.
5. Discuss the application of Monte Carlo simulation.
6. Elaborate on the two-bin inventory control system.
7. Describe the process of choice of (*S, s*) in the optional replenishment system.
8. Explain as to how (*S, s*) policy is a combination of the 'Q' and 'P' systems of inventory control.
9. Demonstrate the need for random members in simulation.
10. How does mathematical approach quicken the simulation search process for optimum solutions.

PART IV

PURCHASING STRATEGIES

43. Procurement Strategies

SALIENT FEATURES OF PURCHASING

The objective of supply management is based on the philosophy that the key to efficient industrial, marketing is not the seller, but the buyer. The terms supply management, purchasing function, procurement strategies or the logistics management signify an activity comprising the following jobs: (i) Acquisition of materials required to support the organisation and government operations. (ii) Procurement and maintenance of capital equipment for development projects. (iii) Purchasing and upkeep of machinery to maintain public utilities. (iv) Material support of public and private organisations. (v) Bulk procurement, storage and distribution of the supplies to support social services such as health, education and housing programmes.

Profit Centre

In all these cases, the objective should be to ensure continuity of supply of uniform quality with lowest cost of the Ultimate product or service. Purchasing accounts for a major portion of the cash outflow or capital outlay in any organisation. In view of this substantial outlay, financial losses arising from mismanagement of supply functions have been high. By the same token, cost reduction can be achieved by better planning and setting realistic objectives and through supply management improvements on better buying procedures especially in capital and materials-intensive organisations.

Deficiencies

Mismanagement of purchase functions is a frequent cause of delays in executing developmental and operational projects. At times, this has led to costly interruptions and sometimes abandonment of important projects. The economic, financial, social and political consequences in these cases have been serious, sometimes causing great embarrassment to the concerned.

Ill-kept public buildings, poorly maintained roads, suspended construction projects, unserviceable machinery, and broken down public facilities, are manifestations of inefficient supply management. Defective purchasing management also leads to disruption in public utility services, electricity breakdown, irregular water supplies, breakdown of public transportations, cut backs in public health, and accidents in the transportation system.

The buyer should realise that he is not a post box of getting the item indented, but contributes significantly to the corporate image and profitability.

CORPORATE IMAGE

The buyer, in any organisation, should be a highly responsible person with sufficient authority and decision-making powers, as the profit made in procurement goes directly in the balance sheet. The supply manager meets various salesmen and can be considered as the public relations officer of the company. He can make or mar the company's image by his relations with the vendors.

The Fourteen Rights

In order to contribute to the corporate objectives, the buyer should plan the procurement managerial strategies in the right way. For his guidance, he has the fourteen rights which are as follows: Right quantity, Right source, Right quality, Right material, Right price, Right time, Right contract and legal aspects, Right place of delivery, Right transport, Right packaging, Right systems, procedures and powers of delegation, Right method of handling, Right materials intelligence covering socio-political, economical, financial, regulatory, technological, natural and inflationary global environments, and Right attitude oriented towards innovative improvement. Right contract or legal aspects should cover all the above purchase parameters.

PURCHASING CYCLE

Need identification is the first step in the purchasing cycle. The need is identified through kardex, bill of materials, material schedule, requirement planning, stock levels or users' special requisitions. Sometimes travelling purchase requisitions are used for reordering repetitive maintenance repair and operating items. The indents are classified according to the source of materials, such as projects, imports, subcontracts, monopoly, tied credit capital, consumable, budgeted, research, bazaar purchases, etc. Budgeted items are segregated from those for which sanction is necessary. The suppliers' directory enables to identify the possible sources. The order is placed according to the buying process, after getting quotation — if necessary by means of an advertised tender, as per the delegation of powers. Clearance by vigilance and pre-auditing purchase orders are insisted by some firms. Suitable follow-up action is initiated to get the item. The purchase order gives complete specification, price, delivery, terms and other conditions. The item is received and taken into the stores after inspection. The supplier is paid as per the terms of contract. However, there is a tendency towards paperless purchasing in progressive organisations.

The ordering cost to carry out the above activities, on an average works out to Rs 750 per order. This cost excludes the inspection charges, materials cost, and freight charges, and is low in the case of locally available material. Simplification of procedures drastically reduces the procurement cost. This cost can be up to 1 per cent of the value of material procured, as it depends on procedures, systems, delegation of authority, source of the item, advertisement, communication and level of integrity in procurement department. Top management must realise that 1 per cent saving in purchasing is equivalent to 10 per cent increase in sales.

Forward Buying

Forward buying or committing an organisation into future is the most common method of buying process. It is the practice of buying materials in a quantity exceeding current requirements, but not beyond actual foreseeable requirements. The time period is usually one year. Most forward buying involves a certain amount of speculation regarding the price. This future commitment depends on the availability of the items, financial policies, economic order quantities, discounts, staggered delivery and stability of market conditions.

Forward buying reduces the risk of inadequate delivery due to an impending material shortage, transportation bottlenecks, or unreliable performance by the supplier. However, forward buying involves the obvious price risk in unstable markets and the additional, carrying charges. Any price advantage resulting from a forward purchase should be evaluated against the above mentioned features to determine the real savings.

Speculative Buying

Industrial buyers, serving in large organisations, do not indulge in speculative buying, as speculation involves buying an item with the intention of profiting by selling at a higher price. The speculation may also be with regard to the availability of the item, and is more prevalent in the commodity market. Sometimes this involves the purchase of material in excess of foreseeable requirements, in anticipation of future need. If the risk is substantial, then the activity should be conducted by high level staff, outside the purchase department, since manufacturing organisations do not make profits from speculation in the materials market.

Hedging

Hedging is associated with speculative buying. The buyer tries to protect himself, by entering into two transactions: a purchase and a sale contract in two markets whose prices fluctuate. Hence the profit or loss sustained by the buying transaction, is compensated by the loss or profit in the selling transaction. This implies that the two markets behave similarly and one can reach the ideal solution of zero loss by a perfect hedge. Such ideal conditions do not exist in actual practice and there will be marginal adjustments of profit or loss due to hedging.

Hand-To-Mouth-Buying

The terms *hand-to-mouth buying stockless purchase, zero stock buying*, or *just-in-time buying* — refer to the practice of buying material to satisfy current operating requirements in quantities smaller than those considered economical. As the term signifies, the production operation literally exists on a hand-to-mouth supply of materials, and this is particularly suited for high value items and capital equipments. This method can be adopted in unstable markets, when prices slump or the demand changes drastically due to design changes in the buyers' plant. This method forces the supplier to carry the stock of items and the buyer can reduce this inventory.

Tender Buying

In public buying, the buyer should not only be honest, but should also appear to be honest. *Tender buying* ensures proper competition amongst the vendors. This method is practised in all governmental buying, and purchases in public sector undertakings. Public purchasing should be handled without any semblance of favouritism, patronage, personal preference and corruption in the choice of sellers. Open tender system, or advertising the requirements in regional/national/global, newspapers, is commonly adopted when the value of purchase exceeds prescribed monetary limits.

The steps involved are as follows:

(a) Formulate the requirements clearly.

(b) Establish bidder's list.

(c) Invite competitive bids through advertisement.

(d) Open the tenders before the representatives on the notified date.

(e) Prepare comparative statements on quality, price, and delivery schedules.

(f) Evaluate the bid.

(g) Award the contract for the *Lowest Responsible* bidder by taking a total view of the final product. *Lowest* implies being below the average standard. *Responsible* means answerable legally and morally.

(h) The quotation should be low, consistent with the quality requirements.

(i) The lowest bidder need not be the best bidder from the buyer's view.

As advertising the bids is costly and increases the lead time, bids are solicited only from reputed limited parties, who have been prequalified earlier. This process is known as *limited tender*. If there is only a single source, he is asked to submit his quotation in writing and this process is known as *single tender*. As far as possible, single tender purchases should be avoided, as it makes every item a monopoly proprietory item and is against the basic objective of purchasing or having at least two suppliers for each item.

Rate/Running Contract

The system of *rate contract* is prevalent in governmental buying, public sector purchase and procurement by direct demanding officers. The rate of an item is fixed usually by the tender system by a central purchase organisation after suitable negotiations with the supplier. But rate contract specifies only the rate and the quantity to be supplied may be zero. Hence, organisations like the Indian Railways use *running contract*, where the minimum quantity is also specified. Rate contract drastically reduces the internal administration lead time. The supplier can also plan his manufacturing activities in a better manner, as the buyers' exact requirements are known to him in advance.

Systems Contracting

Systems contracting is a form of stockless purchasing, under which authorised employees may directly obtain requisition of the material from the suppliers' storeroom. This procedure helps the buyer and seller to reduce the administrative expenses, and at the same time ensures material availability. The contract is simple, covering only delivery period, price and invoicing procedure. This method is particularly useful for low unit price but high consumption value items.

Blanket Order

Blanket order is an incomplete contract with a vendor to purchase certain items. But, it is not an authorisation to ship any material. It becomes effective when the vendor receives a bona fide shipping requisition or other type of release. This reduces the buyer's administrative burden and is useful for low value items. Blanket orders are usually entered for a period of one year. A typical example of blanket order is an annual order for stationery items.

Reciprocity

Reciprocal buying is based on the principle, 'Do unto your customers, as you would have them do unto you'. There is a dispute amongst experts about practising this method, but, it can be used if ethical and economical aspects are sound. This method dislodges competition and hence should be selectively practised.

Bazaar Purchase

Bazaar purchase, also known as *cash purchase* or *imprest purchase*, is confined to urgently needed small items. Besides the buyer, the users also purchase such items from the imprest amount as the items are urgently needed by them. A large number of departments send their own personnel, who waste a lot of time, thereby rendering the cost of purchasing more than the cost of the item. Hence, it is desirable to group such minor items required by different sections, and a single visit to the bazaar may be coordinated.

Subcontracting

Many organisations prefer to buy components, subassemblies, and assemblies instead of making all the items in the plant. The major advantage of *subcontracting* is that the subcontractor is a specialist in his job with lower overheads and hence lower cost of production. The choice of the subcontractor, the infrastructure for manufacturing and his ability to deliver the material are critical parameters in this process.

INDENT CONTROL

Generating a large number of indents increases the cost of procurement and hence, is the need for effective control of indents. Initially, the number of indents can be reduced for high value non-critical items and later on extended to low value non-critical material. A well-known practice amongst the users, is to raise an indent, as soon as materials, usually spares arrive in stores, even though they are intended for some other department. Such grabbing should be discouraged by indent control, which will avoid stockout for the relevant parties.

Indent control results in better planning of inventory levels at the warehouse. The initial resistance of the user department has to be overcome by assuring better service and emphasising the importance of service to all users. Indent control facilitates better planning by the user. However, indent control should not result in the departments substituting the warehouses in carrying at the job of inventory of items.

By adopting a professional procurement strategy, the buyer can significantly contribute to the corporate profitability and liquidity of the organisation. By this process, they can promote ancillary industrial activities, and indigenous development. By encouraging healthy competition amongst the suppliers, the buyer can contribute significantly to the economic development of the country.

TEN POINTS TO PONDER

1. Discuss the objectives of procurement strategy.
2. Examine the profit centre concept of buying.
3. Classify the contribution of the buyer to the corporate image.
4. Describe the purchasing procedure.
5. Distinguish between forward buying and speculative buying.
6. Define stockless purchasing, systems contract, reciprocity and cash purchase.
8. Identify the need for subcontracting.
9. Explain the utility of indent control.
10. Discuss the fourteen R's (rights) of buying strategy.

44. Right Source of Supply

THE RIGHT SOURCE CONCEPT

The modern trend in most organisations is to buy or subcontract, instead of making the item in their own plants. The buyer has to locate, develop and preserve the sources of supply. Since scientific purchasing requires at least two sources for any item, source selection becomes the most important responsibility of a purchase officer. The efficacy of a source depends on its capacity to match the user's high-tech, narrow tolerance, quality requirements and the purchase officer comes to a decision only after detailed discussions with the likely sources.

SOURCE LOCATION

Every buyer has to build a directory of purchases, which has to be updated periodically. Many organisations use the personal computer for storing and updating this vital information. This directory can be a standard catalogue of items, with full description along with up to date prices. It can be built from the following sources of information:

(a) Information received from liaisoning with marketing department.

(b) Periodic visits to industrial units.

(c) The buyer's guide published by the Indian Standard Institute lists a large number of items by the manufacturers who are licensed to use the ISI certification mark. The ISI-marked products are given preference by the Central and State Government departments, as well as some Public Sector Undertakings.

(d) The Directory Catalogue of the member's products published by Industrial Associations like the All India Manufacturers Organisation.

Other information leading to source identification includes the following:

(a) The handouts of the DGTD — Directorate General of Technical Development — and the Supply Directory of the DGS & D — Directorate General of Supplies and Disposals.

(b) The Yellow Pages of the Telephone Directory.

(c) Visiting cards of salesmen.

(d) Advertisements in trade journals.

(e) Announcements of new products in newspapers.

(f) Commercial journals.

(g) Technical magazines.

(h) Industrial exhibitions.
(i) Catalogues from sales departments.
(j) Reports of the National Research Laboratories.
(k) Professional journals.
(l) Exhibiting the requirements in the main reception.
(m) Advertisements calling for source registration.
(n) Competitor's sources obtained by materials intelligence.
(o) Professional association conferences, meetings giving ideas about sources.
(p) Industry survey reports of the Planning Commission and the reports of Ministries for macro-level planning.
(q) Trade Commissioners of foreign countries posted in India.
(r) Indian Trade Consultants in foreign countries.
(s) Information gleaned casually from informal and formal discussion in the market and elsewhere.

In many large organisations, a separate source development officer is responsible for locating and developing new sources. The research survey reveals that the buyers constantly search for new and reliable local/international suppliers.

INFLUENCING A TO Z PARAMETERS

Developing and retaining reliable suppliers, are crucial purchase activities. Before registering the following parameters should be checked, preferably by visiting the company:

(a) Technical capacity
(b) Design skills
(c) Financial background
(d) Bank references
(e) Motivation of workers
(f) Their major customers
(g) Their major suppliers
(h) Technological skills
(i) Manufacturing strategies
(j) Inspection facilities
(k) Testing equipments
(l) Factory conditions
(m) Type of service warranty offered
(n) Facilities for development of drawing
(o) Image of the company in the market
(p) Ability to meet contingencies and emergencies
(q) Previous performance on adherence to deliveries to other units
(r) Willingness for innovations and cost reduction techniques
(s) Capacity for absorbing new technologies

(t) Research and development efforts
(u) Preparation for accepting educational orders
(v) Willingness to share the information on cost of manufacturing
(w) Rejection level and commitment to zero defects
(x) Willingness to accept the buying company's terms for bill settlement
(y) Adherence to the company's contractual terms
(z) Company balance sheets and other publications, in order to assess the size and capabilities of the supplier.

Sources are registered for consideration after obtaining the above information. Visits to supplier's plants are absolutely necessary to check the bona fides as many organisations operate on the basis of fictitious visiting cards and bogus claims.

DEVELOPING RESOURCE

Having thus selected a supplier, a trial order/educational order, may be given after discussing and negotiating the price and other aspects. Future trial orders may be given, if the performance, in terms of quality, quantity, delivery time etc. is reasonably good. By this time, if the supplier is not sincere and has only tried to get an entry into the company, he starts showing indifference in one respect or the other. The trial orders should be continued for at least one year.

If even then, the supplier's performance is above reproach, he can be considered for the development of other requirements. In this process, engineering, manufacturing, technical help and occasional financial help can be given. Large suppliers are not interested in small quantity orders. The efforts devoted to the supplier development should therefore, be proportional to the size of the savings, which in turn is affected by the size of the order.

NUMBER OF SOURCES

The consequences of relying upon wrong suppliers are too well known to be stressed as the profitability and the image of the buying company are affected by stockout conditions. The company will be affected both financially and operationally. Hence reliable vendors should be recruited by gathering data. For efficient procurement operations, it is necessary to maintain up to date records on the performance of suppliers on technical, financial and managerial aspects. Local suppliers are usually preferred.

A question that needs to be answered is whether there should be only one supplier, or more suppliers for the same product. If the entire supply is to be procured from one source, there is a likelihood of getting quantity discounts. But, supply is assured if there is more than one supplier and production need not be held up for want of supplies. A certain amount of competition, and motivation for improving quality results out of two sources. It is necessary to have at least two suppliers for very critical high value production holding items. Having too many vendors for one item is also not feasible. The buyer should also ensure that the sellers do not gang up and form a cartel in supplying the items.

MONOPOLY SITUATIONS

In the case of monopoly supplier or canalised items, where the entire requirements are supplied by a single source, good liaison should exist with the vendor. The buyer should keep himself informed of the difficulties of the monopoly supplier.

Sometimes the buyer becomes monopolistic, by buying the entire product of the supplier. In such cases, the tendency to squeeze the profit level of the supplier is not good ethics. 'Live and let live' should be the policy and the buyer must allow the company to survive by making reasonable profits in the long-term interests of both.

Buyer should decide whether to buy directly from the manufacturer or his authorised agent. The original equipment manufacturers are not interested in small quantity orders for spare parts. The local middleman offers better services, as he has a direct interest in the consumer. Even though the item may be slightly costlier, savings can be effected in lower transportation and inventory charges.

ANCILLARY DEVELOPMENT

In the context of right source, ancillary development plays a crucial role. As a matter of fact, the Government gives price preference up to 20 per cent for small industries, when purchases are made from them. Tax concessions are provided in Central Government budgets every year, for promoting the small scale sector.

Small scale enterprises, whether formal and well-known or informal and unknown, have gained importance in developing countries as outlets for the entrepreneurial energies of people outside the mainstream economy, as providers of jobs for the unemployed and underemployed, and potentially as contributors to national economic growth and economic equality. In India, ancillary development has been an important part of the small industry development programme. The small scale industries number two million units, employ about twelve million people and account for a sizable part of industrial production. It is even claimed that a quarter of the total exports is through this sector.

PROBLEMS OF ANCILLARY UNITS

According to the latest notification of the Union Government, an ancillary unit is an undertaking having investments in fixed assets, in plant and machinery up to Rs 45 lakhs and engaged in the manufacture of parts and components and rendering of services, with a proviso that they should earmark 30 per cent of their production for supply to bigger units. This rupee value needs review as the business expands, and also due to inflationary spiral in India.

Among the important measures to promote ancillaries, the Government has been exhorting the Public Sector enterprises to establish ancillaries progressively. However, in the Public Sector, the number of ancillary units is about 20,000, with an employment of 500,000 people and a total value of procurement amount of nearly Rs 750 crores. Paradoxically, in spite of the directives from Government, only 10 per cent of the total purchases of the Public Sector are made from ancillary units. In the midst of rising costs it is a natural economic proposition to have smaller units with low overheads make the components so that the costs could be kept low. Japanese ancillaries have a share of more than one third of the total supplies as against less than one tenth in India.

The location of ancillaries should be within easy access of the parent industry to be mutually beneficial. Problems faced by ancillary units are in the following areas:

(a) Quality standards and manufacturing methods
(b) Technology changes and upgradation
(c) Payment in time and finance availability.

The parent unit should ensure that quality control parameters are thoroughly explained to the ancillary units before they proceed with the job. Personnel of the smaller unit should be given training by the parent unit. Instances may be cited from the automobile industry which goes in for the maximum percentage of components from ancillaries. A defective spark plug can make an automobile immobile while a defective brake lining can make the automobile crash.

When large enterprises introduce new technology, they should help the smaller units gear up to corresponding modernisation plans. Correct identification of modernisation needs is an important input accompanied by financial backup from banks or financial institutions.

With bigger industries clamouring for 'Just-in-time' system (JIT), whereby they demand supplies on an almost daily basis without their carrying any inventory, the pressure is on the small units to supply progressive batch quantities instead of whole lots. To add to the problem, procedural hassles in the big company result in late payments which can demoralise and even ruin the small units.

Ancillaries should be considered not merely as a supplier of components, but as a supplier of both technical and non-technical support services. For example, there are many engineer entrepreneurs, who have set up machine design units, as well as tool design enterprises. In fact some of them have been able to introduce modern technology of — computer aided design/management—CAD/CAM faster than the bigger units. Similarly, inspection services are offered by small entrepreneurs in big cities.

TEN POINTS TO PONDER

1. Identify the need for right sources in the field of materials management.
2. Examine various ways of locating a new source.
3. Discuss the methodology of developing new source.
4. Elaborate on the mechanism for fixing the number of sources for an item.
5. Explain the methods of tackling monopoly sources.
6. Detail the procedures for retaining good suppliers.
7. Illustrate the role of ancillaries in developing right source.
8. Define the problems of ancillary units.
9. Elucidate the information to be gathered before deciding on a right source.
10. Articulate on the significance of educational orders.

45. Buyer-Vendor Relations

NEED FOR CORDIAL RELATIONS

Successful business is run on cordial relations, in these days of stiff competition. The vendor is often treated as an intangible asset of progressive organisations, and is considered as an extension of the buyer's plant, particularly in organisations which go in for more subcontracting.

A cordial internal as well as external relationship within the organisation and the outside world is necessary to boost business. The pivotal personality in the complex market mechanism is the vendor. The vendor's goodwill fetches a market for the product. Good buyer-vendor relationship enhances mutual motivation and results in better development of the total economy.

The buyer should appreciate the need for a viable producer who can ensure regular supply with constant innovations and improvements. Frequent changing of vendors is detrimental to business. Buying efficiency varies directly with the efficiency of the suppliers. Non-delivery of material affects the companies' profitability adversely. Sellers can solve the buyer's problems, particularly in the area of cost reduction. In view of the above advantages, the buyer should maintain a mutually profitable relations with the suppliers. Supplier management is a major area to the buyer, according to the research survey.

BUYER'S ROLE IN IMPROVING RELATIONS

The purchase department should recognise that in its own interest, it has a vital responsibility in establishing a cordial relationship with vendors. Many buyers organise periodic conferences with the salesmen, supplier seminars and visits to the sellers' plants to foster cordial relations.

The buyer may help the seller, whenever necessary in the following aspects: Obtaining raw materials and licences, rendering technical assistance and suggestions for improving the process, educating the suppliers on the company's requirements, assuring steady orders, underwriting the initial wastage and risk sharing. The buyer should use public relations to maintain as well as enhance the corporate image in strengthening the bond of relationships. The buyer should understand the seller's position in the industry, his marketing strategy, production capacity, technical competence, manufacturing efficiency, and research. The differences and poor performances, should be segregated so that, a case by case approach can be adopted to improve the relationships.

In maintaining cordial relations, the purchase department can take the following steps. Place an order with the supplier with correct specifications and in unambiguous terms. Set out purchase terms clearly with clear instructions on quality, delivery, price, time, terms and all other aspects. Specify unequivocally, the terms of payment in the order itself.

In order to improve the relationships, the buyer should ask the following questions: Is the relationship improving or deteriorating? Are our promises being fulfilled? Are we neglecting anything? How do we stand vis-a-vis our competitors? What steps should be taken to improve the relations further? etc.

The following measures can be taken to improve the relationships: Initiate positive phone calls. Make recommendations for improvement. Be frank in one's views. Communicate more by phone. Show appreciation of good performance. Make service suggestions. Identify real problems. Plan the future methodically. Accept responsibility for faults instead of shifting the blame.

BASIC REQUIREMENTS

In the context of healthy buyer-seller relations, the following points should be considered by the materials department. In the depression years, many purchase managers rebuff vendor's salesmen and discourage their visits. This short term policy is myopic and fails if they are to find the best buys for their organisations. Even at the initial design stages, it is necessary to identify the requirements in order to sell upstream items as far as possible. A salesman calling on a firm is the purchasing department's guest and his visit should be made pleasant and mutually profitable.

Purchase department should fix definite calling hours and make them known to salesmen. Parking space should be allotted for visitors, and salesman should not be kept waiting for a long time. The receptionist should be informed of what areas each buyer handles. It should not be easy for salesmen to spot, how often and when their competitors are calling.

Communicating with the seller in the public reception areas, destroys the effectiveness of the interview. If the salesman is brief and to the point, he is entitled to the purchase managers undivided attention — without distractions of telephone calls or reading the mail.

The salesmen should have faith that the buyer is the ultimate authority on quantity, quality, packaging, lot sizes, price, supplier, shipping point, routing, terms and contracts. This avoids back-door direct selling to the user, over the shoulders of the buyer. The buyer should be familiar with the logistics of business transactions so that he does not make unrealistic demands on the suppliers. The buyer can inform the salesmen about his competitor's terms and this reduces tension in the buyer-seller relations. Definite delivery dates should be indicated on inquiries, requisitions and orders.

A TO Z OF SELLER'S FRUSTRATIONS

The interdependence of the buyer and seller should be analysed from the points of view of their interactions, influence processes and customer service. The seller faces the following problems:

(a) Delays in accounts settlement.

(b) Unwarranted and wavering policies adopted by the buyer in rejecting good items, when not needed.

(c) Leaking sellers' quotations to the competitors.

(d) Inordinate delays in buying-decisions.

(e) Changes in delivery terms, orders and unscheduled delivery demand.

(f) Abrupt cancellation of order without notice.

(g) Periodic modifications in specifications amendment of purchase orders, after issuing orders.

(h) Haggling for the price continuously through negotiation process.

(i) Undue delay in giving an appointment causing the salesmen to wait for a long time in the buyer's office.

(j) Details of requirements, specifications, drawing, etc. are never handed over.

(k) The buyer sometimes insists on unethical and unfair means.

(l) The buyers plant personnel do not adhere to recommended operating and maintenance manuals.

(m) The buyer insists on delivery of spares for outmoded machines installed fifty years ago.

(n) Small uneconomical quantity orders, are difficult to comply with.

(o) The communication on rejections or the advice note is not sent immediately.

(p) The buyer does not believe in joint inspection.

(q) The buyer is always curt and non-cooperative when sellers seek information on future requirements.

(r) The buyer does not encourage new sources as he does not believe in educational orders, and has faith only in established reputed suppliers.

(s) Inadequate physical facilities for meaningful dialogues in the buyer's place.

(t) The buyer does not have sufficient authority to exercise sound judgment in taking decisions, creating embarrassing situations.

(u) The buyer insists on buy-back clause for initial spare parts at the time of buying capital equipment.

(v) Sometimes the buyer forces the seller to take obsolete stores, supplied by other companies.

(w) The buyer insists on practice of reciprocity.

(x) Due to inadequate planning the buyer asks for rush orders in the last minute.

(y) The buyer fails to appreciate supplier's strength and weaknesses and insists on the same price throughout the year without considering factors leading to increased costs.

(z) The buyer never visits the vendors' plants or listens to vendor's difficulties and relies on other vendors' information only. These problems should be sorted out individually in order to improve the relationships.

A TO Z OF BUYER'S PROBLEMS

The buyers' problems with the vendors can be catalogued as follows:

(a) Aggressive salesmanship forcing the vendor to try his new products.

(b) Unsatisfactory long delivery schedules, deviating from the quoted periods.

(c) Delivery of substandard quality, not as per the agreed sample or specifications.

(d) Seller does not remove the rejected materials from the stores, assuming that they will be utilised in times of shortages.

(e) Seller tries to adopt *force majure* clause at the earliest opportunity.

(f) Seller resorts to frequent price increases without proper justification.

(g) Seller bypasses the buyer's authority and directly deals with the user.

(h) Seller frequently reminds the finance department to settle outstanding bills.

(i) The vendor is unable to meet emergency requirements.

(j) The morale of workers at seller's plant is very low, with the workers trying to go slow at the slightest disturbances.

(k) The seller accepts orders on terms beyond his capacities, thereby dislocating many buyers.

(l) The seller always demands advance money, even for procuring regular materials.

(m) The seller is not interested in small orders.

(n) After sales service — ASS — in spite of warranty clauses, and commitments, is always below par.

(o) The seller sends orders in unsealed covers and by name.

(p) The vendor resorts to litigation at the earliest opportunity.

(q) The seller manages three quotations on three letterheads, whenever the buyer's procedure stipulates three minimum quotations.

(r) The seller resorts to unethical corrupt practices to get the order.

(s) The seller sometimes defaults on financial grounds of insolvency.

(t) Sometimes, the seller sends quotations in the last minute, without adhering to the rules.

(u) The seller avoids sending the documents through Bank.

(v) Forgets to give the 'C' form details for purposes of sales tax.

(w) The seller responds to the notice inviting tenders without giving details.

(x) The seller is technically incompetent and inefficient.
(y) Sometimes the seller requests his retail agents to supply the material.
(z) Materials are sent without support of adequate documentation.

Blacklisting

The vendor and vendee should meet if the above problems occur and sort them out on a case by case basis, without being guided by emotions. Blacklisting is only the last step.

The buyer and seller should have complete clarity of communication concerning the need, application and usage of the required item. They should have mutual confidence in the statements and intents of the other party. Professional, ethical, and progressive approach will definitely boost the buyer-seller relationship which is vital for any organisation. A fair price and profit, walk hand in hand and the alliance between the two is the strongest assurance of a strong, free society.

TEN POINTS TO PONDER

1. Identify the need for cordial buyer-seller relations.
2. Examine the prerequisites for good relations.
3. Elaborate on the failures of vendors.
4. Elucidate on the problems created by buyers.
5. Clarify the frustrations of the seller, due to buyer.
6. Illustrate the problem areas of the buyer.
7. Explain the buyer's role in improving the relations.
8. Discuss the seller's role for better relations.
9. Describe the tips to improve the buyer-vendor relations.
10. Articulate on the necessity of mutual plant visits of the buyer and vendor, in the context of buyer-seller relations.

46. Supplier Evaluation

OBJECTIVES FOR RATING THE SUPPLIERS

Purchasing aims at having two suppliers for most of the items. Hence, identifying potential and reliable vendors, as well as maintaining uptodate records on their performance is vital for purchasing operations. The consequences of choosing the wrong type of suppliers are serious both financially and operationally. A supplier's default may sometimes be more serious than a buyer's mistakes. In deciding the market share between suppliers, it is necessary to assess the performance of the suppliers. It is also necessary to educate the supplier with a view to improve his performance. It is desirable to compare one vendor's performance with others in order to improve the overall reliability. Just as the buyer tries to rate the supplier, the seller also rates the buyer with regard to his authority, technical knowledge, professionalism and commitment to promises. Purchase department has the sole responsibility for the choice of the supplier and placing the orders. Hence, it is necessary to assess the vendors' performance on an objective basis, based on price, delivery, quality, and service aspects. In order to fulfil the objective of getting a quality product at minimum costs. Unless the vendor sends material of the proper quality, all the previous efforts and time spent by the buyer become useless and delays, while replacements are secured cause losses. The company also incurs extra cost resulting from inspection and rejecting the substandard items.

FOUR STAGES OF EVALUATION

There are four stages of evaluation of suppliers. In the *survey stage*, all possible sources are explored for their capabilities, based on preliminary information. The *enquiry stage* consists of a detailed analysis of the vendors performance with other consumers. In the *negotiation* and *selection stage*, only those vendors, who pass the enquiry stage, are called in for negotiation on all aspects and the list of approved vendors is drawn up. These three stages are considered as a part of source development and source registration process.

Vendor rating normally refers to the experience stage when the supplier has started sending the material after trial orders. For this purpose, the parameters are quality, price, delivery, schedule and service. In this process quality controls' department's assistance is needed to get information on rejections at incoming materials stage, in process conversion stage and at the outgoing final stage. All the stages of evaluation should be carried out objectively, using scientific methods, without personal biases and prejudices.

A TO Z ISSUES ON RATING

A detailed list of technical, managerial, financial, personnel, and service aspects for vendor rating is given below:

(a) Vendor's reputation as a reliable and financially strong company.
(b) Supplier's proven integrity by past performance.
(c) Supplier's commitment for warranty, guarantee and quality.
(d) Suppliers' cost reduction programmes by using scientific methods.
(e) Labour-management relationship in the seller's plant.
(f) Adequate design and development wing for technological challenges.
(g) Supplier's capability for analytical engineering, installation and commission engineering.
(h) Capacity to train buyer's engineers.
(i) Ability to meet challenging and new tasks by technical leadership.
(j) Research efforts leading to general advances.
(k) Quality of the after-sales-service and spare parts availability.
(l) Dependability and reasonable price of after-sales-service.
(m) Ability to provide maintenance contract needs of the buyer.
(n) Assurance and conformance of minimum delivery time schedules.
(o) Reserve production facilities for emergency requirements.
(p) Supplier diverting to other customer in terms intended for one.
(q) Vendor not becoming insolvent after taking money.
(r) Selling a range of full-time of related items.
(s) Credit availability and financial arrangements of seller.
(t) Local contacts and their technical competence to help.
(u) Suppliers' reputation enhancing product sales of buying firm.
(v) Vendor's capacity to provide selling aids, drawing, etc.
(w) Managerial effectiveness of supplier's organisation, ownership pattern, and professionalisation of personnel.
(x) Status and operational reliability of seller's plants.
(y) Financial status of supplier, certified by the bankers.
(z) Record of labour disputes at the sellers' plant and their impact on the buyer's requirements.

Methodology of Rating

A large number of methods, using quality, quantity, price, delivery schedule, and service factors has been developed for the purpose of rating the vendors. The weightages vary from item to item depending upon cost criticality, availability of the item and market conditions of demand and supply. Vendor rating becomes meaningless if there is only a single source or the item becomes canalised through some governmental agencies.

Since considerable amount of data is needed, a computer is useful in comparing the performances of vendors supplying similar commodities in the same period or the same vendor, over periods of years. These are known as inter-firm and intra-firm comparisons. The defects and the scores may be pointed out to the vendor in order to enable him improve his performance. In this context, the Indian Institute of Technology at Madras, has compiled a vendor rating index using 300 parameters to assess the seller's performance.

Incoming Material Rating Plan

The incoming material rating plan, developed by the General Electric Company is often quoted as a popular vendor rating method. In this scheme, the quality control department complies a list of qualitywise troublesome items, with the drawing numbers. The purchase department compiles a list of items, considered troublesome because of price and service. From the above, a single list of items is used in the rating programme. A total of 100 points is allocated in the following manner. Weightage of 40 points for quality, 35 points for price and 25 points for service. These points are changed if necessary to suit specific items. The quality factor of 40 points is multiplied by the percentage of accepted lots for each supplier. The net price after allowing for discounts and transportation is multiplied by 35 points to get price factor for each supplier. Similarly, the service factor expressed as a percentage of delivery promises, is multiplied by the service factor of 35 for each supplier. These three are parameters are evaluated and added for each supplier in order to get the rating.

Weighted Point Plan

The evaluation criteria is quantified to the above method and points allocated for quality, delivery and price. The number of factors and the points can be altered to suit individual products. An addition of weighted average of the performance parameters, is known as the *weighted point plan*. Norms can be fixed for acceptance/rejection of the vendors.

Categorical Method

The *categorical method* depends primarily on the experience and judgment of the buyer, and hence this may be subjective. The buyer makes out a list of all relevant parameters and annual performance report is prepared. He may hold discussions with inspection, production stores, and other colleagues in formulating the points. Out of a possible total of 100, weightage of 80 and above is considered excellent, 70 to 80 points is very good, 60 to 70 points is good, 50 to 60 points is satisfactory, 40 to 50 points is average and below 40 points is considered poor. By this process, the buyer is forced to think of assessing the suppliers. The cut-off points for low rating, can be determined on the basis of market conditions.

Cost-Ratio-Method

This system involves computing the actual costs incurred in procurement, chasing, transportation, packaging, duties, receiving material, etc. The unit cost incurred by the buyer on the material, when actually received is then computed. The higher the cost, the lower the supplier's comparative rating. In this system, the computation of cost for each item may become tedious.

Costs relating to the quality may include factory visits, approval of samples, inspection, rejection of incoming materials, and losses arising in production, besides the usual purchasing costs. Costs relating to delivery include cost of follow-up, transportation system for speedy deliveries, etc.

Key Questions Approach

Many questions, what, where, how, why, when, what for and who are asked about the item in this method. When these questions are answered, subsidiary questions arise. These questions should include issues on management, labour, finance, technology, design, development and service aspects. In the end a balanced perspective of the supplier emerges.

Matrix Method

Many combinations of systems are possible and appropriate in various sets of circumstances in the sphere of vendor rating. The vendor rating should result in high quality at the lowest possible price. A large number of parameters can be considered, quantified, weightages assigned and analysed with the help of computer. But most of the methods restrict the parameters to quality, price and service.

Quality evaluation is simply the supplier's record with respect to meeting required specifications and is measured as per cent rejections to total material inspected. Consistency of success and integrity in price behaviour provide measurement criteria. Service evaluation includes prompt submittal of data, response to inquiries, delivery performances, special services and other intangible aspects. When there are more number of suppliers and more factors being considered, the matrix approximate of comparison is useful. The vendors are listed horizontally and the factors vertically. Individual marks are entered for each factor and the total entered as a weighted average of the individual factors. This gives the rating in a complex situation and the buyer can take appropriate actions.

TEN POINTS TO PONDER

1. Examine the necessity of vendor rating in materials management.
2. Identify the goals of vendor rating.
3. Discuss the four stages of evaluating vendors.
4. Elucidate the agencies involved in vendor rating method.
5. Describe the vendor rating process.
6. Explain the incoming material rating method.
7. Clarify the concept of weighted point plan.
8. Enumerate the steps involved in cost ratio method.
9. Define the scope of matrix method of vendor rating.
10. Examine the issues in vendor rating method.

47. Right Pricing

STEPS IN PRICING

Price is an opinion, cost is the fact, says the first principle of Financial Management. Price is what you pay according to market conditions, whereas cost is incurred by the vendor. "There is hardly anything in the world, that someone cannot make cheaper, and the man who considers price alone is the lawful prey", says John Ruskin. The price must be technically acceptable. In the context of the common criticism that buyers buy by price and do not know the cost of the item, determination of right price of the item becomes necessary.

Pricing is an area where the top management decision is vital for keeping a project viable. Price goes by many names like fee, fare, rate, premium, rent, tuition, interest, honorarium, bribe, salary, wage, etc. Price has operated as the major determinant of the buyer's choice both in the past and present. The survey has pointed out that the major area of interest of the buyer is setting optimum price targets.

The steps in price setting in a selling organisation include the following:

(a) Selection of the pricing objective as survival/quality leader/market share leader/current profit maximisation.

(b) Demand determination by market surveys.

(c) Estimation of fixed and variable costs.

(d) Analysis of competitor's price and offers.

(e) Choice of a pricing method.

(f) Price modifications in the form of cash/quantity/seasonal discounts.

The buyer should realise that the sellers' firm is not a philanthropic institution, and should aim at discovering the right price of an item by identifying the material content, labour component, depreciation, taxes and overheads.

Price Information

Reliable current price information is important in purchasing activity. The buyer should maintain a current catalogue and price-sheet file for high value items. The second step is to obtain competitive bids. The third method is through negotiation, particularly for non-standard items. Discussions with the buyer often lead to opportunities for price reductions.

In addition to the above, there are commodity market rates, published in newspapers/periodicals and information published by trade associations. London Metal Exchange price is an example.

The reasonableness of the price quoted is to be assessed with the above information, and also, where necessary, by getting data on worker output, labour rates, materials cost, overhead and other factors. Technical and engineering assistance, when required, is to be obtained for a proper evaluation of the price quoted.

A TO Z INFLUENCING FACTORS

The price of an item depends upon a large number of factors affecting market condition. These include demand profile, supply pattern, government policy, social, political, economical, fiscal, monetary, regulatory, macro, micro, global and natural factors. Some of these factors are listed below.

(a) Government taxation policies
(b) Priority industry
(c) Small scale industry
(d) Agro-based industry
(e) Seller located in backward areas getting subsidies
(f) Standard item
(g) Quantity discounts
(h) Delivery schedule (most urgently needed items priced higher)
(i) Payment terms
(j) Precision tolerances and specification
(k) Past experience or reputation
(l) Range of indigenous manufacture
(m) Alternative items
(n) Availability of licence, in the case of imported items
(o) Off the shelf items with huge inventory
(p) Guarantees offered
(q) Assurance of future suppliers
(r) Shelf-life
(s) Rate of obsolescence
(t) Maintenance and repair
(u) Direct materials cost
(v) Conversion cost
(w) Easy availability
(x) Proprietary item
(y) Production capacity
(z) Buyer-seller relations.

These influencing factors determine the pricing strategy in a company. Another crucial factor is the concept of the product life cycle, wherein an item goes through the phases of growth, youth, stability and decay due to competition or changes in fashion. The return on investment on the item denotes the phase in which the item is passing through. Price which influences sales volume, is based upon average full cost per unit, which in turn is partly determined by the underlying volume of sales.

PRICE SETTING

In the past prices were set through negotiations between buyers and sellers. Setting one price for all buyers is a relatively new concept. In small companies, prices are set by top management. In large companies, divisional product line marketing managers set the price. In industries where pricing plays a crucial role, a separate pricing department, headed by a Finance Manager is incharge of product pricing. In government organisations, the financial advisor plays a crucial role in determining the price of an item.

Marginal Pricing

A major consideration in setting the price is that at least the variable cost elements like direct material and direct labour, should be recovered. The margins are fixed depending upon the strategies between 50–70 per cent as contributions to fixed expenses. These margins depend upon market conditions, competition, etc. In the case of spare parts, the margins are much higher compared to other items. This pricing is based on variable costs. It is assumed that fixed cost is not increased for producing the items. Managers are reluctant to employ a contribution approach because of fears that variable costs will be substituted indiscriminately for full costs, and lead to price reduction.

Market Skimming Strategy

Based on the influencing factors, pricing strategies are developed in the selling organisation and the buyer should be aware of these strategies. The *market skimming strategy* enables the buyer to recover his high investment on research, development design, engineering, tooling, etc. This strategy works well till competition arises. This is particularly suited for high-cost, high-price, premium-quality items.

Market Penetration Strategy

For general purpose spares, when buyer has many sources, seller offers lower price to attract the customer. In adopting this strategy, the seller has to consider many other factors, like existing level of competition, after-sales-service, distribution costs, etc. This is suitable for high-quality but medium-priced items.

Price Penetration Strategy

This is usually adapted by non-original equipment manufacturers, and spurious spares suppliers. They pack it as made in USA — Ulhas Nagar Sindhi Association. Here the low price alone determines the market share and hence the quality is likely to be poor.

Target Profit Pricing

This method is based on 'break-even' analysis. The break-even point is determined from the total cost and revenue curves. The *target profit* is applied to determine the corresponding price and sales volume in the break-even analysis chart.

Perceived Value Pricing

This is based on buyer's perception of the value of the product and not sellers' cost. A typical example is spare parts priced by Maruti Udyog, which the buyer feels is exhorbitantly high. Usually in most cases, particularly in spare parts, the buyer perceives the price of the item as very high. Hence he wants the seller to reduce the prices and this is achieved by means of price analysis and negotiations.

Open-End Pricing

When there is an established free market for a material, neither the seller nor the buyer wants to commit himself to a specific price for a long period of time. The purpose of *open-end price contract* is to ensure supply of material for some time in future at the prevalent market rate. But, the buyers normally avoid open-end pricing, since it limits their flexibility.

Cost Plus Pricing

In some cases, the buyer and seller may not exactly know the cost of the contract. In such cases, *cost plus contract* pricing may be the only solution. The cost includes materials cost, labour cost, depreciation and reasonable overheads. The seller under this contract, agrees to do the job either at some fixed price per hour of the work performed, or at his cost plus some extra fee as profit. Sellers have little incentives to keep down the costs and become inefficient and so, these contracts are to be avoided as far as possible.

Going Rate Pricing

This method is based on the competitors' price. Small firms follow the market leader. This method is helpful when details are not available. The small cigarette, manufacturers, (following the prices of I.T.C.) are examples.

Sealed Bids Price

Sealed bid is commonly adopted in procedure-oriented, Government, buying organisations, where the firms give the offer, in a sealed cover under the tender system. They expect that they have quoted less than their competitors. Instead of written quotations, the buyers can also get telephonic quotations from well-known sources in order to have an idea of the prevailing price rate.

Escalation

Escalation clauses in contract permit for adjustment of unforeseen changes in basic costs but otherwise, require the seller to stick to his original quotation. Escalation is due to labour material cost, or currency fluctuations or due to all elements. The formula for escalation is to be made clear to avoid disputes and it is desirable to fix an upper limit for escalation.

Administered Price

The Government fixes the *administered price*, if necessary by subsidising the item, like price of farm inputs, fair price sugar for weaker sections, which are lower than the open market price. For some items like steel and cement, the levy price is adopted in order to boost construction activities.

Price Redetermination

Price redetermination permits periodic negotiation of contract prices. They are broader than escalation clauses. This is useful for contracts that extend over long period for items, for which neither the buyer nor the seller, have had much experience. Sometimes such contracts, include an agreement permitting the buyer to audit the sellers' cost.

In determining the final price, the buyer should also consider (a) the psychological pricing of further orders, (b) selling companies' pricing policy, and (c) impact of price on other parties.

Price Modifications

The prices are usually modified by discounts, which may be in the form of cash discount, quantity discount, seasonal discount, trade discount and cash subsidies. The price can also be discriminatory, depending upon the customer's product, place and time. Prices are increased because of inflation, or because of huge demand. Price cuts can occur due to reduction in market price, excess capacity or genuine decreases in price level. But the buyer requests the seller to reduce the prices, as the sellers voluntarily do not decrease the price. The sellers' price may also decrease, if the competitor reduces the price.

Recently, particularly in the Southern part of India, the power cut on high tension consumers, has forced the manufacturers to use their own diesel generator set for manufacturing items resulting in increased prices. The general inflationary spiral, increased cost of labour, higher volumes of inventory, credit increases taxation, etc. result in a generally increasing level of prices of most items. The buyer and seller should discuss the pricing problems frequently for mutual benefit. Such a step enables the buyer to forecast the price, which is essential to formulate the purchase budget.

TEN POINTS TO PONDER

1. Explain the meaning of right price.
2. Identify the steps in pricing an item.
3. Discuss the requisite information to determine the price of an item.
4. Enumerate the influencing factors on pricing.
5. Describe the role of price setting machinery.
6. Distinguish pricing by market skimming strategy, from market penetration strategy.
7. Define marginal cost pricing.
8. Elucidate the significance of cost plus and open ended price.
9. Articulate on the concept of sealed bid pricing.
10. Examine the relevance of price modifications.

48. Governmental Buying

ROLE OF DGS & D

Governmental buying is buying for the public in public. The basic difference between governmental buying and non-governmental buying, is that the money (rupees) spent in governmental buying, is generated by the collective tax payers. Governmental buying involves buying the required items at the optimum price, in accordance with all applicable laws, regulations, policies, procedures and monitoring systems. The process of defining the need, identifying the source of supply, making the supplier aware of the needs, obtaining offers, arriving at agreed conditions, placing contracts, receiving the goods, settling the supplier's bill etc. is equally applicable in governmental buying.

The Directorate General of Supplies and Disposal (DGS & D) constitutes the central purchasing organisation of the Government of India. It is responsible for the purchase of supplies required by the indenting Government departments, mainly the Defence (excepting military hardware), Railways, Post and Telegraph. DGS & D purchases goods worth about Rs 5000 crores annually for the above three organisations and disposes goods worth about Rs 200 crores per year. DGS & D buys at the lowest rates. Besides its services, mainly rate contracts and registered supply sources — are also utilised by the State Governments, Public Sector Undertakings, quasi-government agencies.

The centralised purchase organisation has enabled a considerable degree of standardisation of procedures and contracting. Over the decades, the DGS & D has steadily increased the value and variety of its purchases. The supply wing is the mainstay of the organisation and has 11 directorates at Headquarters and five regional purchase directorates in other parts of the country.

The planning and development division helps to make the procurement process a success by rationalising the specifications and drawing a comprehensive programme for receipts and booking of indents of common user items. The organisation and methods branch assists the supply wing in devising procedures and systems to obtain optimum efficiency. The statistical branch, which has been computerised, is responsible for maintaining up-to-date statistics and brings out the annual report as well as the Directory of government purchases.

DGS & D consists of supplies wing, inspection wing, progress wing, disposal wing, planning division, coordination division, public relations organisation and regional officers. The supplies has the following 10 Directorates:

(1) Mechanical Engineering Stores Directorate

(2) Workshop Machine Tool Directorate
(3) Railway Stores Directorate
(4) Electrical Stores Directorate
(5) Miscellaneous and Hardware Directorate
(6) Oil Paints and Chemical Directorate
(7) Timber, Wool and Leather Directorate
(8) Vehicle Directorate
(9) Project Directorate
(10) Civil Armament Directorate

The policy for all purchases for stores, quality preference, price preference, registration of firms, removal, blacklisting, banning and suspension of firms, and fees for services rendered, have been laid out clearly.

DISTINCT FEATURES OF PUBLIC BUYING

In the case of public buying, the aim is to ensure the best value of money spent, and to obtain what is needed at the right time and at the most economical price. The difference in industrial and public buying is the concept of public accountability of the latter. Purchases made by Government are financed by the Exchequer and hence Government buying is more organised and executed in such a way, that not only should it be honest, fair and impartial but should also appear to be so.

The concept of public accountability gains still greater significance when the large volume of purchases is considered. Every transaction should be the most economic and appropriate buying. To ensure uniformity in purchase and its applications in such a large volume of work relating to different kinds of material and equipment, certain procedures have been evolved.

CENTRALISED PURCHASING

Centralised purchasing is generally recognised as an important function, particularly in Government. Its objective is to achieve savings by the application of efficient management techniques, especially through volume buying. Under centralised procurement procedures, responsibility rests with an authorised specialist and the procurement is not incidental to other operating functions. If specialised effort is devoted to coordinating the requirements of various agencies or departments, and inviting widespread competition, maximum value can be expected for each rupee spent. By planning purchases, the number of small and emergency orders is reduced. Proper planning and coordination result in less number of tender invitations, less number of purchase orders, and related paper work. Standard procedures are laid down in the form of a purchasing manual to help prevent costly and embarrassing mistakes by ever-changing personnel.

Coordination and communication are improved, practices made routine, and efficiency is increased.

NEED IDENTIFICATION

When the need for goods and services arise from the indentors, the necessary funds are budgeted and specifications drawn up. Quality is an expression of the measured properties, conditions or characteristics of an item usually stated in terms of grade, class or specification and is determined by the application involved. The quality of the items demanded is normally prescribed by way of standard specifications, including National or International standards like Indian standards, British Standard, or in the case of P and T, Railways and Defence by their own specifications. Buying to standard specification frequently induces more suppliers to bid for an order as suppliers know exactly what is wanted and that, their chances of getting an order are as good as those of other suppliers, because they are bidding on identical terms. This generates competition amongst bidders resulting in lower prices.

Depending on the urgency of requirement, the raising of demands can, in general, be classified as follows: (a) *Ad hoc demands* required in the normal course of provisioning: (b) *Operational demands* where the stores are required due to unforeseen circumstances; (c) *Urgent demands* which, if not supplied immediately within the stipulated date, will result in a breakdown of essential services or will cause great inconvenience or loss to the state. In such demands, urgency is determined by the date specified in the indent for completion of supplies.

SOURCE REGISTRATION

Whatever method of purchase is adopted, there should be an awareness of the sources from which procurement can be done. The first step in the process of selecting a supplier is to prepare a complete list of prospective suppliers. It is a standard practice to maintain a list of registered or known suppliers for each commodity. Registration is done after assessment of the technical capability of the firm as determined by the inspection authority, financial standing by way of bankers' reports, and audited financial/accounting statements. Monetary limits for individual orders may also be prescribed. Small scale units registered by the National Small Industries Cooperations under the single point registration scheme, are eligible for consideration as registered firms and no separate registration with the DGS & D is required. Further, small scale industrial development cooperatives of the various states enjoy the facilities available to small scale industrial units. Besides, the offers of unregistered firms, may also be given due consideration for trial orders, when tenders are submitted in accordance with the prescribed procedures.

PURCHASE MECHANISM

The method of purchase is largely dependent on the following factors: (a) The nature of the demand; (b) Likely sources of supply; (c) The urgency of demands; (d) Government policy.

The various methods of purchase that are available are as follows: (a) Purchase by tender, (b) purchase by repeat order, and (c) purchase by negotiation.

Open Tender

Open tender is the usual method of purchase. This involves the issue of an invitation to tender by the buyer. This invitation requests the seller to offer to sell the stores described in the tender schedule on terms and conditions stipulated in the invitation to tender. The seller submits his offer including price. With the issue of the acceptance of tender or offer, the contract is concluded.

The two types of tender enquiries are: Advertised tender enquiry and Limited tender enquiry.

Advertised Tender

The advertised tender is normally followed in Government purchasing. This is resorted to, when the following conditions prevail: when quantity is large or the value of procurement is high; when the delivery schedule allows sufficient lead time for procurement and when source availability is limited and more sources are to be generated.

The advantages of this method are generation of competitive bids resulting in economic procurement and an opportunity to locate and develop new sources. The main disadvantage of this system is the long lead time required due to the process involved. Sometimes advertised tender may result in cartels being formed by the suppliers — causing difficulties, disrupting the supply.

Limited Tender

Limited tender enquiries are issued in cases of urgency, or where all likely sources of supply are known. Such a type of enquiry is normally limited to firms which are registered. However, firms which are known to have the capacity to supply or, are past suppliers, also send the tender enquiry. In cases, where the number of firms is unusually large, the purchase officers can limit the enquiry to a few selected suppliers on the panel, subject to the selection of the firms being by rotation so as to ensure that all firms included in the panel of approved suppliers, get a fair chance to compete for the business. The selective limited enquiry should rarely be issued and generally the enquiry is issued to all the suppliers in the panel.

Purchase of Import Items

On behalf of the indentors, the DGS & D imports goods with about Rs 500 crores per annum from different countries. The import of stores is governed by the Import and Exports Control Act, 1947 and the Import Control Order, 1957. These provide that apart from items on open general licence, clearance from the Directorate General Technical Development from the indigenous angle, and import licence from the Chief Controller of Imports and Exports, should be obtained, prior to the import of any store. In respect of Government Departments, foreign exchange is released by the Ministry of Finance. This foreign exchange may be from free resources, non-convertible rupee and from loans from individual countries or from International Institutions like the World Bank, International Monetary fund, etc. In the case of purchases made against loans, the formalities as laid down in the loan agreements have to be followed in the purchasing procedure.

As in the case of indigenous stores, a list of prospective suppliers has to be prepared. Trade Directories of the various countries as well as assistance of foreign embassies in India are availed of. In the case of orders on foreign suppliers, their capability can be verified by a reference to the Indian Embassy in the country concerned. After placing the contract supply is monitered to ensure that the material is despatched.

Inspection

Normally all stores ordered should be subject to predespatch inspection by the representative of the inspection wing of DGS & D. This eliminates the need for an inspection at site, though provision exists for the consignee's right of rejection after receipt of stores. On the stores being found to be as per the specification in the contract, the Inspecting Officer issues a set of Inspection notes, after which the stores can be despatched in accordance with the instructions given in the contract. Whenever, there are differences, joint inspection is carried out with the indentors. Usually the DGS & D charges up to 1 per cent on indentors as inspection charges.

The advantage of centralised buying is central payment. After the stores are despatched according to the terms of the contract, payment may be claimed by the seller from a centralised paying authority. On obtaining the proof of despatch and inspection, 95 per cent of the payment is usually made and the remaining is settled after the material reaches the user. This reduces the time and effort expended by the seller in chasing consignees for payment.

LIMITATIONS IN PUBLIC BUYING

Public buying in addition to the greater accountability, has also certain other limitations. In the matter of contracts, Law does not make any distinction between the State and in individual i.e., the Government is liable for breach in the same way as an individual. It is now a principle of law that in the process of entering into a contract with a person, the action of the Government should conform to the requirements laid down by the Constitution. In making a contract, fundamental rights of a contractor should not be violated. The public law remedies such as writs can be availed by a contractor to seek redressal of his grievances. Once a contract has been entered into by Government, the Government stands on the same footing as ordinary litigants. No special privileges are available to Government as a party to a concluded contract.

EQUALITY OF VENDORS

Often contractors claim the protection of Articles 14 and 19 of the Indian Constitution, which relate to the principles of "Equality before Law" and "Right to Freedom". Article 14 strikes at the arbitrariness in the actions of the Government and ensures fairness and equality in treatment. Governmental actions should be based on rational, non-discriminatory principles. Under Article 14, grievances have been raised alleging wrongful and arbitrary rejection of lower tenders on grounds of discrimination, arbitrary cancellation of a contract where supplies have not been forthcoming, and arbitrary blacklisting. Whereas a private buyer can choose his sources of supply being accountable to none but his employer, such a provision is not available to public buying. The Government buyer, in the event of cancellation of a contract, should invariably give a chance to the defaulter to quote and accept his offer if the lowest and otherwise acceptable, or in the alternative, forge the extra expenditure incurred, due to this repurchase.

Similarly, liquidated damages on account of delay in supply and forfeiture of security deposit on account of non-performance of the contract cannot be enforced unless the purchaser is in a position to prove the loss. In many instances, firms which have been banned for malpractices have floated new firms with a few facade, but with administration and management in their own hands. The buyer, though fully aware of this situation, has not been able to discriminate against the new firms.

The fear of public accountability and audit queries force the central purchase organisations to adopt time-consuming procedures resulting in inordinate delays. In view of public accountability, the contract entered with the vendors has too many specifications. The DGS & D and other public buying organisations have to follow certain procedures due to the policies of the Government.

These include encouragement to indigenous industry, price preference up to 10 per cent for Public Sector undertakings, and price preference up to 30 per cent for the encouragement to the small scale sector. Obviously, the private sector — medium and large — organisations have been complaining against this discrimination.

In view of the central purchasing process, the total lead time of satisfying the users need, goes up to three years. The indentor invariably complains about the long delivery time. However, the buyer from the central purchase organisations invariably shifts the blame on the indentors not giving the complete set of specifications. In spite of coordination meetings, the buyer rarely takes account of the items causing difficulties in negotiation process and the urgency of requirements. However, during crisis situations, like wars, flood, earthquake, natural calamities, the central purchasing has done wonders by rising to the occasions. Sometimes, various pressure groups like politicians, legislators, foreign countries giving aid, State Governments also interfere in the process of public buying. These problems have restricted the public buyer from fully carrying out his function, of making available the right stores at the right time at the most economical prices.

TEN POINTS TO PONDER

1. Distinguish the differences between Governmental buying and private sector industrial buying.
2. Examine the role of the DGS & D in the context of Governmental buying.
3. Explain the organisational aspects of DGS & D.
4. Discuss the concept of centralised buying in the context of Governmental buying.
5. Describe the need of identification adopted by DGS & D.
6. Clarify the procedure adopted for locating supply sources in Governmental buying.
7. Define open tender and limited tender system as adopted in the Governmental buying.
8. Identify the limitations of public buying in the Indian context.
9. Comment on the advantages of buying by DGS & D.
10. Discuss the volume and nature of transactions in the context of purchasing by DGS & D.

49. Capital Equipment Buying

A TO Z SIGNIFICANT DIFFERENCES

The economic growth of a country depends upon the investment in total stock of new equipments and buildings. The following significant differences are observed in the case of capital equipment buying, as against the purchase of consumption material:

(a) Capital equipments are non-repetitive, purchases occurring once in ten years or so.

(b) The capital equipment purchase is usually differentiated both in procedures and policies from other materials, because of the nature of large investment over a long duration entailing careful budgeting.

(c) Capital expenditure involves committing major outlays to get uncertain returns in the future.

(d) Capital equipment buying involves, project generation, project evaluation, project selection and project execution. Hence network analysis and other planning tools play a significant role.

(e) For accounting purposes, capital equipment is considered as a fixed asset, to be depreciated over the period of its economic life.

(f) The top most level in the organisation is involved in decision-making.

(g) The buyers' job is restricted to identifying possible sources, getting technical literature, getting quotations, etc.

(h) The size of the expenditure is a major consideration requiring the sophisticated methods to evaluate the benefits of the expenditures.

(i) In view of the major expenditure, there are several influencing pressures, arising out of social, political, economical, technological, religious, regional, ethnic, natural, financial, regulatory, and global considerations, all of which cannot be quantified.

(j) The equipment buying, directly influences the manufacturing operations of the organisation and hence is a matter of significant concern to everyone in the company.

(k) The new equipment should be compatible with the local environmental conditions.

(l) Since many requirements of major equipments are specific and typical for different types of industries, the number of suppliers available is very few and the organisation has to decide carefully about the choice of right source.

(m) Purchase of major equipments is a result of careful negotiations, arising out of an analysis of various alternatives available to the company.

(n) In this process, the number of contacts and discussions with the vendors is much greater compared to those for the purchase of consumption materials.

(o) Single bid or two bid system of separating technical from commercial should be evaluated before buying capital equipment.

(p) The legal and contract management, in the case of capital equipment, has to be done in a greater depth, as compared to regular items.

(q) The contract should be unambiguous in all aspects, namely penalty, bonus, price escalation, currency, fluctuation, rated capacity, maximum capacity, concept of *force majure*, validity, resolving disputes, inspection and payment schedule.

(r) The contract should also include clauses of guarantee, warranty, risk purchase, liquidity, damage, basis of price, technical literature, maintenance contract, after-sales-service, commencing installation, operator training, etc.

(s) Suppliers exploit the ignorance of the buying organisation by dumping unwanted spare parts initially.

(t) Such a procedure should be avoided by inserting a buy-back clause in the contract, where unwanted spares can be exchanged for the required spares.

(u) A unique feature in the purchase of capital equipments is the long lead time requirements of finalising the order, suppliers' manufacturing cycle time, transportation time, and inspection lead time.

(v) Since the equipment is custom-built, sufficient lead time should be given to the suppliers for fabricating, manufacturing and assembling. A close follow-up of monitoring the progress is necessary.

(w) The operation characteristics and engineering features should be carefully formulated by the design department and communicated to the supplier for manufacturing. Modifications required in adapting the equipment with the existing operations should be borne in mind while manufacturing and inspecting the features.

(x) Whenever possible, the maintenance and operating personnel should be deputed to the supplier's plant to assess the maintainability of the equipment in the future.

Economic Analysis

(y) Any criterion of measuring benefits from the project should include "bigger the better" principle and prefer the early benefit equipment to the late benefit one.

(z) A thorough evaluation of the economic analysis of the different models should be carried out before buying an equipment. One of the most important considerations is not simply to compare the cost of the equipment that is involved at the time of purchase, but to consider also the cost of modifications or additional fittings required. Hence in the economic analysis, the total cost of the equipment, over its whole life, and the return it brings in the form of profit in its life span are compared.

The materials manager's role in the procurement of capital equipment, is only service-oriented while decision-making is done at the top level. Whereas in the procurement of consumption materials, he makes the decisions. However, he can give ideas on the suppliers and contribute to the planning of capital equipment procurement operations, using network analysis techniques.

ECONOMIC APPRAISAL

Many methods of investment appraisal are available to assess the future performance of capital equipment. The need of the hour, in a capital-scarce economy like India, is to evaluate the equipment at its inception and to monitor the performance of physical assets during their economic life. The traditional criteria for appraising investment decisions include:

(a) Pay back period
(b) Average rate method accounting rate of profit
(c) Internal rate of return
(d) Discounted cash flow method.

The following problems are encountered while using the above methods: The time patterns of costs and benefits — reinvestibility prospects are ignored. Productive life of investment. Operating and variable costs are ignored. These problems result in under over-estimation of profits leading to under/over-investment and consequently to misallocation of investible sources.

Pay Back Period

This approach is the simplest and most widely used method for evaluation of investment. It represents the number of years required to return the original investment, by saving before depreciation and taxes. It is useful in industries characterised by instability, uncertainty and rapid technological change because in such industries as the future is uncertain there is no point in projecting returns on investment beyond a period of, say, five years. Pay back method ignores cash flows from the project after the recovery of the initial investments and assumes uniform cash inflows. The period is obtained by dividing the initial capital outlay of investment, by the average annual cash inflow expected from the interests of depreciation, which is a measure of capital recovery included in the cash inflow.

Average Return Method

Here the average annual after tax profit, as reported in the income statement, is expressed as a percentage of annual investment. Sometimes this is expressed as average investment over the project's life or even total investment. In both the cases, the average annual earning is taken to estimate the return on inventory. This method gives equal weightage to the profit from the project distributed over the project's entire life.

Internal Rate of Return

This is also known as *discounted rate of return, yield of an investment,* etc. It is defined as that rate of discount at which the present cash outflows and the present value of the cash proceeds expected from an investment, equal the present value of the total investment. It is the rate of discount which, when applied to the future cash flows, will equal their sum to the original price of an investment. Let C be the initial capital outlay, R_1, R_2, R_3,..., R_n are the cash inflows over n years, n being the economic life of the investment. S is the salvage value and r the discounting factor. Then, the following equation enables us to identify the cut off part of internal rate of return.

$$C = \frac{R_1}{(1+r)} + \frac{R_2}{(1+r)^2} + \ldots, + \frac{R_n}{(1+r)^n} + \frac{S}{(1+r)^n}$$

Table 49.1 gives the present value table for calculating the value of C and computer programs are readily available to solve the above equation, with varying cash inflows.

Discounted Cash Flow Method

The entire cash inflows — profits after tax plus depreciation in the case of unlowered company — from the project are discounted to the present value, using the appropriate discount rate. The DCF method considers the time value of money or the fact money earns money by compounding method. This takes care of the differences in times at which revenues are generated from an investment and the present value concept is used in the evaluation of investments. Present value simply implies, a rupee earned a year later is the same as 0.909 earned today as 10 per cent interest. In this method, normally the cost of capital, or opportunity cost is taken as the rate of discount.

INVESTMENT ANALYSIS EXAMPLE

Ganapathy Ram Engineering company is contemplating the purchase of a machine for Rs 8200 and Rs 2000 is the profit after tax and depreciation for the next five years. The cost is capital at a low value of tax per cent. Varying cash inflows can also be similarly considered.

Pay back period is investment/cash inflows or [(8200/2000) = 4.1 years]. The company will accept the proposal if its policy is of pay back duration less than five years. To find the internal rate of return, we form the following equation:

$$8200 = \frac{2000}{1+r} + \frac{2000}{(1+r)^2} + \frac{2000}{(1+r)^3} + \frac{2000}{(1+r)^4} + \frac{2000}{(1+r)^5}$$

The value of r satisfying the above equation is 7, which is greater than the cost of capital, viz. 6. Hence the proposal is accepted.

Table 49.1 Present Value of One Rupee

Year	Interest Rate Per cent											
Ahead	1%	2%	4%	6%	8%	10%	12%	14%	15%	16%	18%	20%
1	0.990	0.980	0.962	0.943	0.926	0.909	0.893	0.877	0.870	0.862	0.847	0.833
2	0.980	0.961	0.925	0.890	0.857	0.826	0.797	0.769	0.756	0.743	0.718	0.694
3	0.971	0.942	0.889	0.840	0.794	0.751	0.712	0.675	0.658	0.641	0.609	0.579
4	0.961	0.924	0.855	0.792	0.735	0.683	0.636	0.592	0.572	0.552	0.516	0.482
5	0.951	0.906	0.822	0.747	0.681	0.621	0.567	0.519	0.497	0.476	0.437	0.402
6	0.942	0.888	0.790	0.705	0.630	0.564	0.507	0.456	0.432	0.410	0.370	0.335
7	0.933	0.871	0.760	0.665	0.583	0.513	0.452	0.400	0.376	0.354	0.314	0.279
8	0.923	0.853	0.731	0.627	0.540	0.467	0.404	0.351	0.327	0.305	0.266	0.233
9	0.914	0.837	0.703	0.592	0.500	0.424	0.361	0.308	0.284	0.263	0.225	0.194
10	0.905	0.820	0.676	0.558	0.463	0.386	0.322	0.270	0.247	0.227	0.191	0.162
11	0.896	0.804	0.650	0.527	0.429	0.350	0.287	0.237	0.215	0.195	0.162	0.135
12	0.887	0.788	0.625	0.497	0.397	0.319	0.257	0.208	0.187	0.168	0.137	0.112
13	0.879	0.773	0.601	0.469	0.368	0.290	0.229	0.182	0.163	0.145	0.116	0.093
14	0.870	0.758	0.577	0.442	0.340	0.263	0.205	0.160	0.141	0.125	0.099	0.078
15	0.861	0.743	0.555	0.417	0.315	0.239	0.183	0.140	0.123	0.108	0.084	0.065
16	0.853	0.728	0.534	0.394	0.292	0.218	0.163	0.123	0.107	0.093	0.071	0.054
17	0.844	0.714	0.513	0.371	0.270	0.198	0.146	0.108	0.093	0.080	0.060	0.045
18	0.836	0.700	0.494	0.350	0.250	0.180	0.130	0.095	0.081	0.069	0.051	0.038
19	0.828	0.686	0.475	0.331	0.232	0.164	0.116	0.083	0.070	0.060	0.043	0.031
20	0.820	0.673	0.456	0.312	0.215	0.149	0.104	0.073	0.061	0.051	0.037	0.026
21	0.811	0.660	0.439	0.294	0.199	0.135	0.093	0.064	0.053	0.044	0.031	0.022
22	0.803	0.647	0.422	0.278	0.184	0.123	0.083	0.056	0.046	0.038	0.026	0.018
23	0.795	0.634	0.406	0.262	0.170	0.112	0.074	0.049	0.040	0.033	0.022	0.015
24	0.788	0.622	0.390	0.247	0.158	0.102	0.066	0.043	0.035	0.028	0.019	0.013
25	0.780	0.610	0.375	0.233	0.146	0.092	0.059	0.038	0.030	0.024	0.016	0.010
26	0.772	0.598	0.361	0.220	0.135	0.084	0.053	0.033	0.026	0.021	0.014	0.009
27	0.764	0.586	0.347	0.207	0.125	0.076	0.047	0.029	0.023	0.018	0.011	0.007
28	0.757	0.574	0.333	0.196	0.116	0.069	0.042	0.026	0.020	0.016	0.010	0.006
29	0.749	0.563	0.321	0.185	0.107	0.063	0.037	0.022	0.017	0.014	0.008	0.005
30	0.742	0.552	0.308	0.174	0.099	0.057	0.033	0.020	0.015	0.012	0.007	0.004
40	0.672	0.453	0.208	0.097	0.046	0.022	0.011	0.005	0.004	0.003	0.001	0.001
50	0.608	0.372	0.141	0.054	0.021	0.009	0.003	0.001	0.001	0.001	—	—

The present value factor for an annuity of Re 1/- for five years at 6 per cent is 4.212. Hence the present value of Rs 2000 for 5 years is

(2000) × (4.212) = (8424). The net present value is [(8424 – 8200) = 224] or positive cash outflow. Hence accept the proposal.

Social Cost Benefit and Risk

In the investment analysis, it is possible to introduce sophistication like the concept of uncertainty and risk. Sensitivity analysis through the simulation of alternative scenarios helps identification of internal and external parameters which influence project performance. It also helps to find acceptable range of parameters, variation, estimation of probabilities, probability-weighted performance criteria, distribution parameters, and decision analysis in the context of multiple objectives. It is also possible to bring social benefit-cost analysis, i.e., social profitability concept, price distribution, linkage effects, distortion of costs/benefits, economic efficiency, shadow prices, social efficiency, estimation of multiple social costs, social benefits, etc. These concepts are necessary in the context of investment in socially relevant projects.

MAPI INVESTMENT DECISIONS

The Machinery and Allied Products Institute, in 1967 issued a Business Investment Manual. The first section of this manual deals with the operational analysis and the second section with investment and return calculations. The key policy in MAPI approach is the calculation of the adverse minimum. When a new plant is purchased, a capital outlay is made. As the life of the plant increases, the capital cost of the plant per year decreases. Also, as the life of the plant increases, the amount by which it is inferior, to the latest new plant increases. The sum of the capital and operating costs, after adjusting for time, will decline for a while and then begin to increase as operating inferiority grows more rapidly than the decrease in capital cost per year. MAPI uses the word *adverse minimum* to indicate that both capital cost and operating inferiority are undesirable. The adverse minimum point of the sum, is the point at which replacement has to be made, as per MAPI dynamic conditions. The operating inferiority is judged on how much the plant is operationally inferior to the best alternative.

The basic assumptions in MAPI are that future new machineries will have the same adverse minimum as the present one, and the present new machinery will accumulate operating inferiority at a constant rate over its service life. The details required for application of the MAPI include (a) the service life of plants, (b) terminal salvage value, (c) acquisition costs, and (d) cost of capital.

Basic to the MAPI concept is the distinction between the *relative return* and the *absolute return on investment.* The relative return is based on the improved earnings achieved by replacing obsolete equipments, compared to the situation of continuing with the old equipment. MAPI emphasises that it is impossible to make sound investment decisions, without a clear comprehension of the distinction between the relative and absolute return on the investment. Once MAPI concept is grasped, it is at least possible to develop rational analytical procedures, without which the case is lost from the start.

REPLACING EQUIPMENTS

The replacement decision is concerned with the prediction of replacement cost of an equipment, with the most economical policy. The replacement considerations may be due to inadequacy, incompatibility, inefficiency, obsolescence, excessive maintenance charges, and declining efficiency due to wear and tear. The prediction of costs involves those factors which contribute to: (a) inefficient operations, (b) increased operating cost, (c) forced idle time, (d) increased scrap, (e) increased repair cost, (f) maintenance cost higher than the original cost of the machine, (g) non-availability of reliable spare parts from the original equipment manufacturer due to his technological upgradations, etc. The present asset and its potential replacement have to be evaluated not only on technological grounds, but also by financial mechanisms.

The methods include identifying the cost relationships relevant to the maximum costs, and developing some methods for predicting costs based on probability distribution of life spans. In addition to operating costs, capital costs of procuring the new machine, depreciation policies and tax implications also have to be considered. The measure of efficiency for comparisons of alternative replacement policies, is by the discounted value of all future costs, associated with each policy. These can be computed as average cost per month per year. The average monthly or annual expenses will increase, the longer a replacement is postponed. However, there comes a point of time at which the rate of increase of the operating costs more than offsets savings in average capital costs. At this point of time, the replacement of the old machinery is justified, by buying a new equipment, if there is a provision for the same in the capital budgeting process of the organisation. Since the efficiency of the organisation in measured by profits, replacement should generally occur if it results in an economic advantage.

LIFE CYCLE COSTING — LCC

Evaluation of the working of capital equipment and monitoring the performance of the physical assets, are important managerial activities. The life cycle costing approach spells out the aim of optimising the life cycle cost of asset. It recognises that the costs incurred and decisions made at every stage of life span of the assets, reflect on the future decisions of acquiring, using and owing the fixed assets.

Life cycle cost of physical asset is the total of the following costs: Research, development, specification costs, design and acquisition costs, installation, commissioning, spares provisioning, requirement, training the organisation and methods staff, and auxiliary equipment costs, operating costs, losses due to non-achievement of specified performance, labour charges, indirect materials, support services, a portion of overheads, maintenance costs including labour, disposal costs including demolition, production loss and disposal of the asset less the salvage amount realised. The recent concept of life extension programme should also be included in the calculation of LCC.

The key factors in life cycle costing are the asset specifications, life cycle cost and life span. LCC is trade off between these factors, to arrive at the minimum asset life cycle cost, and to realise the benefits of output quantity, output quality, material utilisation, labour utilisation and overall assets utilisation.

The application of life cycle costing concept is likely to result in lower maintenance and operating costs, and reduced capital investment to achieve a specified quantity and quality of output. The cost of introducing life cycle costing system is high, but a trade off has to be done by comparing the costs with the benefits in specific circumstances.

SECOND-HAND EQUIPMENT

In a capital-scarce economy like India, use of second hand equipment, which involves lower investment is common. Scooters India, Lucknow, is an example. The maintenance cost of such equipment is more, and the buyer is advised to get the machine history. The equipment should be completely evaluated by the technical staff on all aspects including the technological efficiency. It is essential to associate maintenance, design, research and operations departments in this process. Second hand equipment carries no warranty or guarantee. Hence the buyer may run the risk of losing his investments if he is not careful. The second-hand equipment should have been overhauled recently before buying. Since the true condition of the capital equipment is not known, it is difficult to estimate the economic life and the operational life of the equipment from the time of purchase. Usually second hand equipments can be had from the owners, brokers, debtors, auctioneers just like second-hand vehicles. In the case of reconditioned/ rebuilt machines, the original suppliers do not render proper service. Further all transportation, handling, insurance and commissioning cost as well as the risk of breakdown are to be borne by the buyer. Second-hand equipment can be acquired — only if the financial position of the company does not permit huge investment and if the equipment is easily adaptable to the existing operations.

TEN POINTS TO PONDER

1. Enumerate the major peculiarities of capital equipment buying.
2. Identify the critical differences between capital equipment buying and other materials.

3. Discuss the techno-economic analysis of capital equipment buying.
4. Specify the need for economic appraisal in capital equipment buying.
5. Define pay back period method.
6. Distinguish between average rate of return and the internal rate of return.
7. Explain the discounted cash flow method.
8. Examine the relevance of MAPI investment decisions and life cycle costing in capital equipment buying.
9. Elaborate on the replacement decision of equipments.
10. Elucidate the role of buying second-hand equipment in Indian economy.

50. Contract Management

MEANING OF CONTRACT

Contract management is a very crucial aspect in the field of materials management. With every Five Year Plan investing large amounts of money in different sectors, contract management is becoming an important discipline. General Manager (contracts) is a common designation in large companies. The materials manager should have a professional knowledge of contract management, which is very essential for capital equipment and project purchases.

According to Section 2(b) of the Indian Contract Act 1872:

Contract is an agreement between two or more parties, enforceable by law.

Agreement is defined by the contract law as:

Every promise and a set of promises forming the consideration for each other.

The Contract law defined a promise as an accepted proposal. The following conditions have to be satisfied for an agreement to become a contract.

(a) There should be some consideration of the agreement.

(b) There should be offer and acceptance.

(c) The parties entering into agreement, should be legally competent to contract.

(d) There should be no coercion or pressure from any side and the parties should give free consent.

(e) The object of the agreement and of the parties, should be lawful.

(f) There should be mutuality of obligations.

(g) Questions of competency or capacity to contract, may arise under the law for infants, insane person, intoxicated persons, married women, corporations and aliens.

Section 2(a) of the Indian Contract Act 1872 defines a proposal as follows:

When one person signifies to another his willingness to do or abstain from doing anything, with a view to obtaining the assent of that other to such an act or abstinence he is said to make a proposal.

The proposal may be made annually or by writing or by conduct.

VARIOUS ASPECTS OF CONTRACT

Thus, a valid contract, enforceable by law, is formed when the following conditions are fulfilled:

(a) An offer is made
(b) The offer is accepted
(c) Legal relationship between the contracting parties is created
(d) There is a lawful consideration
(e) The contracting parties are competent to enter a contract
(f) The consent to the contract is free and genuine
(g) There is a lawful object
(h) The agreements, created are not declared void by law
(i) The terms of contract are certain and there is no ambiguity or vagueness in terms
(j) Agreement, should conform to the legal requirements.

The contract should make provision for the following aspects: (i) *Force majure* clauses; (ii) penalty, bonus, risk purchase; (iii) disputes/arbitration/ adjudication and jurisdiction; (iv) validity and date of contract; (v) deviation from specifications; (vi) basis for price escalation; (vii) currency conversion; (viii) performance guarantee; (ix) inspection and site tests; (x) mandatory spares; (xi) warranty considerations; (xii) bid guarantees; earnest money deposits; (xiii) packing parameters; (xiv) performance parameters; (xv) payment schedule; and (xvi) delivery and documents.

Offer/Acceptance/Consideration

A contract between two parties matures when an offer is made by one party and the other party accepts the offer. The offer and acceptance should be lawful. The terms of offer and acceptance should be definite. The acceptance of the offer should be communicated to the offeror by the offeree. The acceptance of the offer makes the proposal or offer, a promise and thus enforceable by law. It should be an absolute and unqualified acceptance of all the terms of offer. The acceptance should be made, while the offer is in force and before the offer has been revoked or lapsed.

Section 25 of the Contract Act, declares a contract void if an agreement is without consideration. A consideration is one of the essential elements of a valid agreement, an absence of which renders an agreement void.

Consideration is defined as "Some right, interest, profit, or benefit accruing to one party or some forebearance, detriment, loss or responsibility given, suffered or undertaken by the other."

It is the price for which the promise of the other is bought and the promise thus given for value is enforceable. Section 10 of the Indian Contract Act makes free consent, without coercion or under influence, as an essential requirement of a valid contract.

Fraud and Misrepresentation

Any false representation made intentionally to secure a promise is called a fraud.

The false representation is made knowingly without belief or carelessly. This policy is followed so that the unscrupulous are deterred from designing schemes, skirting or falling outside the precise legal limitations.

Fraud has the following essential elements:

(a) Concealment of a material fact or misrepresentation.

(b) The misrepresentation should be known to be false or there is no concern as to the validity of the statement by the party making it.

(c) Patent infringement, leading to legal complications.

(d) The misrepresentation is made with the intent, that the other party is deceived.

(e) The injured party should act upon the misrepresentation in the belief that it is true.

(f) Reliance upon the validity of a misrepresentation should result in injury to the party deceived.

(g) An intention to deceive by any other action.

(h) Acting fraudulently.

(i) Specially acting against the law.

(j) Misrepresentation may be by word, or act and should relate to past or existing fact and not promises regarding the future.

(k) Misrepresentation should be material so that it is improbable that the contract would have been entered into, under the terms and conditions, if the truth had been known.

Liquidated Damages

When time is a major factor in the execution of a contract, it is implied that completion by a certain date is part of the consideration of the contract. Failure to meet the requirement leaves the offending party liable to the other party for damages. Since such damages may be extremely difficult to estimate, the usual practice is to specify a definite amount per day as liquidated damages in lieu of actual damages suffered. Agreed to by both parties, this amount, enforceable by law, should be justified by the damaged party and the burden of proof rests with the defaulting party.

Penalty and Bonus

An incentive for the contractor to expedite completion when a project is urgent, takes the form of *bonus payment*, if the work is completed ahead of schedule, and *penalty* if the schedule is not met. Both provisions are necessary to be enforceable at law. Courts have consistently ruled that penalties are not enforceable unless, there is an equivalent bonus provision.

The law of carriage of goods determines as to how the seller should fulfil his obligations to deliver the goods to the carrier. For this purpose, the International Chamber of Commerce, has introduced the *inco* terms to be used by the buyer and the seller. Nowadays, goods are usually delivered by the seller, to a carriage before shipment on board takes place.

Performance

Section 37 of the Contract Act provides that the parties to a contract should either perform, or offer to perform their respective promises. The performance should be offered at the place, if a particular place is fixed for such a performance. Performance time should also be adhered to if it has been agreed upon by contracting parties. Within 30 days after the supplier's receipt of notification of award of the contract, the supplier shall furnish performance security to the purchaser in the amount specified in the special conditions of contract. In the event of *force majure* clauses — war, revolution, fire, flood, epidemics, quarantine restrictions, freight embargo, earthquakes, national disasters etc. the supplier shall not be liable for forfeiture of the performance security or liquidated damages.

Inspection

The purchaser shall have the right to inspect and test the goods for their conformity to the contract specifications. The tests may be conducted in the premises of subcontractors or at the point of delivery or at the final destination. The contractor should give due notice to the site engineer, when each stage is ready. No work shall be covered up or put out of view, without the approval of the engineer. Visual checks, dimensional measurements, chemical tests, laboratory analysis and endurance test, will be conducted and reliability and physical aspects will be checked as stipulated in the contract.

Contract Packaging

Major contracts, sometimes, are broken into well-defined compatible packages for better operational control. The packages formed should include such combination of equipments and services that can be advantageously engineered for preparations of specifications for bid documents and development of drawings. The number of packages should be optimised between too few and too many. If the number is too many, then the manpower for monitoring too many packages increases the follow-up costs. On the other hand, too few contract packages may affect the quality of the work because a contractor may not undertake different types of works.

Breach of Contract

Discharge of contract by breach in also provided in the act. A *breach of contract* takes place in the following circumstances:

(a) Any of the contracting party renounces his liability under contract.

(b) Any of the contracting party acts in such a manner as to make it impossible to discharge his liability under the contract.

(c) Any of the contracting party totally or partially fails to perform and fulfil his obligation under the contract.

(d) An anticipatory breach occurs, when there is prior repudiation of the contract by any party.

(e) The injured party is entitled for compensatory damages, *liquidated* damages and penalty.

Disputes

Arbitration is the reference of a dispute by voluntary agreements of the parties to an impartial person, who renders a decision, called an award, after hearing evidence and arguments presented by both the sides. Although private in nature and less formal in procedure than a Court of Law, arbitration is governed by established rules and standards of impartiality. If arbitration is not helpful in sorting out the difficulties, the time-consuming procedure of adjudication has to be initiated by the aggrieved party to redress his grievance.

Termination

Ordinarily, a contract is terminated when both the parties have performed their obligations under the agreement. Specific performance is accomplished when the work is completed exactly to specifications within the specified time limit and the other party makes payment in full. Contracts also may be terminated by (a) waiver, (b) substituted, by a new agreement, (c) a new party added to the substituted agreement, and (d) payment in lieu of performance.

Warranties

A *term* in a contract of purchase/sale may be a condition warranty. A *condition* is a stipulation essential to the main purpose of the contract, breach of which gives rise to rejection of the goods. A *warranty* is a stipulation collateral to the main purpose of the contract, breach of which gives rise to a claim for damages. A *stipulation* may be a condition though it is called warranty. The buyer may elect to choose a condition as a warranty and accept goods and sue for damages for breach of condition. Law implies the following conditions in relation to warranties:

(a) The seller has a right to sell the goods at a time when the property has passed to the buyer.

(b) The buyer shall have peaceful and quiet possession and enjoyment.

(c) The goods shall be free from any charge or encumbrance or third party claim unknown to the buyer.

(d) In a contract for sale of goods by description, goods shall correspond with description.

(e) Where goods are specified by seller as required for any particular purpose and the buyer relies on the skill, judgment, of the seller, and the seller sells goods of such description, goods shall reasonably be fit for that purpose.

(f) In a contract for sale by sample, the bulk shall correspond with the sample.

(g) In sale of goods by patent or trade name, there is no implied condition as to its fitness for any purpose.

(h) Goods are bought from a merchant, who deals in goods of that description, that goods shall be of merchantable quality.

(i) Implied condition as to quality or fitness of goods for any particular purpose, may be annexed by usage or trade.

(j) In a contract for sale by sample, goods shall be free from any defect rendering them unmerchantable and that buyer would have reasonable opportunity to compare bulk with sample.

TYPES OF CONTRACT

According to the Planning Commission's Construction Division Document, "contract" shall mean the notice inviting the tender, and acceptance thereof and the formal agreement if any, executed between the President of India and the Contractor together with the documents referred to therein, including these conditions, specifications, designs, drawings, priced schedules, bill of quantities and schedule of rates.

Works Contract

The buyer often enters into a works contract with construction agencies for construction of roads, buildings, factories, etc. The contractor has to supply the material and the labour for building the immovable property. It cannot be looked upon as a separate contract of sale of goods and another for the supply of labour. The usual practice for such long term contracts is to have a price variation clause, particularly on labour rates.

Bilateral Contract

When there is a mutual exchange of promises, a contract is said to be *bilateral* and most business contracts fall into this category. A promise on one side in return for an act by the other party, is considered to be a *unilateral* contract. It is possible that contracts may be both bilateral and unilateral in that, the consideration on both sides may consist of promises and acts.

Void and Enforceable

Void is the term used to describe certain illegal contracts which are null and generally create no legal obligations. A contract made by a person adjudged insane is void. If a contract can be *avoided*, at the option of one of the parties, it is said to be voidable. Grounds for such actions are insanity, infancy, duress, fraud, etc. The bankruptcy of either party will terminate an offer, as will death or insanity of either. An agreement becomes voidable if one of the parties can avoid it, at his option, but the other party cannot do so. A void agreement cannot be enforced upon any of the parties, while a voidable one can be *resinded* and enforced at the option of the aggrieved party, whose consent was so obtained.

Fixed Contract

In *fixed contracts*, the quantity and price are fixed. The tender relating to the contract indicates the definite number or quantity, time and delivery based on which tenderers can work out the cost and tender, with definite knowledge of the limitations of their responsibilities. The essential feature of this contract is that in the absence of mutual agreement, the contractor can refuse to supply anything above the ordered quantity and can also insist on the purchaser taking the entire quantity ordered.

Rate Contract

Rate contract as the name implies is contract for supply of materials at specified rates during the period of the contract. No quantity is mentioned in the contract and the contractor is bound to accept any order placed during the validity of the contract. But normally, a minimum quantum of offtake is to be guaranteed by the purchaser and even DGS & D orders all the materials which are required to be purchased under rate contract subject to certain reservations regarding price, acceptance and division of orders among contractors. In addition, a minimum offtake is also assured.

Running Contract

A *running contract* is an agreement for supply of an approximate quantity of stores at a specified price during a certain period. In these contracts, the price is fixed but quantity is subject to specified tolerances. In the case of DGS & D running contracts, the approximate requirements of a number of indentors for the period in question are combined and any of the indentors can order his requirements direct on the contractor or on DGS & D. The conditions provide for taking of up to 25 per cent of quantity over or below the approximate quantities indicated in the contract but minimum offtake of 75 per cent is ensured.

SETTLEMENT OF BILLS

The terms of payment fall into two general categories from the buyer's point of view:

(a) The terms which influence the ultimate delivered cost: example — Cash discounts.

(b) The terms which merely designate method of payment.

Limited credit facility, cash against delivery, prepayment discount, consignment terms and progressive payment terms are some terms of payment. In foreign contracts, payment is generally made through Letter of Credit. Whenever advance payment or payment through bank against despatch documents is agreed, a provision should be made in the order for the seller to take back the rejected goods, if any, with despatch documents through bank.

In the case of foreign contracts, full payment by Letter of Credit is generally agreed only after obtaining a performance bank guarantee for certain value. The standing and reputation of the seller is to be verified before agreeing to any terms which are not advantageous to the buyer. Advance with the order is to be avoided as far as possible.

TEN POINTS TO PONDER

1. Define the significance of contract management.
2. Elucidate the statement "An agreement enforceable by law is a contract"
3. Comment upon the statement — "Before entering into a contract, you have to ascertain the competence of the party, you propose to enter into contract with".
4. Identify the relevance of offer, acceptance and consideration in the context of contract management.
5. Explain the statement an agreement without consideration is null and void.
6. Distinguish void, voidable and valid agreements, by giving illustrations.
7. Critically analyse the statement that impossibility of performance renders the agreement void.
8. Enumerate the A to Z contents of contract.
9. Describe the relevance of fraud, mistake, misrepresentation, breach, liquidated damages, penalty, bonus, and disputes in the context of contract management.
10. Articulate upon the types of contracts.

51. Legal Aspects

NEED FOR LOGISTICS

"Law is an ass. My word is my bond! Caveat Emptor or buyer beware." "Law is a dark room, with the lawyer providing the much-needed focus light." "Law is only for the strong and wealthy not for the weak and the poor." "Interpretations of specific laws undergo sea-changes, as a result of some judicial pronouncements. Hence, highly legalistic approach to purchasing is both unnecessary and unprofitable." These are the comments heard in the business meetings between buyers and sellers.

Purchase is the legal act viewed from the act of the buyer and sale is the act viewed from the point of the seller. Both have the basis on contracts. Each transaction of purchase invokes the provisions of several laws and regulations. The transactions should not be illegal or opposed to the public policy of the country. According to Section 4 of the Sale of Goods Act, *a contract of sale of goods is a contract whereby the seller transfers or agrees to transfer the property in goods to the buyer for a price*. The purchase or sale of goods is governed by several laws in the country and the buyer should be familiar with the same.

In particular, he should know the relevant provisions of the Indian Companies Act, contract laws, law of agency, Sale of Goods Act, insurance schemes, claims procedures, Insolvency Act, Negotiable Instrument Act, excise duty, customs levies, foreign exchange regulations, sales tax, local taxes, octroi duty, law of carriage of goods, etc. The materials manager cannot be an expert on all the commercial laws, but he should have access to legal expertise within the company or from external sources. He should realise that every one of his transactions of purchase binds the company legally. The buyer functions in the legal capacity of an agent. In this capacity, he is liable for his actions both to his employer and to those with whom he deals.

A TO Z OF PURCHASE ORDER

The purchase order is of primary importance, while considering the legal aspects. A purchase order is a legal instrument which expresses the buyer's contract of sale at a certain price.

(a) No material can be purchased without a duly authorised purchase requisition.

(b) The purchase order contains the following aspects: Date of issue of purchase order with reference number; suppliers name and complete address; consignee's name and address.

(c) Sometimes the computer is used to trigger a purchase order.

(d) A set of terms and conditions, is usually printed on the back of the order, but this should be indicated on the order.

(e) Description of the quality, specification, brand name and code number.

(f) Quantity with the pricing unit.

(g) The reference number and date of supplier's quotation.

(h) Price terms, unit price, discount, sales tax and other taxes.

(i) The total value of the purchase order including the breakdown of staggered delivery values.

(j) The definite date or rate of delivery avoiding phrases like "immediate", "urgent", as soon as possible.

(k) Packing size and packing charges.

(l) The purchase order should specify if there are any special markings to be carried out.

(m) Detailed shipping instructions on the method of despatch, name of port/airport/railway station or place of delivery.

(n) Freight charges, as to whether it has been paid to be paid.

(o) Inspection methods, agency, sample size, joint inspection, stagewise inspection.

(p) Removal of rejected materials and substitution of these by quality materials is also mentioned.

(q) Insurance details such as name of the insurance company if the supplier arranges, or to be arranged by the buyer.

(r) Invoice details, number of copies, attaching purchase order copy for verification.

(s) Method of payment — as 95 per cent on proof of despatch and inspection — or any other method.

(t) Warranty clauses, if desired by the buyer, should be sought for.

(u) Order number, with the requisition details or work order number.

(v) Resolving disputes, by arbitration adjudication, jurisdiction of the court, machinery for resolving differences, etc. are mentioned particularly for the orders of capital equipment.

(w) The list of recommended spares approved by the buyer, for capital equipments.

(x) The loading, unloading mechanism at the receiving warehouse may be indicated.

(y) Provisioning for escalation and application of *force majure* clauses.

(z) Aspects such as liquidated damages, transit damages, excess supply, short supply, etc.

The purchase order may be in a standard form, computer printout, travelling requisition, letter, contracts, agreement form, multicopy snapout form, service contract, works contract, letter confirming the verbal order, etc.

Purchase orders should be signed only by those authorised to do so and the buyer has to use 'on behalf of his company' with his signature.

The buyer should inform the supplier if changes in any of the above conditions, including changes in quality, quantity, cancellation, etc. take place.

BID EVALUATION CRITERIA

Usually any purchase order is subject to certain terms and conditions, which should be explicitly explained for major purchases. Ideally the seller returns the buyer's acknowledgement copy of the purchase order duly signed indicating his acceptance. While the buyer insists on adhering to his terms and conditions, some sellers send the material subject to the terms stipulated on the invoice, which may not tally with those specified by the buyer.

Bid evaluation is an important prerequisite before purchase order placing. The analysis of the investment and return on major equipment, has slightly different ingredients from regularly purchased items. The technical engineer in collaboration with the buyer should evaluate the bids on all technical, non-technical, and commercial aspects. This is done in a phased fashion and the following are eleven phases for major bid evaluation, before placing the purchase order:

(a) Pre-select vendors
(b) Prepare detailed inquiry
(c) Receive bids
(d) Make preliminary evaluation
(e) Make technical evaluation
(f) Make commercial evaluation sometimes technical and commercial are done simultaneously
(g) Conduct pre-award meetings
(h) Condition bids by considering intangible and service factors
(i) Select prospective vendors
(j) Conduct pre-committee meetings
(k) Award the order to the lowest responsible bidder and review by following up the vendor.

TERM OF PURCHASE

Contracts for sale/purchase should be in conformity with terms of offer and acceptance. Stipulations/specifications in the contract should be performed according to such terms. Parties by mutual consent may relax contractual terms; however, unilateral change of terms and conditions of contract are not valid. Contract being resultant of offer and acceptance, concluded contracts cannot be varied or altered by one party to the contract. Extension of time for delivery/supply of goods by the seller beyond the period fixed in contract does not bind the seller, unless he agrees to and accepts the extension and agrees to perform during extended time. Extension of time once does not mean or imply that further extensions of time should be granted to seller. The computation of date of breach and damages in contract can be performed, under the contract or mutually extended date.

INCO TERMS

The supply of imported stores and other purchases is effected according to the INCO terms, developed by the International Chamber of Commerce. Recognising the considerable diversity and plurality of commercial laws and practices in different countries, the ICC has developed the following terms.

(a) In *FOB contracts* the stores are delivered Free On Board the vessel at the port of shipment. The prices include all charges from the contractor's works to the delivery point, viz. on board the ship at port of shipment.

(b) *C & F contracts* are those in which the prices include, the cost of the stores and the Freight up to port of entry of the importing country. Insurance has to be paid by the buyer.

(c) In *CIF contracts*, the insurance charges are also included in the prices apart from C and F. Insurance, can be done from the firm's warehouse abroad to buyer's warehouse in India and can be for a period of 30/60 days from the date of receipt by buyer.

(d) *CIF & C* includes Cost, Insurance, Freight, and Commission type of contracts.

(e) *C & F* is used for goods which are to be carried by ship, the freight or carriage paid to, is used for inland transport only, including national and international area. Ex-quay means the seller makes the goods available to the buyer on the wharf.

(f) *FAW contracts* imply that the material is put alongside wharf. This is also known as *FAS* (Free Alongside Ship). Goods are placed along the side of the ship by the seller. Loading into the ship is the responsibility of the carrier/buyer.

(g) FAW contracts stipulate for fair average quality. The usual quality supplied in the trade at the place of supply abroad at the relevant season is the average quality.

(h) *FOR contracts* mean that the seller takes on the responsibility, of importing the stores on behalf of the purchaser, clearing the stores from the port of entry and delivery, on board rail in India. This is done by the supplier's Indian agent.

(i) *FOT contract* means Free On Truck to be used for road transport similar to FOR.

(j) *Exship* implies that the seller shall make the goods available to the buyer in the ship at the destination named in the sales contract. The seller has to bear the full cost and risk involved in bringing goods there.

(k) *DAF* or *Delivery At the Frontier* means sellers obligations are fulfilled, when the goods have arrived at the frontier, but before the customs border of the country.

(l) *DRP* or *Delivered Duty Paid* denotes the sellers' maxim obligation, as opposed to Exworks.

(m) Ad or free at airport of the departure makes the seller, fulfil his obligations by delivering the goods to the air carrier.

(n) In order to minimise the payment of ocean freight in foreign exchange, it is the policy of the Government that orders for imported stores should be placed on Free On Board or Free Aside the Shore basis.

(o) Post contract management is an essential requirement of a good purchase organisation. It is not necessary that once an order is placed, the supplies will be automatically received. Efforts are required to monitor supplies and expedite suppliers where necessary. Measures like liquidated damages for delay in supply, or cancellation and repurchase at the risk and expense of the supplier in case supplies do not materialise, or are not likely to materialise have to be taken.

LAW OF AGENCY

Agent is a person who acts for another in all lawful transactions. There can be no agency in criminal, illegal, unlawful acts, acts opposed to public policy. *A facit alium facit per se* — is the maxim — he who does an act through another does it himself. Agency is a matter of contract or conduct implying a contractual relationship.

True test of agency is whether a person acts for himself, or for another with the consent of, or at the instance of another. Agency may be express or implied. Employees and servants are agents; but agent is not a servant or employee; the distinction is that employee or servant is controlled by principal in the manner of doing his work; an agent is not. Principal tells the agent *what* to do; but instructs servant/employee *what* to do and *how* to do it. Agent is in the position of principal in respect of acts done in legitimate sphere of agency.

Commission agents, buy and sell, upto the point they act as agents they are in the position of active confidence and beyond that, they are in the position of principals. Brokers are agents for very limited purposes and they may be agents for both buyer and seller and receive commission from both, depending on the terms of brokerage agreement.

Purchase agents-procuring articles/commodities for Government or for public use under licences issued, are not servants of Government, they are only licencees and not agents in the true sense of the term. They are not in any position of trust or active confidence vis-a-vis the Government, they are liable for infraction of terms of licence. Power to buy could be granted to an employee or an agent. The terms of power would fix the extent of authority. The buyers are agents of their companies and are not personally liable especially if they are employees.

NEGLIGENCE

Mistakes, loss/damage or errors would not absolve the principal of liability vis-a-vis third parties. However, the concerned officers would be subject to penalties by the employer (a) where the purchasing officer acted in a dishonest manner, (b) where he acted without authority or contrary to authority, (c) where he acted in a grossly reckless or careless manner though without dishonesty. He will not be liable for errors of judgment provided it is bona fide and free of dishonesty. Fraudulent purchases, secret commission for the officer, inferior quality accepted against contract specifications even without dishonesty, patent dishonesty in purchase, refusal to consider objections by other staff members in the said purchases, unusually a hasty transaction resulting in loss to employer, purchases when not required, or when the company carried a heavy inventory of stocks, are all circumstances where the officer concerned would be liable for disciplinary action.

The principal whether a company/firm or proprietorship would be liable on the contracts of its officers provided they have been acting in such transactions in the usual discharge of their duties as employees. Contracts made by officer/employees in the usual course of business without collusion, fraud or dishonesty, would bind the principal.

Agent's negligence in the discharge of duties of principal will not absolve the principal of liability. Negligence does not include acts outside authority; it means work done in a careless manner, without due care and caution. Contractual obligations are not avoided by plea of negligence of agent. They could be avoided on plea of collusion between agent and third party, fraud on principal to the knowledge of other party to the contract.

TEN POINTS TO PONDER

1. Comment on the statement "A pure legalistic approach to buying is both unnecessary and unprofitable".
2. Elucidate the concept of *caveat emptor*.
3. Discuss the legal responsibilities of a purchasing agent to his firm.
4. Distinguish the authority possessed by buyers and salesmen who act as agents to their respective organisations.
5. Examine the extent of liability of the buyer as the agent of his company.

6. Explain the legal status of a purchase order, which has been accepted by a vendor, who has used an acceptance form, containing terms conflicting with those of the purchase order.

7. Describe the buyer's legal rights, when he cancels a purchase order, with no delivery date specified on the purchase order.

8. Enumerate the details of bid evaluation criteria for capital projects.

9. Discuss the applicability of Inco terms and the best suited to India.

10. Explain the legalities of the law of agency.

52. Foreign Purchase

NEED FOR IMPORT CONTROL

In this interdependent world no country can be self-sufficient and self-reliant, meeting all its requirements only with internal transactions. A developing country like India, to catch up with advanced nations should import sophisticated equipments. Import means bringing into and export means taking out of India. In this context, Ricardo's economic principle is—countries should concentrate on items in which they specialise and export these items, but import other essential items. Even the most advanced nations like America and Japan import their requirements. India imports goods work about Rs 50,000 crores per annum, mainly, fuel, fertiliser, equipments and defence needs. In spite of the tremendous export incentive given to traditional and non-traditional items, Indian exports amount only to about Rs 25000 crores. The gap between imports and exports has been increasing day by day, necessitating stricter control in the imports. In spite of growing volume of business, the country hardly accounts for about 0.7 per cent of the total global international trade.

The difficulty arising from the shortage of foreign exchange has been exacerbated by steeply rising prices. The depleting foreign exchange reserves cannot service even two months of imports. This has resulted in the lowering of parity rupee value in international markets. To tackle this problem, India has imposed rigid controls on industries to induce increased production of import substitutes. Despite these measures, imports are likely to remain a significant factor in the foreseeable future. The establishment of local industries frequently creates a need for foreign exchange, since many materials and services are imported. The import trade control policy depends upon a large number of factors like social, economical, natural, fiscal, foreign exchange reserves, trade credits, aids, tied credits, global pressures, etc.

OBJECTIVE OF IMPORT CONTROL

Import control was first introduced as a war-time measure in the early stages of Second World War. The object was not only to conserve foreign exchange reserves, but also make the best use of the available limited shipping space. The import trade control was kept alive by emergency provisions and was ultimately replaced by the Imports and Exports Control Act 1947. This act was amended in 1955 and again in 1977. A large number of notifications has been introduced and included in the Handbook of import-export procedures, updated and published every year. The import-export policy is usually announced on April 1, for a period of three years.

The main objectives of the import-export control are as follows:

(a) Provide easy access to import essential industries.

(b) Safeguard the reasonable interest of domestic industries.

(c) Discourage excessive import of items manufactured in India.

(d) Encourage export of commodities which are not of strategic importance, sensitive in nature or required for mass consumption.

(e) Improve the foreign exchange by balancing the need for import and export production. The policy has been liberalised recently.

The policy for various commodities—capital goods, raw materials, components, spares, etc. is reviewed in consultation with the technical authorities of Directorate General of Technical Development, Administrative Ministries, representative bodies and trade associations. Such review enables the indigenous industries to have access to raw material inputs at the most competitive price.

The principal objectives of the recently announced import-export policies are as follows:

(a) The validity period is for three years.

(b) Encourage rapid and sustained export growth including export of services, with special emphasis on exports which generate higher net foreign exchange earnings.

(c) Facilitate availability of necessary imported inputs for sustaining industrial growth, including essential imported capital goods for modernisation and technological upgradation.

(d) Simplify and streamline procedures for import licensing and duty exemption scheme for exports.

(e) Support recognised indigenous research and development institutions for building up their scientific and technological capability for technology absorption and development.

(f) Promote efficient import substitution and self-reliance.

The policy has also to be seen in the light of super 301 clause imposed by the United States of America against the practice of Indian Protectionism; and the import-export policy has to be reviewed.

ITEMS COVERED IN IMPORTS

At present, import control covers practically all articles included in the schedule to the Import Control Order 1955. Their import is prohibited except with an import licence or a customs clearance permit or an open general licence issued by the Government clause.

Import of gold, silver, currency notes, bank notes, and coins is controlled by the Reserve Bank of India under the Foreign Exchange Regulations Act. Every importer shall have to get the importer's code number as per para 231 of the Import Policy Book.

CATEGORIES OF IMPORTERS

Established importers are those who are actually engaged in the import trade of the articles, comprised in any serial or sub-serial number of the import schedules during the last financial year, falling within the basic period specified for the said serial number.

The categories of actual users are as follows:

(a) Industrial
(b) Non-industrial
(c) Hospitals and institutions
(d) Registered exporters importing goods
(e) Cooperative societies
(f) Canalised agencies like State Trading Corporation, Metal Mineral Trading Corporation etc. who import for others
(g) Import/export trading houses
(h) New industries.

Quota certificate is issued by licensing authority to an established importer as a token of acceptance of a particular commodity falling under a serial or subserial number of import trade control. The licensing fee for the established importers is very limited and has become very nominal.

Industrial users are such units who need imported items for their manufacturing process. These include: (i) Scheduled industries on the registers of the Directorate General of Technical Development; (ii) Scheduled industries not on the registers of DGTD and non-scheduled industries like hospitals, etc. and (iii) Small scale industries with a capital investment of not more than Rs 35 lakhs.

Non-industrial users include the following: (i) Any commercial establishment which holds a certificate for at least 3 years under the local law; (ii) Any establishment holding licence for at least 3 years under local law relating to exhibiting films to public; (iii) Any laboratory or scientific institution; (iv) Any service industry; (v) Public sector non-industrial undertaking; (vi) Any local authority.

The import policy also defines the other following classifications: capital goods, components, spare parts, consumable/part, consumer goods, accessories.

Replenishment Licences

Registered exporters are entitled to special import facilities by way of replenishment of materials required in the manufacture of products exported and these are called Replenishment (REP)-licences. REP-licences are issued to manufacturer, exporters, and manufacturing export houses. Import of items mentioned in Appendices 3 and 6 will be allowed only up to 20 per cent of the face value of the REP-licence. Imports of canalised items mentioned in Appendix 8, will be permitted up to 20 per cent of value of REP-licence. Import of tools will be allowed under this policy. Import of items in Appendices 5 and 7 will be allowed under this facility.

SPONSORING AUTHORITY

The sponsoring authority for large scale industries is the DGTD, which ensures that the items are not indigenously available. The other sponsoring authorities as per Appendix 15, for the relevant industries include the following:

(a) State Director of Industries
(b) Commissioner for Small Scale Industries
(c) Trade Associations
(d) Commissioner of Textiles
(e) Chairman, Tea Board
(f) Chairman, Coffee Board
(g) Technical Director, Sugar
(h) Chairman Rubber Board
(i) Ministry of Petroleum and Chemicals
(j) Iron and Steel Controller
(k) Coal Controller
(l) Central Water and Power Commission
(m) Central Silk Board
(n) Chairman, Coir Board
(o) Director General, Shipping
(p) Jute Commissioner, Calcutta
(q) Agricultural and Marketing Advisor
(r) State Director of Handlooms.

Sponsoring authorities recommend applications for licences, in accordance with the procedures. In the matter of assessment of requirement, they ensure the compliance of the general directives issued by various agencies. They obtain clearance from indigenous angle wherever necessary, before recommending the licences. They ensure that the recommendations do not exceed the ceiling allotted to them. They send an abstract of recommendations to the controlling authority.

LICENSING AUTHORITY

The Office of the Chief Controller of Imports and Exports, Udyog Bhavan, with its regional offices, working under the Ministry of Commerce, is the licensing authority. Licences are issued on the basis of recommendations of the sponsoring authority within their jurisdiction. The reasons for rejections of licences are communicated to the parties and penal action is taken against the licensees for violations.

DOCUMENTATION

Several documents are necessary for importing and customs clearance. These are:

(a) *The bill of lading* gives title to the importer. This is an undertaking by the carrier of having received from the consignor, the goods under description from origin to be delivered at the port of destination. The bill of lading is issued by or on behalf of the carrier, and is evidence of a contract of carriage as well as proof of delivery of goods on board the vessel. A bill of lading may be either freight pre-paid or freight payable at destination. Sometimes, non-negotiable documents such as sea-way bills, liner-way bills, freight receipts, etc., are used.

(b) The seller, at his own expense furnishes without delay a clean bill negotiable bill of lading for the agreed part of destination, as well as the invoice of the goods shipped and the insurance policy.

(c) It enables the importer to claim the goods at the destination. It is a contract between shipper, carrier and consignee.

(d) The bill of lading should cover the contract of goods, be dated within the period agreed for shipment, and provided by endorsement or otherwise for delivery to the order of the buyer or buyer's agreed representative.

(e) It should contain date, place of shipment, names of carrying vessel, consignee, and consignor, port of destination, number, contents, identification marks, freight details, insurance, etc. It is a negotiable instrument. Such bill of lading should be a full set of on board or shipped bills of lading or a "received for shipment" bill to the effect that the goods are on board.

(f) These endorsements should be dated within the period agreed for shipment.

(g) A *clean* bill of lading is one which bears no superimposed clauses expressively declaring defective condition of the goods or packaging.

(h) The following conditions do not convert a clean into an unclean bill of lading:

(i) Clauses which do not expressly state that the goods or packaging are unsatisfactory like second-hand cases or used drums

(ii) Clauses which emphasise the carrier's non-liability for risks arising through the nature of goods or packaging

(iii) Clauses which disclaim on the part of the carrier knowledge of contents, weight, measurement, quality or technical specification of the goods

(iv) Foul bill of lading signifies damaged cases and is not acceptable.

The other documents include the following: Commercial invoice, certificate of origin, packing list, detailed description of goods, country of origin, price/ value, name and address of the buyer, name and address of the seller, vessel details, port of discharge, import permit numbers, letter of credit details, trade discounts, packing details, marine-insurance information, credit details, terms of payment, rate of exchange in trade contracts, export and import permit numbers, etc.

IMPORT PROCEDURE

The first act of a prospective importer is to ensure that the item is not indigenously available. The buyer should then approach the sponsoring authority. The last date for filling applications is 15th December of the licensing year to which the application pertains and 31st January duly recommended by the sponsoring authority. The validity period of licence is as given below: The licences issued to actual users for raw material, components. For emergency spares, it is 12 months.

The requirement of the industrial actual user will be met through: Open general licence, Supplementary licence, and Allotment through canalising agencies as per the policy book. The actual user can import raw materials, components, consumables, and samples, other than those included in Appendices 2, 3, 5, and 8 under the open general licence. The application form for all actual users is Form A, which appears in Appendix five D of the Handbook of procedures. Sufficient number of copies, as indicated should be submitted. Every application has to be accompanied with fees as prescribed in Schedule III of the Import Control Order 1955 vide Appendix 2, of Handbook of rules and procedures. The fee is Rs 50 if the goods specified in the application do not exceed Rs 50000. Re 1 per 1000 or part thereof will be the fee for value lying between Rs 50000 and Rs one crore. When the value of the goods exceeds Rs one crore Rs 1000 up to one crore and fifty paise per thousand subject to a maximum of Rs 25000.

Letter of Credit

Automatic licensing has been abolished and the majority of items allowed so far has been shifted to Open General Licence (OGL) in Appendix 6. Applications for supplementary items may be submitted as per Chapter 5, of the import/ export procedures. These will be considered only on the recommendations of the sponsoring authority. Repeat operation of supplementary licences of the select products is detailed under Appendix 16.

The application should be to the respective licensing authority, under whose jurisdiction the import is located. The past licences, export efforts and consumption data should be certified by a chartered accountant. The letter of credit is an arrangement, whereby the obligation to pay an exporter is undertaken by a bank. The bank's credit is available, to the importer. The amount for which the credit is issued is fully stated with the name of the account. The letter of credit lists the documents to be attached to it with validity period. Import licences have a definite validity period. Beyond this, a grace period of 60 days is available. The letter of credit can cover a further period of 15 days, in order to facilitate negotiations of documents. In the case of revolving letter of credit, i.e. where the letter of credit is renewed automatically on the draft being honoured, the bank will ensure that the aggregate drawings do not exceed the value of import licence.

A TO Z PROBLEMS OF INTERNATIONAL BUYING

The problems of international buying are mentioned below:

(a) Outflow of the precious limited foreign exchanges.
(b) Unrealiable delivery schedules.
(c) Long procurement lead times.
(d) High transportation charges.
(e) Communication difficulties.
(f) Political uncertainties.
(g) Fluctuating foreign currency.
(h) Restricted shipping and cargo-carrying capacity.
(i) Inadequate and congested port facilities.
(j) International terrorism including narco terrorism.
(k) Problems in customs clearance.
(l) Difficulties in insurance claims settlement.

(m) The need to cope with different commercial practices and unfamiliar commodity specifications.

(n) Problems of representation in various foreign countries to provide effective on-site technical and quality control inspections.

(o) Lack of experience and technical skills.

(p) Devaluation of rupee and global inflationary trends.

(q) As import entitlements are based on previous years' consumption, organisations more often tend to exaggerate their consumption pattern, to get higher entitlements.

(r) In an attempt to secure renewals, organisations make hasty imports to show utilisation of existing licences.

(s) The government's policy of imposing sales tax and other levies on purchases made through the canalised agencies, leads organisations to lift the entire year's requirement through mid-sea transfers, resulting in increased inventories.

(t) Flash strike on minor differences in major ports like Cochin, Calcutta, Bombay.

(u) Importing all components and assembling in India or the *screw-driver technology* is harmful.

(v) Import negotiations should be on TRIPS — Trade Related Intelligent

Property Rights and focus attention on trade related aspects instead of the attempt to concentrate on substantive norms or standards of the parts.

(w) Different types of legal systems operate in different countries, leading to difficulty.

(x) Rackets in imports leading to international maritime frauds.

(y) The buyer can pretend the quality of imported item with a manufacture's guarantee certificate or by an independent inspection agency.

(z) The contract with the foreign supplier should involve a clear understanding of the quality, price, and other contractual obligations.

TRADE REFORMS

The following are the highlights of the 13-point major trade reforms announced by the Commerce Ministry recently.

(a) Cash compensatory scheme (CCS) suspended with effect from July 3, 1991.

(b) Replenishment Licensing Scheme (REP) to become the principal instrument for export-related imports. It will now be called Exim scrip and can be freely traded.

(c) All exports to have a uniform REP rate of 30 per cent of the Freight on Board (FOB) value from the present rates ranging from five to 20 per cent of the FOB value. Special rates for gems and jewellery certain metal-based handicrafts and books and magazines will, however, continue.

(d) The new REP scheme to give maximum incentive to exporters whose import intensity is low, such as agricultural exports.

(e) All supplementary licences stand abolished except in the case of the small scale sector and for producers of life-saving drugs/equipment. These two categories would be entitled to import both under Open General Licence (OGL) or through supplementary licenses.

(f) All additional licences granted to export houses stand abolished. However, they will enjoy a REP rate of 30 per cent of FOB value and will be granted an additional REP of five per cent of FOB value.

(g) All items which figure in lists like the limited permissible list now to be imported through the REP route.

(h) Unlisted OGL contained in the Exim policy stands abolished and all items under this category to be imported only through the REP scheme.

(i) Advance licensing for obtaining imports for exporters will remain open. However, the REP rate for advance licence exports is being increased from 10 per cent of Net Foreign Exchange (NFE) to 20 per cent of NFE.

(j) All import licensing for capital goods and raw materials except for a small 'negative list' to be removed in three years.

(k) All items to be decanalised except those that are essential and major changes in the entire canalisation policy to be announced.

(l) Financial institutions would also be allowed to trade in Exim scripts. In due course, the Exim scrip to be replaced with foreign exchange certificates which will be more easily tradable.

(m) The rupee is expected to be fully convertible on the trade account in three to five years.

OBJECTIVE OF CANALISATION

The concept of canalised imports, has been introduced to achieve the twin objectives of optimal utilisation of the scarce foreign exchange and to bulk purchase efficiently in the maze of international arena. It aims at minimising the cost, not only for the individual importer, but for the economy as a whole. It is possible to achieve this after a thorough study of the markets, credits, aids, demand, supply, price, availability, currency fluctuations of the global situation and general agreements on trade tariff (GATT).

The provision for submitting applications directly to the canalised agencies for items which are canalised and required by actual industrial users in the manufacturing activities have been detailed in Chapter IV of the Import Policy Book. The items which are canalised are given in Appendices II Part A, Part B of the policy book.

The canalised agencies include the State Trading Corporation, Metals, Minerals and Trading Corporations and several public sector organisations. The canalised agencies account for an annual import value of about Rs 7000 crores. There is a monitoring Price Committee, with representatives from different Ministries to review the import programmes of the canalised agencies. However, the functioning of the canalised agencies has been criticised by the users, as they charge the users, up to five per cent as mark up/commission.

TEN POINTS TO PONDER

1. Examine the necessity of import trade control policy.
2. State the objective of import policy.
3. Enumerate the steps involved in import buying — from the point of source selection to the arrival of the item in the factory.
4. Discuss the customs procedures while importing items to India.
5. Explain the port formalities involved in imports.
6. Describe the advantages and disadvantages of canalisation.
7. Elucidate the problems involved in imports.
8. Identify the documents needed for imports.
9. Articulate on the role of sponsoring authorities.
10. Categorise the importers and define their roles.

53. Indigenisation

RELEVANCE OF IMPORT SUBSTITUTION

In a developing country like India, the policy of import substitution should be enshrined as a major national objective. It is the major tool, with which we can achieve our goal of self-reliance. Substitution means local alternatives for items, which are not presently available. Beside individual company's research efforts, the policies of several organisations — like the Planning Commission, different ministries, Department of Science and Technology, National Research Laboratories, Council of Scientific and Industrial Research (CSIR), Directorate General of Technical Development, and other agencies should be oriented toward import substitution. Self-sufficiency and self-reliance have to be achieved in selective areas, where proven competence is available over a period of time.

Due to increased economic development and the global inflationary spiral, Indian imports are increasing day by day and are likely to reach the Rs 50,000 crores mark in the near future. The yawning trade deficit or the excess of imports over exports has been around Rs 10,000 crores in the recent past. The foreign exchange reserves have been diminishing at a fast rate and are below Rs 5,000 crores, which implies only less than ten months imports will be serviced. Hence, is the necessity for everyone to adopt the slogan 'Be Indian and buy Indian' and substitute imports by local items. The major imported items include fuel, fertiliser, machinery, spare parts, defence equipment, transport items, edible oil, electronic items, dry fruits, wool, food, chemicals, manufactured ferrous and non-ferrous items, etc.

BENEFITS OF INDIGENISATION

Import substitution aims at the following advantages.

(a) Conserves the precious foreign exchange reserves of the country.

(b) Better cooperation between the buyer and seller is possible because of close physical proximity.

(c) Delivery dates are more certain as transportation is only a minor factor in delivery.

(d) Lower prices can result from consolidated transportation and insurance charges.

(e) Local talents, expertise, research, etc. can be developed to boost employment in India.

(f) Shorter lead time can frequently result in reduced inventories. Fuller utilisation of capacity and material conservation is possible by import substitution.

(g) Expedited orders are likely to be complied faster.

(h) Implied social responsibility to the community is fulfilled.

(i) Indigenous sources, as a result of the economies of scale, can offer higher quality with reduced price.

(j) Local companies can often provide superior technical assistance.

(k) Encourage indigenous industry, by price preference.

(l) With greater production capacity, the local industries can handle fluctuating demands.

(m) Shortages are less likely because of their broader markets.

(n) Local companies know the local manufacturing conditions as well as skill-mix of the operators.

(o) Establishment of technology transfer centres and technology parks will improve the thinking power of Indians to meet the challenging task of substitution.

(p) Import substitution is the only answer when the world powers threaten economic blockade and embargo like super 301 clause due to confrontation with our hostile neighbours.

(q) Restricted shipping space and cargo-handling capacity at the national level necessitate import substitution.

(r) Inadequate port facilities, congestion in the ports, periodic dock labour unrest, necessitate import substitution for the smooth running of the industry.

(s) Different commercial practices in various countries.

(t) Currency fluctuations.

(u) Communication problems.

(v) Difficulties in inspection of foreign goods.

(w) Problems of rejected imported items, etc., add to the urgency of import substitution.

(x) The foreign specifications and tolerances may not suit the Indian weather conditions and workers' knowledge.

(y) Sometimes the high image and professional perfection of foreign suppliers remain only theoretical.

PROBLEMS IN INDIGENISATION

Organisations face a large number of problems in the field of import substitution and these are mentioned below:

(a) Inability to meet the precise and high-tech nature of the demands of the industry according to production, design, maintenance and safety standards.

(b) Using inferior quality raw material in the absence of imported special steel to manufacture spare parts.

(c) Lack of technical know-how on the part of vendors, particularly in the field of reliability estimates and adherence to warranty clauses.

(d) Foreign collaborators not permitting to switch over to indigenisation due to fear of safety and quality considerations.

(e) Vendors not interested in small quantity of requirements of spare parts and do not want to spend money on educational orders.

(f) The suppliers produce good quality samples according to the given specifications, but not adhering to the specifications of approved samples.

(g) No one is interested in investment of developing drawings of intricate spare parts, in the absence of firm orders.

(h) Inadequate infrastructure facilities and technical as well as entrepreneurial skills.

(i) Improper testing equipments and inspection facilities.

Price Increases

(j) Insisting on venture capital from the buyer's organisation with a few ancillaries reporting sick later on.

(k) Law morale of employees due to poor wages resulting in frequent strikes at vendor's premises.

(l) Increasing the price after obtaining the order, by more than 100 per cent at the time of supply.

(m) Increased cost of manufacture with diesel generator sets due to power cut.

(n) Resorting to *force majure* clauses and delaying the supplier.

(o) Cost overruns and time overruns in most cases.

(p) Delays in replacement of rejected material.

(q) Non-removal of rejected parts from stores, even after six months.

(r) Reliability of spare parts not specified.

(s) Poor after-sales-service even in warranty period.

(t) Insisting on advance for material every time or requesting for supply of raw materials and the industry reporting sick with the industrialists becoming rich.

(u) Resorting to psychological pricing mechanism.

(v) Repeated failures of the same part.

(w) Poor organisation at vendor's premises.

(x) Not applying well-known cost reduction techniques like learning curve and value engineering for repeat orders.

(y) Too much time spent on price haggling during negotiations.

(z) Lack of professional approach for quick response, prompt repair and least repeat failures by the vendors.

INTELLECTUAL PROPERTY RIGHTS TRIPS

Developed countries do not want to transfer the high technology as the international trade, technology transfer, patents, trade secrecy, aid, investment's etc., are governed more by political rather than humane considerations. The focus is only on *screw driver technology* — or selling all components to be assembled in India. With relaxation of global tension, the problem of technological mercenaries, flowing out of industrialised world to rich developing countries, offering their skills in weapon design and developments, may be solved. A core issue in this context, is that of intellectual property rights. Certain dominant technological nations have adopted an approach which not only affects technology transfer and development of weaker nations, but also inhibits research and development all over the world. TRIPS Trade Related Intellectuals Property Rights should be available to the weaker, poor and developing nations. However, ultimately, it is a question of collective wisdom for collective management on international peace and sustainable development.

STRATEGY OF TECHNOLOGY

Technological innovation is essential for the birth and the growth of industrial enterprises as well as socio-political and economic health of the countries. The stock of technology on which innovation depends, can be increased by research and development — both at macro and micro levels. It is necessary to adpat the bought-out technology to the local conditions before absorbing the same. The strategy for achieving technological self-reliance should include the following phases: (a) surveying the alternative technologies available in the globe, (b) selective buying of technology, (c) import substitution, (d) development of threshold technology, (e) development of essentially indigenous technology.

All countries have to import technology. But this has to be dovetailed to and close knit, with the existing practices in India. Having amassed imported technology, the quickest pay-back to our country is from adapting technology or import substitution. Development of essentially indigenous and threshold technology, has to be the bulk of our effort in achieving self-reliance. For better assimilation of imported technology, it should be purchased outright or a dynamic research institution has to be associated with every import of know-how. In this processing it is necessary to involve and coordinate the efforts of interested parties, like the Government, Ministries, trade associations, the Planning Commission, national laboratories, universities, freelance inventors, and corporate researchers. It is also necessary to ensure that, compromises are not made in the realms of quality, safety and pollution while discussing import substitution.

The indigenously manufactured item should be of equivalent quality and compare favourably with the landed cost of the item in the long run. There are a large number of techniques available for import substitution:

(a) Efforts at all levels should be properly channelised and coordinated towards indigenisation.

(b) Establishment of technology transfer centres.

(c) Critical review of the existing tolerances without affecting safety.

(d) Maintenance of technological parks, in the cities.

(e) Application of cost reduction techniques like value analysis, standardisation, brain storming, supplier seminars, etc.

(f) Training technical personnel abroad to study the latest technology.

Role of R & D

Research and development — both at macro and micro level — are essential for technological innovation and indigenisation. Despite having the third largest number of scientists in the world, the scientific development in India has not come of age. The reason is that, the existing scientific establishments are basically for development techniques for doing well-defined tasks, geared to imitating and not for innovative tasks. The scientists who study abroad are dissatisfied with the environments of the existing scientific institutions. These do not have a healthy ambience for the growth of young minds and their scientific potential. Some of the unfortunate ones commit suicide due to frustration.

Normally, the screw driver technology of importing semi-finished items and assembling in India is practised. This is in spite of the fact that the government spends about Rs 1000 crores in the national laboratories like — Council of Scientific and Industrial Research, Indian Agricultural Research Institute, Medical Research Laboratories, Nuclear Research, Defence Research Laboratories, etc. Individual organisations also spend up to 1 per cent of their turnover for the in-house research. But the benefit accrued from these efforts is not commensurate with the investment. The theoretical work done in Universities and the research laboratories should be translated to commercial shopfloor level. Closer liaison is necessary between scientific institutions, freelance inventors, universities and technological institutions with the commercial organisations, in order to get optimum results.

Import substitution on a selective scale should not be a major hurdle in India, particularly when the country is able to successfully build missiles, satellites, nuclear power plants, war planes and other sophisticated items. Proper environment, motivation, recognition, national awards, financial incentives and other measures should be instituted in this regard. Licensing procedures should be liberalised at all levels to encourage indigenisation.

TEN POINTS TO PONDER

1. Identify the need for import substitution.
2. Explain the role of DGTD and other government agencies in the process of import substitution.

3. Comment on the statement that "indigenous items are of poor quality at higher cost".
4. Give your suggestions to reduce the adverse trade gap.
5. Enumerate the advantages of import substitution.
6. Critically analyse the difficulties of indigenisation.
7. Examine the relevance of intellectual property rights in the context of indigenisation.
8. Describe the strategy for import substitution.
9. Discuss the role of R & D in indigenisation.
10. Define the problems in the area of technology transfer.

54. Art of Negotiations

In the industry, the top management spends a substantial amount of time in negotiating with the labour. The buyer always tries to negotiate with the vendor on quality, price and delivery schedule. Even procedure-oriented Government organisations talk about the sanctity of tender on the one hand, and negotiate for price reduction with the lowest tenderer on the other hand. There is nothing wrong in negotiating on a selective basis, if it is advantageous to the buyer and the seller.

DEFINITION OF NEGOTIATION

Negotiation, a word from Latin Civil Law, means trading deliberation leading to an agreement. According to the Oxford Dictionary, it means, conference with another, with a view to compromising on agreement. It attempts to find a formula, which gives each party the best possible value for money in a given situation. It aims to achieve the best value for money spent by the organisation, taking into consideration the overall requirement of quality, quantity and price according to the market conditions.

The negotiation skill is inborn in every human being even though the intensity of the negotiation skill varies from individual to individual.

Everyone does buy or sell something. So, everyone takes negotiation for granted as a routine practice of reducing prices by 10 per cent. Negotiation is not haggling, bickering horse trading but should be based on facts. But negotiation is not bargaining for lower prices. For the professional buyer, negotiation means a tool of getting items at right price and other contractual terms. Negotiation is an art as well as science. It is a constant creative search for alternatives, directed at established objectives.

NEGOTIATION PARAMETERS

It is one of the most interesting functions of purchasing. Negotiation reviews, analyses and results in a mutual agreement on the following materials management aspects:

(a) Price discounts
(b) Specifications
(c) Quality tolerances

(d) Quantity requirements
(e) Delivery terms
(f) Legal terms
(g) Contractual obligations
(h) Despatch terms
(i) Insurance management
(j) Payment conditions
(k) Inspection standards
(l) Rejections
(m) After sales service
(n) Spare parts
(o) Bonus/penalty
(p) Guarantee
(q) Warranty
(r) Freight
(s) Buy-back clauses
(t) Liquidity damage clauses
(u) Currency fluctuations.

However, price, quality and delivery schedule play a more crucial role, out of the above factors.

The professional buyer seeks to optimise the relationship between quality and price. Sometimes he may prefer one vendor's product over all others, although the price is out of line. Substitutions call for negotiation to find a realistic price which will be equitable to both parties. In some cases, where both the buyer and the seller are eager to do business, this may be extended to analysis and negotiation of the vendors' cost of producing the product.

SITUATIONS TO NEGOTIATE

If a buyer resorts to negotiation every time and wants to reduce the price by 10 per cent, he will be called a 10 per cent buyer. Negotiation has to be done selectively and is resorted to in the following cases:

(a) Negotiation is useful when competitive bidding is missing.

(b) When the buyer feels competitive bidding has not resulted in fair and reasonable price.

(c) When purchase is stretched over a long period of time.
(d) When long time is required to manufacture items.
(e) When buying of non-standard products is involved.
(f) When specifications are vague and indefinite:

(i) when order quantity increases due to production changes, (ii) when purchases are for a large volume of the same item, (iii) when tooling or development money is required to be invested by the seller:

(g) When additional high volume requirements are anticipated for high cost multiple source items.

(h) When there are changes in drawings.

(i) In case of supply of spares.

(j) When there are changes in packing and delivery points.

(k) When no acceptable bids have been offered by seller.

(l) When additional work or existing project has to be carried out.

(m) When disputes cannot be solved by arbitration or adjudication.

(n) When all tenders have quoted high price and retendering is presumed to take a long time.

IMPORTANT ASPECTS OF PREPARATION

The most important, but neglected aspect is the preparation for negotiation meeting. It is unfortunate that high level policy-making executives run from one meeting to another without any preparation. Inadequate preparation due to very short notice is commonly observed when governmental purchases are held in foreign soil. Hence the importance of adequate preparations for negotiations cannot be overemphasised.

Some of the steps to be followed in the preparation process for negotiations are:

Establish objectives of the negotiation process. Gather knowledge of vendor's capabilities like manufacturing process, material sources, etc. Allocate areas for negotiations after knowing facts, objectives and limitations. Fix the meaning of fair and reasonable price. Ensure that contract assures fulfilment of all terms on delivery, quantity, payment quality, service etc. Establish a sound basis for understanding between the buyer and the seller. Estimate sellers' possible approaches, strengths and weaknesses. (How badly the seller needs the order? How certain does seller feel about his chance for getting it? How much time is available to reach an agreement regarding terms and conditions?)

The buyer is the leader of the negotiation team. The team's composition varies with the type of item, situation and organisation. Team includes personnel from purchasing, production, quality, finance and engineering departments. The team should hold a meeting, prior to negotiations with sellers. The team should develop the negotiation strategy and tactics after discussions and complete agreement on the terms of the contract. Adopt the approach of offering alternatives. It is necessary to have a mock negotiation exercise. Factors like variations in quantity and specification, price revisions under escalations, facilities to be provided in the form of tools, materials, money and unforeseen contingencies/ repairs, etc. must be taken into consideration during negotiations.

A TO Z APPROACH TO NEGOTIATION

The attitude to be adopted by the team and the approaches to negotiation should be clear to the members. The following are some guidelines relating to good approaches for the negotiation process:

(a) Purchasers and sellers appreciate that their facts must be contradictory.

(b) Listen to the supplier's view.

(c) Be prepared to discuss any point.

(d) Learn about the supplier's strategy and him as an individual, from his speech and discussions.

(e) Do not debate.

(f) Concentrate on the positive points while acknowledging the negative.

(g) Do not bluff — do not make false statements.

(h) Understand the buyer's limitations and make only the promises which could be carried out.

(i) Ask questions relevant to the subject as this may bring good information.

(j) Do not agree in the beginning, to keep back something to offer in the last as concession.

(k) Do not try to prove that the seller is wrong, as you may not have any advantage.

(l) Negotiate at purchaser's place as far as possible.

(m) Negotiate with people who can give concessions.

(n) Use new techniques/approaches to convince the seller.

(o) If negotiation is not proceeding well, reschedule the meeting or give minor concessions at the right time.

(p) Aim at an integrated solution, in which both the parties get the best possible out of their objectives.

(q) Once the agreement is reached, put it in writing.

(r) Ensure that copies of the agreement are available to all concerned.

(s) Facts, rather than guess work should be the basis for negotiation.

(t) Quantity discount analysis for complex comparisons, cost analysis for simple comparisons and break even analysis for price structure should be used.

(u) Theory of bargaining gives an idea of how to arrive at an optimum price which is agreeable both to the buyer and the seller.

(v) The competitor's figures need not be divulged and misleading information should not be given.

(w) It is not necessary to solicit proposals, where there is no intention of buying.

(x) The learning curve gives an idea of how to modify labour costs and how to account for improvements in learning.

(y) Strategy deals with planning or directing negotiations.

(z) Tactics, defines the mode for gaining advantage or success, covering the actual manoeuvres to implement the strategy.

Negotiations Meeting

Once the formalities and preparations are complete the buyer is ready for negotiations with the vendor at the stipulated time and place. The meeting reviews specifications, classifies points, and understands price-cost analysis of both the parties, in order to arrive at a negotiated price. After determining the supplier's workload, the delivery time is fixed. An integrated solution acceptable to both the parties is arrived at. The result of the agreements is put in writing.

Negotiation should result in better overall understanding, smoother buyer-seller relationship, mutual respect, confidence, knowledge, responsibility and give and take. Negotiation should lead to acceptance of contractual terms and specifications and establishment of right price and right sources. Negotiation enables fruitful discussion and to get over bottlenecks in the buyer-seller relations.

QUALITIES OF NEGOTIATOR

The negotiation process should emulate the qualities of Lord Rama's emissory and negotiator Lord Hanuman, who is considered as an ideal negotiator. Today's another typical example is the working housewife, who successfully manages her home. The qualities of negotiator are given below:

(a) The buyer should realise that a key trait of successful purchase executive is his skill in the art of negotiation.

(b) He should be perceptive and tactful.

(c) He should be a good listener.

(d) He should understand the supplier's suggestions and points of view.

(e) He must objectively assess the suppliers' views.

(f) He should possess the analytical ability and should be able to convince the supplier about these facts.

(g) He should have a complete command of the situation and knowledge about the items.

(h) He should be humble, but should not allow to be browbeaten.

(i) He should lead the negotiating team to a proper direction to achieve the set objectives.

(j) He should express his ideas through simple language.

(k) He should be fully responsible for what he does.

(l) He should adopt ethical practices.

(m) He should believe in the basic principles of scientific materials management.

(n) He should be quick to change his stand depending upon the trend of negotiations to the advantage of his concern.

(o) He should be impersonal and cool-headed.

(p) He must have a sense of humour to break the monotony.

Knowledge of Suppliers

He must have a complete knowledge of the following factors:

(q) Market demand and supply condition
(r) Suppliers facilities
(s) Suppliers workload
(t) Financial status of the supplier
(u) Items to be manufactured
(v) Previous price history
(w) Cost-price analysis
(x) Learning curve and value analysis for cost reduction
(y) Different types of discounts, contracts and terms.

It is preferable that the negotiator should possess a satisfactory IQ level. The qualities ascribed to an ideal negotiator are as follows:

(a) toughness, (b) analytical bent of mind, (c) simplicity, (d) flexibility and firmness, (e) rapid thinking, (f) ability to listen patiently, (g) tactfulness, (h) versatility, (i) patience, (j) persistence, (k) fairness in judgment, (l) professionalism, (m) ability to argue convincingly, (n) creative leadership, (o) decency and resilience, (p) dependability, (q) innovation, (r) intelligence, (s) optimistic, committed and compassionate outlook, (t) courteousness, (u) impersonal outlook, (v) fairness in dealings, (w) efficiency, (x) cooperative attitude, and (y) quest for excellence.

TEN POINTS TO PONDER

1. Define the meaning of negotiation.
2. Discuss the objectives of negotiation.
3. Give examples of negotiation from day to day life.
4. Identify the parameters of negotiation.
5. Explain the situations best suited for negotiation.
6. Emphasise the relevance of preparation in negotiation.
7. Detail the approach to negotiation process.
8. Examine the significance of the theory of bargaining and learning curve in negotiation.
9. Illustrate the qualities of an ideal negotiator.
10. Enumerate the steps involved in the process of negotiation.

PART V

Materials Management Policy

55. Materials-User Interface

CONCEPT OF INTEGRATED MATERIALS MANAGEMENT (MM)

The functions — procurement, purchasing, materials management, business logistics, inventory control, and warehousing — are being carried out in many organisations under the scope of integrated materials management. All activities — of organising, planning, scheduling, implementing and controlling in connection with the functions store keeping, distribution, purchase, disposal and transportation — are included in the purview of the integrated approach to materials management. Thus we can define, materials management as a coordinated approach responsible to plan, acquire, store, move and control all types of input materials.

ORGANISATION OF MM

The materials function is on par with the company's other cross-sectional functions like production, finance, marketing and personnel management and is a partner to production, sales, and development in all aspects of materials supply. The materials manager usually reports to the Chief Executive of the organisation in most Indian companies according to the research study. The impact of the materials department's efforts on corporate profits depends on:

(a) The materials cost;

(b) The cross-sectional approach of penetrating all the financial aspects and over laying it with resource-oriented thinking;

(c) The degree to which materials management can control inventories, raw materials, semi-finished items as well as finished goods;

(d) The personnel assigned to the materials function.

The organisation of materials management can be done by activities, products, stages of manufacture and by geographical location. The concept of centralisation and decentralisation, depends upon several factors. Generally, the purchase of high value items can be centralised while buying low value items and the stores location must be decentralised. The organisational requirements of integrated materials management must include linkages with neighbour functions and should be directed towards the corporate goals.

A TO Z OF DECENTRALISED ORGANISATION

In most cases, the user units clamour for the requisite authority which would enable them to meet expeditiously their material requirements.

(a) Under the decentralised system, users are better able to plan and organise their supply services better because of the immediate concern.

(b) It is possible to delegate direct responsibility for getting programmes executed without delays.

(c) The central organised set-up is advantageous because of less inventory with less working capital.

(d) But, it has proved inadequate because those responsible for procurement lacked sufficient appreciation of local problems and priorities.

(e) Enough experienced high calibre staff cannot be made available to man the decentralised system and the extensive network of supply effectively.

(f) The decentralised system also generates fragmentation and wasteful duplication.

(g) All these result in unnecessary dissipation of resources, standard procedures throughout the organisation due to divergent interests.

Decentralised Activities

The following activities are generally performed at the decentralised level:

(h) The stores and tool stores have to be completely decentralised and located nearest to the user.

(i) Supply policies and instructions should be adopted to the local operating conditions.

(j) Short-term planning and forecasting of immediate operational requirements.

(k) Definition and detailed specification of requirements.

(l) Market surveys and purchases to meet specialised *ad hoc* emergency needs.

(m) Store-keeping, materials handling and distribution services.

(n) Inventory and stock control of raw material, spares and semi-finished goods.

(o) Preservation and protection of stored items against damages.

(p) Management of disposal of surplus, obsolete, scrap item.

(q) Upkeep and maintenance of infrastructure facilities, including warehouses, tool stores, vehicles, and testing equipments.

(r) Personnel action, at lower levels on selection, recruitment and conditions of service.

(s) Public and customer relations, including vendors.

However, the following are centralised in many organisations:

(t) The policy formulation manual and coordination.

(u) Long-term materials planning.

(v) Bulk purchasing to take advantage of discounts of common items and helpful for informational exchange.

(w) Consolidated global materials intelligence, to introduce competitive source of supply.

(x) Uniform policy for negotiations of purchasing.

(y) Development of consolidated inventory system, particularly for high value items.

(z) Provision of shared facilities and public relations service.

PROFIT CENTRE GOAL OF MM

The integrated materials management, which is conceived as a comprehensive supply system, can substantially contribute to a company's profits, through its systematic influence on materials cost and overhead expenses. In most of the industries, the input materials cost — including fuel and utilities, constitute over 60 per cent of the final product. Japanese companies which have reduced materials cost by over 5 per cent in a duration of 18 months, have a common slogan — "We live by sales, but make our profit from materials".

To understand the profit centre concept of materials, let us consider an organisation having an annual turnover of Rs 100 crores. The input materials cost is 60 crores, the labour cost is about 15 crores. Overheads, depreciation and taxes account for about Rs 15 crores and the profit about Rs 10 crores. The aim of any organisation is to improve its profit, say, by 30 per cent; this can be achieved in a variety of ways. One way is to improve the turnover by 30 per cent by increasing the price and or quantity, so that the gross profitability can be increased by 30 per cent, but the net profitability will be much less. Another way is to examine the feasibility of reducing the materials cost by 5 per cent, which will reflect in an increase of Rs 3 crores in profits. Thus, profitability is increased by either reducing the materials cost by 5 per cent or sales by over 30 per cent. This ratio of 1:6 goes up to 1:10 in actual practice due to increased advertisement and additional manpower needed in sales.

Return on Investment

Another major objective of organisations is to improve the return on investment. This is obtained as a ratio of return to the investment on fixed and working capital. The return on investment can be increased either by increasing the return and/or decreasing the working capital, bulk of which is in the form of inventory.

MATERIALS RELATIONSHIP WITH OTHER DEPARTMENTS

In any organisation, a large number of departments strive to attain the organisational final objectives. All these parameters have varied duties and responsibilities. In order that they pull together, there should be good interrelationship and excellent coordination of efforts.

Within the warehousing itself, there should be coordination between receipt, inspection, documentation, disposal and issue to different departments. Similarly, the purchasing activities of source development, ordering system, buyer-seller relations, follow-up, purchase intelligence, should be synchronised. The materials function procures the items, while the production manufactures the product and the sales markets the product.

Within the materials department, the manager must ensure smooth cooperation between the functions of purchasing, stores, inventory control, information flow, transportation etc. The role of the user is very critical in the cycle of supply activities. Their full cooperation and proper use of resources are essential to ensure effective application of scientific materials management techniques. The materials manager exists in the organisation to serve the user, and must have maximum interaction with the production executives for prompt supply of quality materials. The materials manager can have a clear idea of the spares requirements from the maintenance engineer. The materials function should have knowledge, in answering the queries of the finance manager, who arranges the working capital finance of the inventories. The materials manager also interacts with the marketing manager with regard to customer service, the market intelligence, price trends of commodities, forecasts, finished goods inventory level, and market research. The materials department helps the sales department in selling at as low a cost as possible, so that the company can maintain a competitive selling position. The personnel manager can help the materials function on legal aspects, fidelity, guarantee, manpower planning and in motivation. The computer department enables him to get the required information in the quickest possible time.

Externally, the materials function interacts with transporters, carriers, railways, insurance companies, bankers, suppliers, subcontractors, excise departments, sales tax officers, customs departments, import licensing authorities, vigilence officers, inspectors and other government agencies. The materials department should organise itself to meet the challenging needs arising from the above, and to sort out the differences with other departments.

The production is usually the most important department served by the materials function. The purchase department identifies the need and procures the needed items for production. The production department must, in turn, live up to its individual role responsibilities for the total result to be economical, from the company's point of view. In this process, it is desirable that the following information must be interchanged between production and materials department.

A TO Z INFORMATION EXCHANGES

(a) Obtaining technical information and advice on materials

(b) Establishing right specification and proper tolerances for the materials

(c) Scheduling orders for materials and fixing delivery schedule

(d) Specifying methods of delivery and transportation

(e) Inspection of materials for quality

(f) Materials accounting in each work order

(g) Fixing levels of different components of inventory

(h) Receiving, warehousing and storage of items

(i) Executing construction contracts, agreements and service contracts

(j) Realisation of scraps, surplus and salvaged items

(k) Materials handling and traffic management

(l) Considerations on make-buy-lease criteria

(m) Following up with suppliers, for urgent relations

(n) Ensuring good buyer-seller relations

(o) Rejections at final stage, manufacturing and incoming stage

(p) Market research for purchasing

(q) Ideas on cost reduction techniques

(r) Schedule of overhauling and its requirements

(s) Shutdown, stockout, shortages

(t) Production planning, bill of materials, and MRP

(u) Resolving ambiguity of drawings for machinery spares

(v) The greater the accuracy of an item, the higher is the cost and difficulty in procurement of the item

(w) Preparation for technological upgradation

(x) Discouraging the tendency to erect mental barriers around new thinking and an inclination not to cross such boundaries, must be discouraged

(y) Both must appreciate that, if profits are to be maximised, the specified materials must be economical to procure and to fabricate, and they should be easily available, from more than one efficient, low-cost producer.

(z) To tackle the above information requirements, the materials manager must play the role of a business economist and a peer of the management at the top echelon.

The materials management can contribute significantly to the corporate goals of increased profitability and improved return on investment and hence is treated as a profit centre. The objective of materials management should be to ensure continuity of supply and at the same time reduce the cost of the final product. The goal conflict of high level of customer service, versus low working capital locked up in inventories, forces materials management to do an optimising function, comparable to a balancing act on a tight rope.

A TO Z ASPECTS OF PROBLEM AREAS WITH USER

(a) In performing the balancing act of optimising high service level against low inventory, the materials department comes into conflict with the user in the following areas:

(b) The user, normally the production department, complains about the inferior quality of service in the form of incoming materials. The user often has to follow up with vendors, which is a materials management activity.

(c) The user always complains about the procurement delays resulting in production stoppages.

(d) He also complains against the materials function for the accumulation of work in progress.

(e) The materials manager, on the other hand, complains about the bulk of the users' indents market most immediately needed, without caring for the organizations' system and procedures.

(f) It is the duty of the production manager to advise his materials counterpart, well in advance about the anticipated changes in production plans.

(g) An analysis of the indents in several companies reveals that a substantial proportion of the indents are incomplete with regard to the specifications, resulting in increased delays.

(h) A major area, in which production and materials can coordinate, is documentation or avoiding removal of material without indents.

(i) The user should not indent items by suppliers' part number as this process makes every item a proprietory item.

(j) He should not increase the proportion of non-stock items and resort to temporary codes.

(k) The prerogative of declaration of an item as absolete lies with the user, as the materials department can only compile a movement analysis of the items. The user should not categorise non-moving item as a critical or insurance part.

(l) The user is formally and informally associated in the procurement of materials, as he contributes significantly to the technical deliberations in the negotiation prices and the tender committee.

(m) The user has to be informed of the long lead items and fluctuations in the lead times, particularly for critical components.

(n) Without the user's cooperation, it is difficult to introduce consumption norms and scientific inventory control concepts.

(o) The user's active involvement is necessary in providing technical know-how for ancillary development and to the new vendors.

(p) Feedback on rejections from the various production shops influences the vendor rating process.

(q) The well-known cost reduction techniques like standardisation, simplification, codification, import substitution, variety reduction, value analysis, material requirement planning, just in time inventory, etc. can be applied only after getting the whole-hearted collaboration of the user departments.

(r) Norms for acceptance/rejection/rectification of the incoming materials can be developed only with the help of the user.

(s) Decision-on make-buy-lease has to be taken only by the user.

(t) The user is associated with the development of drawings for critical spares, after-sales-service contracts, capital equipment purchase, and determination of initial quantum of machinery spares.

Need for Collaboration

(u) Coordination with the user is a must in the areas of material requirement planning, materials schedule, production planning and production control.

(v) Close cooperation is necessary in deciding about the reuse of waste, in the plant.

(w) Cooperation between the materials and development functions, is necessary for product development and in updating the bill of materials.

(x) The user's acceptance is necessary for altering the service levels, arising due to budgetary constraints.

(y) The user's cooperation is required in solving bottleneck problems by subcontracting measures.

(z) By means of effective cooperation between the materials and the user, one can succeed in reducing the level of capital tied up in inventories as well as decrease the input cost. This sort of collaboration is imperative for total control of flow of materials from the suppliers to the users within the company, and via the processing points in production to allocation for marketing and thus for ultimate consumer satisfaction.

TEN POINTS TO PONDER

1. Define the concept of integrated materials manager.
2. Discuss the advantages of centralised and decentralised aspects of materials management function.
3. Comment on organisation of materials management by activities, by stages of manufacturing, and by products.
4. Justify the relevance of profit centre approach to the materials function.
5. Examine the internal interrelations of materials function.
6. Identify the relations with external agencies.
7. Detail the materials information requirements from production function.
8. Elucidate the problems created by production for the materials function.
9. Crystallise the areas of collaboration between the materials and user departments.
10. Specify the improvements to be effected by the materials function for the user.

56. Materials — Finance Interface

A TO Z ROLE OF FINANCE

(a) It is not a mere coincidence that the first letter for both materials and money is *M*, as materials is nothing but money in the form of goods. This necessitates the finance executive to be associated with the policy-making process of the materials function.

(b) The finance executive is often short of funds.

(c) Shortage of funds results in post-ponement of settling the supplier's bills, forcing the supplier increase the price. This vicious circle goes on.

(d) Materials department can contribute significantly to the corporate profitability and liquidity and hence to the goals of finance function, necessitating a close coordination between the two functions.

(e) Materials Department has to plan justifiable levels of inventory investment for the purpose of annual budgeting.

(f) Controlling the different components of inventory is a key result area, in which both the functions are interested.

(g) The finance function has to furnish estimates on costs such as, ordering charges, order processing charges, inventory holding charges, set-up costs, shortage costs, over-stocking costs, materials handling costs, freight charges, insurance premium etc.

(h) Finance function renders a helping hand in all the cost reduction projects of the materials function and also in reducing the inventory.

(i) Without the involvement of the finance, it is difficult to make make-buy-lease-decisions.

(j) Finance plays a key role in tender committees and in negotiation process with the suppliers.

(k) Useful tips can be had from the finance function for investment in additional storage and transport facilities.

(l) Finance is involved in routine coordination between receiving, invoicing and credit control.

(m) Finance plays a crucial role in the development of manuals, delegation of power, charging off an item, and minimisation of records.

Difficulties with Finance

(n) Finance department often can be nasty, insisting on book value of non-moving obsolete stock, allowing it to accumulate in the warehouse.

(o) For financial propriety, the finance sometimes, insists on three quotations, for a single source item, perhaps from the same typewriter.

(p) A major source of irritant the materials function in every organization faces, is the year end stock verification process, where the materials accounting questions intensively on the differences between the book value and the physical value.

(q) The finance insists on financial concurrence even for budgeted items, causing delays in procurement.

(r) The finance executive is interested in the methods of inventory valuation and pricing the issues, as the company's profitability depends to some extent on the above.

(s) The finance can introduce policy changes for capitalising insurance spare parts to take advantage of taxation policies.

(t) Since the materials function is a major expendor the company's resources, different types of audits and questions are asked on the materials system, to ensure that the transactions are not only honest, but appear to be honest. This can be converted into an opportunity for overall improvement.

(u) Since the Reserve Bank of India is interested in the company's using the bill market scheme, the finance manager can help the materials in procurement by demand bills, usance bills and accommodation bills.

(v) Financial techniques like break-even analysis, marginal costing, price versus cost, quantity discounts, etc. can be useful to the materials executive in deciding the quantum of purchase.

(w) Finance plays a crucial role in capital expenditure purchases incurred in connection with the project purchases.

(x) Liaison with finance is necessary in the areas of import purchases as a detailed study of the currency fluctuations and the implications of floating currencies, is necessary at the time of opening the letter of credit.

(y) The costs and consequences of the role of penalty clauses, risk purchase clauses, liquidity damages, demurrages, insurance premium, etc. can be objectively assessed by the finance department.

(z) The materials department accounts for a major share of the company's cash outflow and has a considerable influence on the level of its current assets. It substantially contributes to the corporate profitability and liquidity, which are the key result areas of the finance function. With the slogan that 'uncontrolled inventory is industry's cancer', the finance and materials function significantly contribute to the corporate goals of increased return on investment.

INFLATION AND MATERIALS

In developing countries like India, the inflationary spiral has often been suggested as a major reason for hoarding inventories. Inflation, defined cynically as confiscation without compensation, depends on several factors. These include social, ethnic, religious, political, economical, financial, monetary, budgetary and natural factors. Money supply has been increasing at a fast rate and about Rs three lakh crores is available with the banking system in the form of time and demand liabilities, with about one third being tied in working capital financing. The parallel economy with a guestimated circulation of over Rs 50,000 crores, affects the price levels. The revenue deficit of over 10,000 crores during the recent years and increased defence expenditure, are obviously a major influence on the price levels.

Since about 50 per cent of the people are living below poverty line, the Government has to adopt populist measures like waiving loans upto Rs 10,000 for farmers subsidised rice schemes, right to work, low agricultural tax, subsidised farm inputs, etc. Populist slogans like, "Work for every one" and "Water for every field" contribute to the inflation. Every budget — Central, Railways, States, Local bodies, etc. — imposes heavy taxes resulting in inflation.

The trade deficit of about Rs 10,000 crores and the precariously low foreign exchange reserves have forced us to borrow from several foreign and internal resources. Approximately Rs 1500 per head for the 85 crores of Indian population is the amount of foreign borrowing, with an equal amount of internal borrowing as well. Economists fear that the above may land us in a debt/death trap. A runway price increase is witnessed due to increased budgetary revenue deficits, deficits in international trade, enormous non-productive election expenses, subsidies for agricultural inputs, concealed subsidies, investment in non-productive gold asset, increased defence expenditure, decreased foreign exchange reserves, revenue losses in public sector undertakings, overheated economy, falling rupee value, corruption in high places, adopting pupulist schemes for short term gains, monsoon failures, higher administrative expenses, etc. The finance and the materials manager have to weigh the pros and cons of inflation versus inventory carrying charges to decide on the materials budgeting policies.

MATERIALS BUDGET

Budget is derived from the French word 'bougette' which means a small bag. 'Budgeting' as applied to business situation generally refers to budget planning and preparation, budgetary control and related procedures. It is essentially a managerial process. Business budget is a plan covering all phases of operations for a definite period in the future. It is a formal expression of policies, plans, objectives, and goals laid down in advance by top management. A budget translates the requirements into financial terms. The short term annual budget deals with repetitive purchases, while the five year long-term budget includes capital equipment purchases. In this process, materials function has to interact with other functions. The quantity estimates and price recasts have to be obtained from the users and finance, in formulating the budget. Budgetary control involves the use of budget and budgetary reports throughout the period to coordinate, evaluate and control day-to-day operation in accordance with the goals specified by the budgets.

Pre-requisites for Success of Budgetary Control

In order to ensure success for budgetary control system, the following essentials are to be observed. The entire programme must have the active and wholehearted support of top management. The principles of formal organisation must be applied at all levels of operations. Responsibility and authority for budget preparation, operation and supervision must be definitely determined and formally laid down. Delegation of budgetary responsibility must be consistent with delegation of operational responsibility. The relationship between budgeting and accounting should be clearly defined. Concerned personnel should be educated in regard to the implementation of budgets. The uses and limitations of budgeting must be thoroughly appreciated and understood, by finance and materials personnel.

Advantages of Budgeting

The advantages of budgeting may be described as follows: It forces early consideration of basic policies. It creates adequate and proper organisation by assigning of responsibility for each function. It compels all members of management from top to bottom to participate in the establishment of goals. It compels all members of management to make plans in harmony with plans of other departments. It forces management to put down in black and white, what is necessary for satisfactory results. It enables management to plan for the most economical use of resources. It instills into all levels of management the habit of timely, careful and adequate consideration of all factors before reaching at decisions. It removes uncertainties relating to basic policies and objectives. It pinpoints efficiency or its lack. It forces attention of management to the effect of the expected trend of general business conditions. It aids in obtaining optimum bank credit. It checks progress or its lack, towards the objectives of profitability and corporate.

The budget plan is based on estimate, which may go wrong or change under pressure or circumstances. The budget must be kept continually adapted to fit changing circumstances; the execution of budget plans will not occur automatically. The budget will not take place without active participation in management and administration.

The best interest of any organisation can be served, only if the functional managers are able to see eye to eye and are prepared to subordinate their views to the overall interests of the company. Their harmonious relationship, which is very necessary for achieving the corporate goals, depends upon mutual confidence, trust, faith, and cooperation between the different functions. Successful interactions with the other functional heads will enable the materials manager to be a leader in providing an adequate supply of desired quality goods in the organisation.

TEN POINTS TO PONDER

1. Discuss the meaning of money and materials.
2. Deliberate the relationship between materials and finance.
3. Elucidate the process of finance helping materials in achieving profit centre approach.
4. Identify the type of questions asked by the financial executive.
5. Examine the scope of smooth auditing in materials function.
6. Deliberate on the cost reduction role by the function.
7. Explain the materials-finance interface.
8. Discuss the relevance of inflation on materials management.
9. Detail the mechanics of the materials budget.
10. Elaborate the steps to be taken when budget norms are exceeded in the case of capital items, raw material inventory, machine spare parts, inventory, semi-finished goods inventory and finished goods inventory.

57. Make-Buy Decisions

IMPORTANT CONSIDERATIONS OF MAKE-BUY DECISIONS

One of the crucial questions to be faced by the materials manager is whether to *own* an equipment or *lease* it. If the decision is to own, then, the question arises as to whether to make the item in the factory's workshop facilities or buy the same from the vendors. The tendency all over the developed countries is to buy/subcontract as much as possible.

The influencing factors in formulating make/buy policy include the following:

1. Availability of the plant facilities
2. Cost of manufacturing items
3. Time for manufacturing
4. Price of buying
5. Total lead time for buying
6. Quality level
7. Possibility of exposing process trade secrets
8. Attitude of work force
9. Procedures for buying
10. Procedures for fabricating
11. Anticipated discounted cash flow of funds to be invested
12. Anticipated useful life of the equipment
13. Specialised skills available in the plant
14. Production capacity
15. Technological innovation
16. Seasonal/cyclical sales
17. Employee goodwill
18. Product guarantee/liability
19. Value of the component as compared to the volume of business
20. Excess factory overhead or creeping overhead
21. Manufacturing stagnation
22. Competitive market
23. Reliability of suppliers
24. Quest for challenges and professional excellence.

These factors have to be evaluated by the cost accounting and purchasing department.

REASONS FOR MANUFACTURING

1. The existing capacity is grossly underutilised, resulting in huge overheads.

2. Manufacturing fits the know-how, tradition, equipments and expertise available in the organisation.

3. There is only one source capable of manufacturing outside and dependence on a single source may be dangerous.

4. The design of the part or its processing is confidential.

5. It is difficult to transport the equipment as it contains fragile parts and is over-dimensional consignment.

6. Making in the plant will facilitate control of part changes, inventories and delivery schedules.

7. Direct supervision is needed to make the complex part and assume its control.

8. Idle capacity is available to absorb the excessive overhead.

Supplier Failures

9. The suppliers are unreliable and do not adhere to their commitment.

10. The company needs special quality of materials, but only in small quantities. In practice, no supplier is willing to accept an order with such conditions. If he accepts, he fails to deliver the goods or charges exorbitantly.

11. When the organisation is in competitive market, it has to work on right production schedules and the production programme to meet the exigencies of competition. The parts, for which the organisation is dependent upon outside suppliers, may not be supplied at short notice, as the suppliers in turn have their own unknown lead time problems.

12. Tax considerations can often be a major factor when materials are purchased: Often, excise duties or sales tax or octroi can make a considerable difference to the ultimate cost of a product.

13. There might be a desire to maintain a stable work force during periods when sales have declined.

14. There might be a desire to integrate plant operations like vertical integration or downstream integration.

15. The item may be closely allied to some other components made by the company. Its manufacture may closely fit and suit the company. Its manufacture may also suit the know-how available in the company. The available skill-mix of the company can be fruitfully deployed in making the product.

16. The raw material required for producing the items is readily available at economic rates.

17. No additional tools, jigs, fixtures, dies, paints, etc. are required to make the item.

18. If the item is purchased, instead of being made, it would render people surplus.

19. The genuine costs of the supplier are not known to verify his quotation.

20. Where complex materials requirements are hard to produce, a firm may choose to assume producing the same.

21. A hostile union may seize the opportunity to irritate the management as a result of the decision to buy an item previously made in the organisation.

PREFERENCE FOR BUYING

1. The trend all over the world is to buy the components and concentrate on the main business.

2. The statement that "it is always cheaper to manufacture than to buy", is highly erroneous particularly in large undertakings.

3. If the plant capacity is used to excess, overcrowding and operating inefficiencies may lead to additional variable cost.

4. Rerouting of materials through the production process may disrupt load levels and scheduling on the machines.

5. Long efficient production runs may be replaced by shorter changing cycles, increasing the set-up costs and overheads.

6. Detailed cost studies indicate, it is cheaper to buy than to make.

7. Space, equipment, and time are not readily available, at competitive rates, to commence the production operations.

8. The investment in infrastructure facilities of inspection and tooling is not attractive due to small quantity requirements.

9. The risk involved in the fluctuating demand due to seasonal and cyclic factors must be faced by the suppliers.

10. The need for special techniques, skills, and facilities, make buying more logical.

11. The top management wants to concentrate on its own line of speciality.

12. Patents as well as buyer-seller relations favour buying.

13. If a decision to manufacture the items is made, chances of buying the same items from outside become remote as suppliers lose interest. To avoid such a situation, many firms favour buying the item and developing more sources of supply to keep up the competition.

14. The company favours standardisation, whenever possible. Such standard items can be purchased profitably than making them. Usually there are more than one supplier for such items and in that situation, subtle negotiating techniques of the experienced buyers will be of great help.

15. On account of manufacturing certain items required by Industry, some suppliers develop expertise. Indeed, with a view to keep ahead of their competitors, they also undertake research to improve their production and specialise in manufacturing those items. The companies requiring those items may well consider the advantages of this research, expertise and cost against a decision to make it.

16. All Government departments in our country have a decided policy to promote small entrepreneurs. They encourage development of ancillary and small scale industries. They expect large industries to sub-contract their various requirements instead of making most of the items themselves. In this climate, it is necessary to think twice, before launching production of a part or material required. Even the permission to make such manufacturing programme may be denied.

17. The existing plant facilities are capable of being used more profitably to make other components.

18. Patents or legal barriers prevent the company from making the components.

19. It is possible to have a wide selection of the item from the supplier.

20. The need for special techniques or equipment makes buying logical. The suppliers may help to produce a better-engineered product, as he is a specialist in the manufacture of items.

21. Credit regulations, taxation policies, withdrawal of investment allowances, etc., force the company to buy from outside.

22. Equipment as well as personnel, that can perform a variety of different jobs, provide the internal flexibility required to consolidate or split work among different production areas as business fluctuates. External suppliers can be used as buffers to absorb the shock of production fluctuations and absenteeism of workers.

ILLUSTRATION ON MAKE/BUY WITH COSTS

A company is currently manufacturing Electricity Meters and Radios. It has its own fabrication shop where there are four coil-winding machines among other machines. The company has capacity to produce (or assemble) 360,000 single phase electricity meters, and 40,000 radio receivers per annum. But the four-coil winding machines are not enough to handle the load of winding coils for the electricity meters and radios. Other details are as follows:

If 360,000 electricity meter coils are manufactured by the company, then only 10,000 coils can be wound for radios, compelling purchases of 30,000 coils for radio manufacture.

If 40,000 radio transformer coils are wound by the company then only 350,000 electricity meter coils can be wound and the balance of 10,000 coils for electricity meters have to be purchased.

Cost data

Cost of coil winding for electricity meters per unit:

Raw materials costs (variable)	Rs 1.50
Direct labour costs (variable)	Rs 1.00
Overhead costs (variable)	Rs 2.00
Overhead costs (fixed)	Rs 0.75

Price that has to be paid if this unit is purchased is Rs 5.00 per unit. Costs of coil winding for transformers used in radio:

Raw material costs (variable)	Re 0.40
Direct labour costs (variable)	Re 0.20
Overhead costs (variable)	Re 0.10
Overhead costs (fixed)	Re 0.10

Price that has to be paid for this unit if purchase is Re 0.75 per unit. Now the alterantives are:

(a) Make 10,000 coils for electricity meters and buy 30,000 coils for radio manufacture.

(b) Make 30,000 coils for radios and buy 10,000 coils required for the manufacture of electricity meters.

The variable costs of alternative (a) are:

Variable costs of making 10,000 Electricity meter coils @ Rs 4.50 per coil would be	Rs 45,000
Plus variable cost-purchasing 30,000 coils for radio manufacture @ Re 0.75 per coil	Rs 22,500
Total variable cost of alternative (a)	Rs 67,500

The variable costs of alternative (b) are:

Variable costs of making 30,000 coils for radio manufacture @ Re 0.70 per coil	Rs 21,000
Plus variable costs of purchasing 10,000 coils for electricity meter manufacture @ Rs 5/- per coil	Rs 50,000
Total variable costs of alternative (b)	Rs 71,000

The alternative which will cost less is obviously alternative (a) of making 10,000 electricity meter coils and buying 30,000 coils for radios which saves Rs 3,500 (Rs 71,000 – Rs 67,500).

Continuous Review

The above illustration is an oversimplification of the issue. Make or buy decisions are far more complicated and involve various other factors and considerations. The top management, in coordination with materials, costing, design and industrial engineering functions — must evaluate the quality, quantity, price, service and other aspects in the make/buy decision process. The process is a continuous one, as the effect of each decision on the corporate profitability, has to be studied. Delivery failures, improved quality, new products, price levels and general economic development have to be monitored to arrive at make/buy decisions. The application of make/buy principles should not be allowed to obscure the analysis and application of other management techniques like market analysis etc.

TEN POINTS TO PONDER

1. Identify the considerations involved in make or buy decisions.
2. Enumerate the influencing factors in make/buy policies.
3. Elaborate on the statement "In make/buy decisions the main six considerations relate only to cost, quality, quantity, service, skill mix and capacity".
4. Discuss the conditions conducive for making.
5. Elucidate the cases essential for buying.
6. The quality of half the number of suppliers for the purchased components, is below par. Suggest suitable action.
7. Examine the role of materials manager in make/buy decisions.
8. Comment on the statement "Most of the make/buy decisions are volatile in nature".
9. Criticize the statement "Make/buy decisions are applicable in assembly oriented-engineering organisations only".
10. "Some top management personnel prefer vertical integration of making all items in the company, while others prefer subcontracting to the maximum extent". Explain the pros and cons of the above.

58. Equipment Leasing

BACKGROUND

Leasing has become very popular in India, in the recent past. Hire purchasing of transport vehicles, cars, two-wheelers, helicopters, residential flats, office accommodation, computers, copying equipments, charter hiring of ships, television, refrigerators, luxury items, generating sets, equipment machinery, etc. are examples. This type of financing has had a steadily accelerating growth in recent years, due to the fact that the user has begun to recognise that an asset produces the same profit or cost reduction, whether it is owned or leased.

Leasing as a financial instrument, was evolved in the United States and the first leasing company set up in 1952, was called the U.S. Leasing Corporation. Since then there has been a mushrooming of leasing companies all over the world including India. Today, about 300 leasing/hire purchase, finance companies, advancing about Rs 10,000 crores are operating in India.

Nationalised banks, non-banking finance companies, private firms and partnership companies collect deposits from the public exclusively for the purpose of hire purchasing business, leasing of assets and instalment buying of consumer durables. They also ensure a high rate of recovery of about 95 per cent and also the recycling of repayment against earlier loans.

CONCEPT OF LEASING

In recent years, the word 'lease' signifies a transaction of several years duration, which frequently approximates the depreciable life of the equipment. The ownership of the equipment being financed is vested with the lessor, but the right of use rests with the lessee. The payment for the lease is in the form of an initial *down-payment* or *front end fee*. The remaining amount is payable in the form of lease rentals payable at regular intervals, in advance, in terms of monthly/ quarterly or annual instalments. As such, the lessee does not have to put up a stake in the acquisition of the asset. After the fixed period of the lease, the lessee is given an option to renew the lease from time to time, at a nominal or a *pepper corn rental* for the next period of the lease. The lessee would have recovered his entire investment and profit thereon, i.e. if full payment has been done.

Once the terms of the lease have been negotiated to the satisfaction of both the parties, the lease agreement drawn up, giving the detailed mode of payment, frequency of rentals, etc.

Leases are usually for a basic term of an agreed number of months or years. Renewal or purchase options can be negotiated by the lessee.

Most leasing contracts provide that the lessee maintain the equipment and insure the equipment in favour of the lessor as well as the lessee. The applicable taxes on the based equipment must be paid by the lessee.

TYPES OF LEASES

There are various types of leases:

(a) In the *financial lease*, the lessor acts as a financing agency only. The lessee acting as the lessor's agent, selects the equipment, places the order, takes delivery, and maintains the same. During the agreed lease period, the rentals cover the cost of the equipment, interest charges and lessor's profit. These normally would be full payout leases.

(b) On the other hand, in *the operating lease*, the lessor company selects the equipment, pays for it and gives it on lease. As the lease rentals do not amount to full payment, the lessor company indulges in repeated leasing until the costs and expenses are covered.

(c) The *payout lease* implies that the lessor will collect payment which will be equal to his cost of purchasing the equipment so that it may be leased to the lessee.

(d) The *non-payout lease* means that the lessor will collect less than his equipment cost over the term and take the equipment back to release it.

(e) The concept of leasing can also be suitably extended to *second-hand equipment leasing* under some special conditions.

(f) When international trade is involved in leasing, the type of leasing resorted to is called *cross border leasing*.

A TO Z ADVANTAGES IN LEASING

(a) Leasing is a superior form of financing compared to outright purchase on account and has a number of advantages.

(b) It finances cent per cent of the acquisition cost and the lessee does not have to pay any capital acquisition cost.

(c) Where the lessee organisation is in a non-tax paying position, the cost of funding the asset, if it were owned by it would be significantly higher, as it would not be in a position to take advantage of the tax shield on interest payment. The implicit cost of funds is lower as a part of the tax benefit passes on to the lessee in the form of lower lease rental.

(d) The lease rentals can be so devised to ensure that they suit the cash flow of the lessee company. The payment can be made front-ended or back-ended accordingly.

(e) The lease agreement being standardised, obviates the necessity of rigorous loan agreements. It is made easier by the fact that new leases can be hooked on to the old.

(f) Apart from the invoice cost, various other costs including inspection, delivery, installation software, etc. could be added to the capital cost and amortised over the lease period. The higher equipment cost can be funded.

(g) As the ownership title vests in the lessor and does not pass on to the lessee, the title vests in the lessor himself. The lessee does not have to incur the burden of obsolescence.

(h) The lessee does not have to invest funds in capital acquisition and divert the funds generated for internal resources or from other avenues to more pressing problems of working capital.

(i) When the requisite resources cannot be fully funded by the user industries, leasing is useful.

(j) The profitability of the enterprise is increased through the leverage of additional debt for increased production, resulting in overall increased profitability.

(k) Sometimes, the manufacturers offer a lease plan to their customers directly or through an associated finance company and the seller often reduces his profit margin in this type of transactions.

(l) At times, lease brokers act as intermediaries between lenders, leasing companies, and leasing users. The broker determines the need of the prospective lessee and solicits bids from financing companies or banks.

(m) The entry of banks, in the leasing business, decreases the cost of leasing to the user.

(n) Those offering leasing service, may have access to sources of financing which may not be available to the user.

(o) It may be difficult for individual user of leasing to negotiate directly with investor trusts.

Contractual Terms

(p) A particular lease contract may permit the lessee to charge lease payments against business operating expenses for an amount, which is more than the depreciation deduction for purchased assets.

(q) At times when the Government allows investment tax credit, the lessor may pass a portion of the credit to the lessee.

(r) Government regulations may require a company to group several different assets in one group for which the guideline life is specified. The asset user may find that his particular machine has a life expectancy, less than the stipulated guidelines. Under these circumstances, the asset may be leased on a contract, tailored to actual experiences, for the purpose of tax benefits.

(s) The leased assets can be a part of, off-balance sheet financing, with the lease contract appearing only as footnotes. These companies acquire the assets without enforcement of capital budgeting requirements.

(t) Many organisations offering leasing services are willing to appraise the assets already purchased in the organisations and lease them back to the user.

(u) As lease contracts are being written for longer terms, an equipment user may be able to insure that his equipment usage cost is stabilised for a period of many years, as a hedge against inflation.

(v) If an user is contemplating leasing many different items of a plant, a master lease contract covering the entire series may be entered, decreasing the paper work.

(w) The credit worthy lessee can usually convince his lessor that no initial rental need be made, until the equipment is commissioned.

(x) If the lessor offers his lessee, a contract which permits cancellation of the lease on short notice without penalty, the lessee can hedge against obsolescence.

(y) Any manufacturer can boost his sales, by offering a leasing plan to his customers, without investing his own capital, in these long-term receivables.

(z) The leasing company's marketing department has an opportunity to approach the lessee towards the end of the basic lease, to suggest that an improved machine be leased to replace the existing machine, thereby facilitating the introduction of new products.

A TO Z PROBLEM AREAS IN LEASING

Let us now discuss the problem areas in the realm of leasing:

(a) Even with the scope of accepting deposits from the public and securing credit/term load from financial institutions, up to ten times of net owned funds, there is inadequacy of resources to meet the ever growing demand for lease financing.

(b) The interest or cost of borrowing to the using organisation has been rising unduly. The outstanding deposits constitute on an average about five times the net owned funds.

(c) The growth in deposits has been affected by the decision of the Reserve Bank of India that the leasing companies can accept deposits for periods of 25 to 60 months.

(d) The increase in the duration has inflated the cost of funds and decrease in deposits as the options before the investing public are numerous.

(e) The banks pay 8 per cent on six months deposits and those receiving interest income from this source can also take advantage of exemption provisions in the income tax.

(f) According to the hire purchase agencies, the existing procedures need modification, for granting refinance facilities atleast to the extent of 25 per cent of fresh advances — against hire purchase documents.

(g) It has also been suggested that the commercial paper can be issued for stipulated amounts after securing the required credit, as is being done by select industrial units.

(h) Many non-banking finance companies have been refunding deposits at the request of the constituents earlier than the maturity dates and interest is paid on a slab rate basis, but the monetary authorities are not happy over this type of pre-closure.

(i) Bank credit too must be extended more liberally, to the hire purchase companies, with a more liberal ratio than the present three times of net owned funds.

(j) Due to intensified competition from other leasing companies, many companies have not been able to secure the desired return on investment, through offering leasing services.

(k) There has been high mortality among supplier leasing companies, which were unable to secure the needed capital for equipment purchasing as their marketing activities expanded.

(l) Banks are becoming more aggressive in leasing. As the smaller leasing companies secure their finances from banks, the banks control the money cost to the lessor: and may eliminate the small leasing companies in the long run.

(m) Rate competition among the lessors is increasingly observed and this has decreased the profitability of small leasing companies.

(n) The aggressive marketing staff must bring not only new leases, but also secure the money required to purchase equipment needed by the lessee.

(o) It may be economical to lease the equipment when it is needed only for short period. On the other hand, most leases are for definite terms and terminable on short notice, it may result in depriving the leases of a needed facility.

(p) In general, in a lease, the lessee bears the dismantling loading, unloading, transportation, installation, and commissioning charges both ways from lessor's location to the lessee's and return. But in case of buying the equipment, it is sold at a delivered price.

(q) The rent of leased equipment is an item of expense for tax purposes while, purchasing is an investment to be depreciated over a number of years. Only when lease payments exceed the value of allowable depreciation and additional tax, then only benefit is accrued.

(r) The most important disadvantage is the lessor retains control of the equipment and not the lessee. This loss of ownership places restrictions in operations, and freedom for lessor to inspect the equipment.

(s) The lessor sometimes may insist on using the equipment with the materials supplied by his associated vendors, whereas the buyer would like to have his own prerogatives in the matter of suppliers.

Credit Appraisal for Leasing

(t) The lessor has to appraise the lessee's capacity to service the lease as well as its loan obligations for which purpose the debt service coverage ratio in a modified form to cover lease principal and interest repayment is inspected. Even though leasing is an off the balance sheet source of financing, the banker must appraise the future rentals on lease net of tax as included in the modified debt service coverage ratio. Lease rentals are always quoted on the basis of per 1000 of the cost of the based equipment per period.

(u) Where the lessee is a tax payer, then alternatives under the lease or loan option must be simultaneously considered. The lessee must consider the net present value of the outflows under both the options. From the lessor's point of view, the internal rate of return post tax value should be used to evaluate the after tax return to the lessor.

(v) In general, lease contract can be arranged for any asset, which has sufficient value to justify the paper work involved and sufficient durability to permit the lessor to realise salvage value/if the lessee fails to fulfil his contract.

(w) The potential lessee must exercise caution in selecting his lessor, so that he can be sure that the lessor will continue in business during the lease term.

Leasing Second-Hand Equipments

(x) Leasing of second-hand equipment of a company can be done on a sale and lease back basis. Such items can comprise equipments on which investment allowance has already been claimed by the leasing company and hence the equipments are not eligible for further investment allowance. The equipments will have to be necessarily not more than eight years old with a reasonable balance of useful life. Since the question of investment allowance does not arise on such second hand equipments, lease rental plan could be restructured to a period of five years across the board.

(y) The major issue in such sale and lease back is the sale price to the lessor and repurchase price to the lessee at the end of the lease period. Such sale and lease back arrangement has the advantage of reducing the net block, without affecting the productive use of the asset.

(z) As a general rule, the primary disadvantage of leasing is its cost. Interest rates are typically higher on leases than on direct loans, because the lessor has to bear all the risks, besides covering the financing charges, obsolescence, inflation, etc.

FUTURE OF LEASING

There has been a rapid growth of leasing cum hire purchase companies in India during the 1980's. Apart from companies like Sundaram Finance, 20th Century Finance and First Leasing in the private sector we have leasing companies started as subsidiaries or sister concerns of nationalised banks including SBI, Canara Bank, Bank of Baroda, etc. IDBI regulations prescribe that only those companies whose annual turnover exceeds Rs 10 crore, present Return on Investment, i.e., profits exceed Rs 1 crore and who have been in business for the past five years, can be allowed to start a leasing company.

Leasing laws in India relating to sales tax levy on lease and other aspects need to be better-structured so that leasing industry can make further rapid strides for contributing to the economic development of the country.

The resources of non-banking financial companies must be fruitfully employed, without any diversion of funds on a large scale, for creating long-term assets or diversifying activities. Then, the required funds can be mobilised in various ways with correct understanding of outstanding issues.

In the final analysis, the decision to lease/buy/make should be based on the evaluation of costs, consequences, funds availability, returns of alternative methods, technological upgradation and qualitative factors.

The Indian leasing industry has been contributing only 2 per cent of the total industrial investment as against 25 per cent in advanced economies. With the relief in the Tax Laws and incentives announced in the recent budgets, it is hoped that leasing will assume its full stature in contributing to the development of Indian economy.

TEN POINTS TO PONDER

1. Explain the concept of leasing.
2. Give examples of leasing business in India.
3. Examine the contractual aspects of leasing.
4. Identify the types of leasing methods.
5. Enumerate the advantages of leasing.
6. Describe the disadvantages of leasing.
7. Relate leasing with financial aspects of a company.
8. Define the concept of credit appraisal for leasing.
9. Elucidate the significance of second hand equipment leasing.
10. Discuss the role of leasing in future in India.

59. Insurance Management

NEED FOR INSURANCE

The major aim of any industry/business is to make profits and therefore it has to take some risks. The compensation is realised on potential profit. These commercial risks are self assumed and are not transferable to a professional risk carreer. The warehouse manager, who is incharge of transporting and storing the largest current asset of the organisation has to ensure the safety of the assets in order to protect their earning potential, particularly in today's context of increasing activities of lawlessness, arson, loot, pilferage, sabotage, fire, ethnic riots, communal troubles and terrorism. Insurance comes in handy in such risky situations. Even though the finance function is concerned with insurance, the warehouse manager should have a knowledge of the costs and consequences of Insurance Management.

A TO Z OF INSURANCE POLICIES

(a) In life-insurance, the subject-matter insured is human beings and as such there is no difficulty either in underwriting or claim settlement of such business. The sum insured is an agreed value, which is generally decided, based on the premium paying capacity of the insured.

(b) But most of the general insurance policies are contracts of indemnity covering different types of properties, commodities cash or statutory liabilities against certain perils.

(c) Most of the Government departments like DGS & D do not believe in insurance as, according to them, the loss is negligible, even though general insurance is handled by another Government undertaking.

Computation of Costs

(d) The basis of insurance is the transfer of risk, at a predetermined cost, related to the risk to the insurer or the under-writer, who undertakes to assume the risks. The insurant has to pay the premium or the cost related to the risk to the insurer. There are many types of risks covered, but the common risks insured are losses due to fire, pilferage, in-transit damages and other liabilities such as personal/ vehicle accidents.

(e) Insurance contracts are contracts of indemnity and premiums are based on the law of averages, as stipulated by the Tariff Advisory Committee. The average cost of insurance is about three per cent.

Loss Minimisation

(f) The computation of the cost of the risks involved is based on statistical and acturial principles. The average total of insurance cost of transit and warehouse in industry is about three per cent of the average value of the stocks, while its range fluctuates from 1.5–8 per cent.

(g) By suitable precautionary measures, the warehouse manager can reduce the total loss. The nature of packing, the mode of transport, installation of fire extinguishers, the extent of care exercised in materials movement and the storage of materials, have a bearing on the loss minimisation. Whenever there is an improvement in any of these, the insurers should be persuaded to reduce the premium.

(h) From the point of cost reduction, self insurance can be progressively introduced, by first excluding claims for less than a specified amount.

Preventive Steps

(i) Prevention is better than cure. Hence insurer must take the following steps:

(i) Prevent documentary frauds.
(ii) Prevent losses during inland transit and storage points.
(iii) Provide proper warehousing and godown facilities.
(iv) Ensure prompt delivery at ports.
(v) Insist on good markings.
(vi) Avoid old vessels and prefer liners.
(vii) Appoint reputed clearing agents.
(viii) Verify the financial stability of the charters.
(ix) Avoid ports with chronic congestions.
(x) Ensure that the charters do not sell your cargo to avoid demurrages.
(xi) Keep documents on letter of credit, bill of lading, commercial invoice, insurance certificate, and certification of original packing.

Indemnification

(j) In the event of loss or damages, the insurers will indemnify the insured for the expenses necessary to restore the damaged machinery to its condition immediately before the occurrence of the damage. The expenditure also comprises, disassembly, freight, and reassembly costs. The value of scrap or parts, which are still usable is deducted from the amount of indemnity.

(k) The insurer has to bear a certain amount of loss from his own account. This deductible relievés the insurers from minor losses where the administration expenses required would not be in an adequate proportion to indemnity. The deductible ensures the insurer's interest in loss prevention and to reduce the insured's administrative costs.

(l) By opting for a higher deductible than normal, the insured can generally get a reduction in the premium. Large industrial houses and Public Sector Undertakings opt for higher than normal excess and secure substantial premium reduction.

(m) By indicating alternative quotations with different deductibles and corresponding premium rates, the insurers are able to offer cover which is geared to the respective conditions and allows the insured to select the extent, to which he is willing to become involved in bearing the losses.

Fire Insurance

(n) Fire insurance covers fire, lightning, water damage, firebrigade damage, falling of walls, damage during removal and damage due to rain after removal. The basis of rating depends on the type of construction, occupation, commodity, warranties, and fire extinguishing appliances.

(o) The items covered include the following in the case of fire insurance: (i) Fixed assets in the form of buildings, fixtures, fittings, pipelines and compound wall; (ii) Movable and immovable machinery, electrical installations, boilers, generating sets, air conditioners, cooling systems etc.; (iii) Stock and trade in the form of raw materials, finished goods, in-process material; (iv) Goods held in trust or commission, bullion, works of art exceeding a certain value, rupee notes, stamps and explosives are excluded from this cover; (v) Policy A provides cover for simple risk, Policy B is restricted to dwelling houses, shops, and Policy C gives an extensive cover.

(p) The basis of rating depends on the construction classification of building as provided by the Tariff Committee and also the combustible nature of the item and open warehousing. As per the principle of indemnity, the market value of the property should be the sum insured for building and machinery and the cover can be taken on the reinstatement value basis.

(q) Fire policies are not valued policies.

(r) Usually *declaration policies* can be adopted for fluctuating values of stocks. This policy is for the maximum possible amount but the amounts are declared monthly. A provisional premium at 75 per cent is paid and adjustments are made on the basis of average declaration. Minimum retention is fifty per cent. Claims are settled for the cost of rebuilding or replacement and hence the sum assured must be related to these costs. Payment is based on actual reinstatement.

Consequential Loss Policy

(s) Properties at more than one location, but in the same town, can be covered under *floating policies* for a single sum insured. The rate is according to the highest rated property. Policies can also be taken for loss of profit due to material damages. Loss of profits is more substantial than material damage, because it affects the earning power of the business and hence is necessary for complete protection. But this policy is issued only if material damage policy has also been taken.

(t) Fire policy C — covers perils like lightning, explosion, implosion, impact, riot, storm and flood. The policy proves indemnity to the insured in respect of the pecuniary losses suffered by him due to loss/damage to the insured property due to insured perils. The after effects of a serious fire or any other perils covered under fire policy are far-reaching and the consequential loss due to the fire are more than the material damages. Such consequential losses are due to the partial or total interruption of the industrial activity. Such interruption causes reduction in turnover, resulting in not only loss of net profits, but expenditure unnecessarily incurred to meet standing charges which continue despite the reduction or cessation of the activity carried on.

(u) Apart from insurance protection against the loss of net profit, the fire may result in the insured being unable to utilise a section of employees. Such compensation can also be covered under the CL — *Consequential Loss insurance* scheme. The terms *gross profit*, *net profit*, and *consequential loss* are defined in the policy itself.

Under Insurance

(v) A specimen list of usual standing charges, helps to ascertain whether any particular charge will be reduced in proportion to the reduction in trade. The specimen list includes (i) Interest on loans, (ii) Bank charges, (iii) Rent, rates and taxes, (iv) Duties, licences and patent fees, (v) Director's fees, (vi) Pensions, (vii) Legal audit fees, (viii) Insurance premia, (ix) Advertising, (x) Travelling expenses, (xi) Dividends, (xii) Heating, lighting and power charges, (xiii) Depreciation, (xiv) Repairs and renewals, (xv) Bad debt service, (xvi) Donations.

The indemnity provided will be commensurate with the sum insured under the policy. The clients declare the estimated annual profit while taking insurance. While ascertaining losses, the adequacy of the sum insured is examined by considering the turnover maintained for the 12 months preceding the date of loss and multiplying the gross profit rate. The amount thus arrived at is the gross profit which should have been insured; but if the sum insured falls short of this then, *under-insurance* is applicable.

Marine Insurance

(w) Commodities from surplus areas flow to deficit areas within the country or in the world. The basic principles of indemnity contribution, good faith etc. are applicable to sea-borne cargoes.

Marine insurance is divided into hull and cargo insurance *hull* deals with insurance of ships and their liabilities, *cargo* deals with the transit of materials of all types, both internal and international. The rates are fixed on the basis of actual total loss averages in marine insurance. Due to the peculiarities of goods in transit, the general principles of insurable interest at the time of claim have been modified. The marine policy is a valued policy where the principle of indemnity is allowed to be modified on an agreed basis, i.e. the policy can cover the invoice value of the cargo, the amount of the import duty and a reasonable amount of profit and incidental expenses.

The insurance of duty and profits is available only for the holders of the replacement licence. An *open insurance scheme* is available where a total sum is assurance for a particular period and as the despatches are made, the value is debited. As and when the total despatches equal the sum assured, a new contract is made. Adequate care has to be taken in the choice of transport ships in view of the large number of maritime frauds taking place in the world.

The standard policy has provisions for durations and the periods covered. The duration period is the time spent on board up to safe landing, i.e. port to port. The perils covered are the perils of the sea such as fire, pirates, war, jettison, marine risks, heavy weather, collision and stranding.

Internal Transit

(x) In the case of internal transit, rail, road and air transit of materials can be insured from loading time to seven days at the destination. It is always desirable to insure all transit materials, including imports and exports with Indian insurers as this not only allows speedy claims settlement but also conserves the scared foreign exchange. The transit insurance provides insurance cover from the time the goods leave continues during the ordinary course of transit and terminates either at the place of storage or on the expiry of 60 days after completion of discharge from the overseas vessel at the fiscal port of discharge.

Motor Insurance

(y) Motor insurance is done for private cars, commercial vehicles, motor cycles, specialised vehicles. The three policies are ACT, third party cover and comprehensive cover. *ACT* covers body injury or death of the third person and is essential for driving the vehicle. The third party cover includes damages of the property of the third party. The *comprehensive policy* covers the damages to vehicles and the legal dues to an employed driver. Other passengers are not covered by this insurance. Discounts in premium are offered for fleets and members of automobile associations. No claims bonus is a speciality of motor insurance.

Burglary Insurance

(z) Burglary insurance is available for business premises and is relevant to warehousing. Loss of cash in transit, cash in safe, and declared materials can be covered under this. Only forcible violent entry is covered under this. A fidelity guarantee policy is usually taken to cover pilferages by employees. Items like heavy castings which cannot be pilfered need not be insured.

Storage Cum Erection Insurance

Engineering insurance is available for machinery erection, testing, breakdowns, boiler explosions and loss of profit. Specialised risks in manufacturing industries are covered under this scheme. The basic concept of storage cum erection insurance, is to offer comprehensive and adequate insurance protection against all the risks involved in erection of plant and steel structures. This policy is a child of the process of industrialisation and is of recent origin. Often the conclusion of such an insurance is rendered compulsory by financial institutions granting funds for the project so that the financial institutions are assured of a measure of property. It is imperative that both marine/transit and storage cum erection covers should be taken so that a comprehensive cover from supplier/ warehouses to final commissioning of the plant is available. If the supply contract for project erection to the insured provides for a maintenance period, it is possible to grant maintenance cover for this period; under erection risk insurance policy includes maintenance visits cover extended maintenance cover and maintenance guarantee cover.

TEN POINTS TO PONDER

1. Elaborate on the need for insurance to the warehouse manager.
2. Identify the factors relevant to computation of insurance premium.
3. Explain the concept of loss minimisation.
4. Discuss the preventive steps to be taken by a company.
5. Describe the meaning of indemnification.
6. Examine the relevance of fire policy in the warehouse.
7. Articulate on the term consequential losses.
8. Give the details of marine insurance.
9. What is meant by the term *under-insurance*?
10. Amplify the terms inland transit insurance, motor insurance, burglary insurance, storage cum erection insurance.

60. Claims Management

INSURANCE CLAIMS

The warehouse manager has sometimes the responsibility to make claims on the insurance companies or transport companies for the cargo lost/damaged in the storehouse or in transit.

Burglary Claims

An immediate notice of the fire or burglary must be followed by details. Either the insurer or the insured may appoint qualified surveyors who will verify the loss, observance of warranties and the claim. The payment will be only for the value of the assessed losses on a pro rata of the sum assured. This is to protect the insurance companies against under-insurance. In burglary insurance, the sum insured is reduced on payment of claims. Reinstatement is possible by payment of pro rata premium. In the case of motor insurance, the repairs have to be authorised and a satisfaction voucher must be given. Third party claims are not negotiated, rather they are paid according to judicial claims.

Marine Cargo Claims

The surveyor determines the cause and quantum of loss in the case of marine insurance. The surveyor has the right to ask for all relevant data, records, reports, etc. for early submission of survey report. The survey fees are reimbursed if the claims are admitted. The basis of payment is the gross damaged value as a percentage of the gross insured value. The salvage/damaged property cannot be abandoned to the insurer. A new exclusion clause, has been added recently by London Underwriters, which is as follows: "Loss or damage or expense arising from insolvency or financial default of the owners, managers, charters or operators of the vessel is on excluded peril." It is advisable to consider insuring the cargo under cargo clause 'A' policy which gives full protection.

Contracts and Claims Settlement

Imports and exports depend on the types of contracts, customary trade practices of different countries. Political, economic, social, regional, ethnic and religious environment of the countries play an important role in this process. The other factors in marine claims settlement include port conditions, labour relations, warehousing facilities, packing standards, efficiency of shipping, role of clearing agents, financial honesty of exporters and shipowners, availability of loss assessors, claim settling agents in foreign exchange, bilateral relations between the two countries and political conditions on the route of the voyage. Foreign exchange fluctuations, shifting of insurable interest from seller to buyer, statutory port regulations, customs and inland transport facilities, at times may create problems for prompt claims settlement.

CLAIMS MISMANAGEMENT

Many claims are lost because of non-clarity of information, delays in furnishing claims and a misunderstanding of the warranty clauses. The claims cell should facilitate the timely intimation of the damages/shortages to the concerned agencies. The overall objective of claims management should be to reduce the total insurance bill, and improve the insurance cover. Self insurance of small items and restricting insurance only to catastrophic risks may be considered.

MARITIME FRAUDS

The transport and warehouse manager should be aware of the frauds which can occur in the international maritime business. These are due to (a) documentary frauds, (b) packing frauds, (c) fraud by carriers, and (d) fraud by shipping as well as clearing agents. The documentary frauds can be due to letter of credit, invoice, clean bill of lading and loss of receipt. The packing fraud arises because of inferior materials, short packing and despatch of damaged goods.

Fraud by Carriers and Agents

The problems attributable to carriers arise due to the following:

(a) Unscrupulous ship owners, charterers and operators.

(b) Diversion of the ship to certain ports where goods are purchased without supporting documents.

(c) Scuttling of the ship.

(d) Selling of the cargo by the master and crew of the ship.

(e) Pilferage during inland and sea transit.

(f) At times, due to negligence, the short landing and damages are not noted in the log book.

(g) The unloaded cargo, due to negligence by the clearing agents, is damaged or at times not traced.

(h) Bad warehousing or storing in open without tarpaulin covers results in major losses at the ports.

(i) Pilferages at the port, customs, warehouses, are common features.

(j) A pre-despatch survey for interior transportation may protect the rights of subrogation and claiming refunds, etc.

(k) In the case of exports, the losses are to be established and assessed in a foreign country and the goods are authorised by the ports, if prompt actions are not taken.

A TO Z OF CLAIMS SETTLEMENT

(a) Proper planning, documentation and follow-up of the movement of goods till their destination can reduce losses.

Settlement of Marine Claims

(b) Prompt notice must be given in writing, furnishing the policy declaration number, description of goods or short delivered to the insurance agents, before making a claim.

(c) The place, address and the person to be contacted for assessing the loss must be given.

Documentation

(d) The name of the shipper, name of the steamer, address of the clearing agent, and date of arrival of goods at the port of discharge.

(e) Date of delivery of goods, whether open delivery, if taken with short landing or damaged certificates, must be notified.

(f) Notice must be given to carriers within 72 hours about the ship survey status.

(g) The details of log book particulars must be kept ready. The information as to whether a claim is made on the carriers by a registered notice of quantifying the loss within such days, must be provided.

(h) The other documents include bill of lading, shipping bills and protest letter to the master of the ship.

(i) The claim forms with duly discharged original policy along with the survey report and surveyors bills, must be given.

(j) Copies of the letters of the claims, made against the vessel, railways or other carriers must be forwarded.

(k) Letters of subrogation and power of attorney must be prepared.

(l) If customs duty is paid in advance, bill of entry duly signed by the customs officer and port officer, must be presented.

(m) Exchange rate on the date of loss should be shown.

(n) The salvage value of the goods may be indicated.

(o) Even if the claim is pending for settlement, the warehouse manager can take appropriate steps for recovery from the carriers.

Settlement of Fire Claims

(p) Written notice with full details and policy particulars must be given within 15 days.

(q) Take all possible precautions to minimise the loss by means of fire protection, segregating the sound goods from the partly damaged/fully damaged goods. If possible, try to separate the water damaged goods from the sound ones immediately and dry them.

(r) Before the loss assessor reaches the premises, do not disturb the debris. Take photograph of the scene of accident and cooperate with insurers, surveyors, fire brigade, police, customs and other statutory authorities.

(s) Check up the policy particulars itemwise in order to ensure that the cause of loss is due to a peril covered in the policy and the sum assured is adequate.

(t) Verify whether the damages are covered by any other policies.

(u) Write to the financiers the details of loss and send a copy to the insurers.

(v) At the time of the survey, involve all the concerned heads of departments to furnish the relevant details to the surveyors and officials.

(w) Fill up the claim forms at the earliest and furnish all supporting dates, stock books, plans, invoices, bank accounts, sales registers, excise details and customs records.

(x) If necessary, request the surveyor to send a preliminary report and ask for an 'on account payment' based on market value.

(y) A regular follow-up action up to the time of settlement in a fair and reasonable manner is necessary. In case of dispute on quantum only, take advantage of the arbitration clause.

(z) For any reason, if the claim is not settled within 12 months, the limitation period should be extended by the insurer in order to prevent time barring of the claims.

Settlement of Railway Claims

The railways have laid out a series of contract clauses: S73, S74, S75, S76, S77, S78, S80, S82, S151, S152, and S161. These clauses outline the care to be taken by the bailee, bailees responsibility, responsibility of railway administration, responsibility of delays, deviation of routes, wrong delivery responsibility of railways in case of damaged goods, exoneration of claims, notification of claims, application of compensation, jurisdiction of claims, tribunal rights to legal representation and appeals.

Section 152 states that the bailee, in the absence of any special contract is not responsible for the loss, destruction or deterioration of the thing bailed, if he has taken care relating to bulk, quality, value, packing and the goods bailed. The Railway administration will be responsible for the loss of goods except in the operation of *force majure* clauses.

About 5 lakh claims are made every year and the railways settle about Rs 5 crores on 0.1 per cent of the total value of freight carried. The procedure and process of documentation are the same as in fire and marine transit policies.

TEN POINTS TO PONDER

1. Define the concept of marine cargo claims.
2. Discuss the relationship between contract and claims settlement.
3. Evolve steps to reduce claims mismanagement.
4. Identify the maritime frauds due to documentation.
5. Detail the frauds by carriers and agents.
6. Enumerate the steps for settlement of marine cargo claims.
7. Explain the procedures of settlement of fire claims.
8. Describe the settlement of railway claims.
9. Elaborate on the precautions to be taken for speedy settlement of claims.
10. Examine the role of warehouse managers in the settlement of claims.

61. Pricing of Issues and Stock Valuation

IMPORTANCE OF VALUATION

Materials account for about 65 per cent of the final product cost and inventories form the single largest asset in any organisation. Hence, the pricing of materials and stock valuation affect not only the product pricing, but also the profitability of the organisation. Stock valuation enables an organisation to know the financial strengths and weaknesses. Valuation of the inventories converts the physical quantities into monetary figures and becomes very crucial in judging the organisation's performance.

The stores personnel are normally not burdened with the valuation of stock or pricing the issues, as they are responsible only for the physical numerical count. But it will be to their advantage if they know the methods so that they can have an idea of how much funds they are responsible. Valuation of stocks has to be done on two occasions: (a) once, while issuing to the consumers to enable the cost of production, and (b) secondly, when preparing the periodic performance statements to value the stocks that are on hand. Moreover, by suitably altering the methods adopted, an organisation can paint a rosy picture of its position. For instance, gross profit is obtained by deducting the cost of materials consumed from the net sales; in a bad year, a higher profit can be returned by adopting a lower value for the cost of materials consumed.

VALUATION STRATEGY

The strategies of valuation of the materials range from the conservative practices such as market price or cost of procurement whichever is lower on one hand, to modern developments, such as the replacement costs on the other. Obviously, the method of valuation of stock will affect the pricing of issues as well. The methods of valuation adopted by the organisations include the following: (a) FIFO — First in first out; (b) LIFO — Last in first out, (c) NIFO — Next in first out; (d) HIFO — Highest in first out; (e) Base stock, (f) Simple average, (g) Periodic simple average, (h) Weighted average, (i) Periodic weighted average; (j) Standard cost, (k) Replacement price. Each method has its own advantage and disadvantage.

For income tax purposes, only one of the following methods, i.e. the FIFO — First in first out system — or the standard price with a reconciliation statement, or the actual price or the market value whichever is lower is accepted. Under the Companies Act, 1956 a change in the method of valuation has to be informed along with the effect on the reported profit, as a result of the change. This is done by means of a footnote in the profit and loss account.

We shall evaluate some of the methods by means of a simple transaction of an item in a month available in Table 61.1. The most common methods adopted in practice are FIFO, LIFO, weighted average and standard cost. We will confine our attention to the raw materials only, whereas the same logic holds good for semifinished items and finished goods.

Table 61.1 Transactions in a Warehouse

Date	Transaction	Price/Unit (Rs)	Quantity Units
10 Jan	Receipt	8	1000
20 Jan	Receipt	9	1000
23 Jan	Issue	—	1500
28 Jan	Receipt	10	1000
31 Jan	Issue	—	1000

The FIFO Method

The FIFO (First In First Out) method assumes that whatever may be the physical practice, materials that come first are issued out first. Hence the valuation method values these according to the latest prices paid. The working is explained in Table 61.2.

Table 61.2 Entries in FIFO Method

Date	Receipts			Issues			Stock	
	Qty.	Rate	Value	Qty.	Rate	Value	Qty.	Value
10 Jan	1000	8	8000	—	—	—	1000	8000
20 Jan	1000	9	9000	—	—	—	2000	17,000
23 Jan	—	—	—	1000	8	8000	500	4500
28 Jan	1000	10	10,000	500	9	4500	1500	14500
31 Jan	—	—	—	500	9	4500	—	—
	—	—	—	500	10	5000	500	5000
Total	3000	—	27,000	2500	—	22,000	5,500	49,000

The final stock is 500 units with an inventory value of Rs 5000. This method ensures that the materials are issued at actual cost and hence there will be no profit or loss intrinsic to the method. The operation is simple, as long as fluctuation in price is not much. The number of entries in the issue column will increase as the prices vary more, or the transactions are increased, resulting in arithmetical mistakes. In a fluctuating market, every batch or monthly production will have different materials cost, making comparisons of performance, a difficult process.

In a falling price market, the materials costs of production will be shown higher and in an inflationary situation, the materials cost will be lower showing increased profits. The costs charged to production lag the actual prices by this method, thus impeding quick decision-making on price changes in a competitive environment. But the value of materials in stock does not vary significantly from the market. It is also difficult to issue materials like heavy plates, stacked one over the other, on first in first out basis, as it will result in double handling of the material.

LIFO Method

The LIFO (Last In First Out) method assumes that the latest material that has entered is issued first. This implies that the stocks on hand are valued at older prices. The working of LIFO method is shown in Table 61.3.

Table 61.3 Entries in LIFO

Date	Receipts			Issues			Stock	
	Qty.	Rate	Value	Qty.	Rate	Value	Qty.	Value
10 Jan	1000	8	8000	—	—	—	1000	8000
20 Jan	1000	9	9000	—	—	—	2000	17,000
23 Jan	—	—	—	1000	9	9000	500	4000
28 Jan	1000	10	10,000	500	8	4000	1500	14,000
31 Jan	—	—	—	1000	10	10,000	500	4000

The resultant stock is 500 units with an inventory value of Rs 4000. In this method, the production is charged at the latest price, so the production costs reflect the latest price and hence the product cost reflects market conditions so long as the receipts are recent. Hence, price changing decisions can be taken more realistically. In an inflationary situation, LIFO charges on production are high and inventory values are low resulting in a higher book profit. On the contrary, when prices are falling, the effects will be just the opposite. It is difficult to apply LIFO method for perishable items and material with limited shelf life, as it is necessary to consume first, the lots received first. Here also the entries are more if the transactions are more, resulting in arithmetical errors.

Simple Average Method

In the simple average method, the prices paid for the purchases from which the issues are made, are averaged to price the entire issue. In the example, the average price is taken as (9 + 8 + 10/3) = 9. To give an even weightage to the actual price paid, the time of purchase and the existing market conditions, the average price method of valuation can be used. In this method, the prices paid for purchases, from which the issues are made, are averaged to as Rs 9 to price entire issues. This method is very simple to operate and give reasonably valid figures when the prices do not fluctuate much.

In periodic average pricing method, the prices paid during a period, say, a year are taken and the calculations change from one period to another. This method can be used advantageously for batch type of production by suitably adjusting the periods. It should be mentioned that errors are possible due to approximations in calculations. Since the valuation is not at cost, the closing stock value will be only an approximate value. For example, FIFO shows the closing stock as Rs 5000, whereas the average price will show (50 × 9) = Rs 4500 as the closing stock value.

Weighted Average Method

The simple average methods do not consider the quantities purchased and this can give rise to differences which have to be reconciled. It is also reasonable to expect the price paid for a large quantity of items to have a comparatively greater weightage in determining the valuation rate than the price paid for a smaller quantity. To meet this requirement, weighted averages of the prices are taken as the basis, the weights being the quantities that are bought at these prices.

In this case, the differences are negligible and the price follows the market quite closely. If there is a large number of items, computerisation may be necessary to have an updated issue price. Many Indian organisations have adopted this method. Weighted average price can also be calculated periodically, once in a year or so, and this method is then known as periodically weighted average. The strategy should be to calculate uptodate price for high value items and periodic prices for low value items. The working of weighted average method is given in Table 61.4.

Table 61.4 Weighted Average Entries

Date	Receipts			Issues			Stock		
	Qty.	Rate	Value	Qty.	Rate	Value	Qty.	Rate	Value
10 Jan	1000	8	8000	—	—	—	1000	8.0	8000
20 Jan	1000	9	9000	—	—	—	2000	8.5	1700
23 Jan	—	—	—	1500	8.5	12,750	500	8.5	4250
28 Jan	1000	10	10,000	—	—	—	1500	9.5	14,250
31 Jan	—	—	—	1000	9.5	9500	500	9.5	4750

The closing stock of 500 units is calculated at the rate of Rs 9.50 and the value is Rs 4750. Thus from Tables 61.2–61.4 we note the differences in the closing stock values.

Standard Cost

Usually the performance of purchasing is judged against the standard price. The entries in the standard cost method are simple, where all the issues and stocks are valued at the standard cost. The standard cost is estimated at the beginning of the budget period. This takes into account the market conditions, need, demand, supply availability, storage facilities, handling, etc. Thus in the previous example if Rs 9 is the standard estimated, then the finished stock is Rs 4500, with the total purchase as Rs 27,000 and the total material issues will be Rs 22,500. This method is simple to operate and pricing reflects the market conditions and hence this is a popular method.

Other Methods

Variations of FIFO and LIFO result in HIFO highest in first out and NIFO next in first out. These methods are only of academic interest and not practised. The replacement price method has a multiple time forecast as its basis. Here each issue is priced at the cost that would be incurred to replace it in stock. However, this is very difficult to implement and can be only used for the very high value super A items for purposes of internal control. This method follows the market price without lag. NIFO can be a close approximation to the replacement price method.

Base stock method is similar to FIFO except that a minimum stock is held at its basic price. This minimum quantity depends on factors such as cost, criticality, availability and market conditions. This minimum stock is dipped into only in case of emergencies, and hence the value of base stock will always be available in the warehouse.

Pricing Issues and Pricing

Data must be collected for the proper costing of production and the right source to collect the data is the place of origin namely the point of issue of materials. To enable this, the issues must be accounted for the job order or cost centre. This facilitates easy computation of the total issues of each item to be respective cost centre. We have seen the mechanism of pricing the issues in the above methods. Once the quantity and the basis for their valuation are known, then the total consumption value can be computed by finding their product and summing up over all the material issues. In a computerised accounting system, such jobs can be accomplished automatically in the update run. Another advantage of such data collection is its utility in budgeting exercises.

For pricing any finished goods or for quoting in tenders the material cost per unit of product manufactured will have to be computed. The more accurate this computation is, the more effective the company will be in a competitive market situation.

The material requirements can be budgeted using the previous consumption figures as a basis and applying appropriate correction factor for development and growth of the business.

TEN POINTS TO PONDER

1. Discuss the importance of stock valuation in an organisation.
2. Identify the methods of valuation of stock.
3. Describe the FIFO method and its role in profitability of the company.
4. Enumerate the FIFO method process and its role in profitability of the company.
5. Explain the differences between simple average, weighted average and periodic weighted average methods.
6. How do you arrive at standard cost of an item?
7. Discuss the relevance of replacement pricing and base stock method in any organisation.
8. How do you cost the issues?
9. Link the methodology of materials budget formulation with costing the issues.
10. Explain which method of valuation is most suitable to Indian conditions.

62. Materials Audit

A TO Z TYPES OF AUDIT

The word *audit*, has been derived from the Latin word AUDIRE which means to hear. Audit practice existed even during the Egyptian civilisation and dates back to 3500 BC.

(a) In order to ensure that the financial and performance statements depict the true state auditing of accounts and transactions becomes a necessity. Today the auditor is a qualified executive/consultant and he has specialised skills and judgment in examining the financial statements and the facts based on which such statements are made

(b) Whoever performs the audit function, whether he is an internal agent or external agent, employed on a statutory or governmental basis, carries a great responsibility on his shoulders. He is the expert who certifies that the transactions are valid and that the decisions are unbiased, honest, and were taken in the best interests of the company. The auditor has to ascertain the reliability of the records, the validity of the basis of valuation and the propriety of the decisions taken in the overall hierarchical structure of the organisation.

(c) During the 20th century, the role of audit has added new dimensions and become more consultative and participative. The audit function is concerned with the evaluation of performance, including non-financial areas ensuring internal controls on efficient and effective appraising of the decision-making process. Different types of audit are required for the evaluation of the corporate business performance.

(d) After verifying the adequacy and effectiveness of controls, the auditor evaluates and submits his reports by giving various suggestions.

(e) Materials account for over 60 per cent of the cash outflow of many organisations, and all the four components of inventory — namely raw material, machinery spares, semi-finished and finished goods — account for 90 per cent of the working capital. Inventory is also the single largest asset in many organisations. Hence the auditors are interested in the material transactions.

Work in Progress

(f) Work-in-progress (WIP) and audit function usually concentrates on the input side only and sometimes on the performance of finished goods under marketing functions.

(g) Rarely, auditing department checks the quantity/value or control of the semi-finished work-in-progress inventory, which could be sizable and laying in the shop floor for many months in assembly-oriented engineering industries. In this context, any raw material is known as semi-finished goods as soon as it arrives in the shopfloor. Similarly, any ready-to-despatch material, lying in the shopfloor is also known as semi-finished goods, even though the value added may be tremendous. Hence there is a need for the auditor to thoroughly examine the status of control of the semi-finished work-in-progress inventory.

(h) *Financial Audit* has been well known for quite some time in India. It has been designed to check the fairness of the accounting statements as prepared in conformity with generally accepted accounting principles; also carried out by external auditors as part of the legal requirement under Sec. 224–232 of the Indian Companies Act 1956 and placed before the shareholders of the company at the annual general body meeting.

(i) *Operations Audit* aims at ensuring that specific tasks are carried out effectively and efficiently. It is utilised to review and appraise the efficiency and economy of operating methods and procedures of the organisation.

(j) *Management Audit* is concerned with assessing the management's accomplishment of organisation's goal in the strategic decision-making areas. It is concerned with the evaluation of how well management is accomplishing its objectives; performing the managerial functions of planning, organising, directing and controlling; and reaching effective decisions in accomplishing the organisation's vision and objectives. The adequacy of management's decisions and the manager's ability to move towards the goal are reflected in the management audit.

(k) *Performance Audit* is usually undertaken by the members of the Board of Directors. It is employed to determine the quality of top management team that makes the key decisions in the organisation, as well as the quality of its cooperation in accomplishing the organisation's objectives.

(l) *Propriety Audit* is intended to function as an independent assessment of managerial decisions and actions but, with an added objective of verifying the use of authority for such decisions and actions and ensuring that cannons of financial propriety are followed in taking such decisions.

(m) *Cost Audit* has been ordered under Sec. 233B of the Indian Companies Act 1956 under which provision, the cost auditor who is a qualified Cost Accountant under the Institute of Cost and Works Accountants Act 1959, has to give an independent opinion whether cost records as prescribed by law have been maintained or not; the cost statements give a true and fair view of the cost of production/processing and marketing of the product under audit. Cost audit does not involve mere certification of data. It is much more comprehensive, involving efficiency, performance and propriety.

(n) *Internal Audit* is an independent appraisal activity within an organisation for the review of the operations. It can be considered as a managerial control which functions by measuring and evaluating the effectiveness of other controls. The objective is to assist all members of management in the effective discharge of their responsibilities by furnishing them with analysis, appraisals and recommendations of the activities reviewed. What was once regarded as a watch dog function and an extension of the finance function, has today come to be recognised as a general management function in India. The internal auditing profession has been accorded a legal status through Section 227-4A of the Indian Companies Act 1956. This has been further enhanced by an amendment, in 1988, extending the list of companies which should have an internal audit system commensurate with the size and nature of business. Companies should receive qualified reports from statutory auditors, if, in their opinion, the internal audit section does not comply with this requirement.

(o) *Social Audit* is concerned not only with reporting the organisation's involvement in socially oriented activities, but also with determining whether or not, it has met its objective for each activity. The social audit leads to a social performance report for management and perhaps awareness for outsiders too. The social assets are subtotalled by different suppliers of social resources, namely, staff, labour, organisation, general public, community and stockholders. The social costs are, social or economic resources consumed by company operations and are a cost sacrifice or determinent to the society. There is wide agreement that business corporations should refrain from capital crimes in this context.

(p) *C & AG Audit.* The Comptroller and Auditor General of India audit is applicable to all companies, where the Government holds share capital of more than 50 per cent, under Section 619 and 620 of the Indian Companies Act 1956. The scope of the audit includes propriety, social, financial and efficiency audits. The final audit report is placed before the Parliament for Central Public Undertakings and before the Legislature, for state Public Undertakings. This audit should ensure that all transactions conform to the accepted standards propriety and are unlikely to affect the efficiency of the organisation, besides conforming to the laws of the land. Public accountability to the citizens is an important consideration of this audit. The quality and quantum of accountability for public sector or government departments are quite different from the accepted norms of accountability in private sector. The CAG is thinking of getting constitutional immunity for its findings.

(q) *EDP Audit* has become relevant with the advent of computers in almost all sectors of the economy, as India is catching up with the rest of the world in information technology. It is an independent evaluation of the data security, software and hardware security, virus attacks, etc. of a computer installation with a view to ensure the reliability and effectiveness of operating controls to enhance the reliability of the computer system. This is also known as *Information Systems Auditing*.

(r) *Safety Audit* has assumed tremendous significance in the Indian context, particularly after the Bhopal gas tragedy. It covers a technical audit of the safety of plant and equipment of an industrial organisation, including environmental safety, pollution control measures, adherence to safety norms, etc.

(s) *Security Audit* is concerned with auditing the security of fixed assets and current assets, i.e. plant, machinery, warehouse, building, in transit material, stored items, outgoing products.

(t) *Compliance Audit* concerns itself with ensuring the adherence to procedures, policies, systems, rules, regulations, precedences, and statutory requirements.

(u) *Efficiency Audit* is concerned with evaluation the quality of performance against predetermined standards. This is covered in operational audit and performance audit. Sometimes this is also known as cross-section audit, wherein a reasonable percentage of items are taken for the audit cycle, commencing from the receipt of purchase requisition to consumption of materials to produce the finished goods.

(v) *Effectiveness Audit* is concerned with the effectiveness in achieving results including the strategies such as management audit and social audit.

(w) *Strategy Audit* has been developed by Wheelan and Hunger. This model could form an ideal approach to management audit of the corporate strategy. The steps include evaluation of current performance, examination of the corporation's strategic managers, scanning the corporation's external environment, scanning of internal corporate climate, analysis of strategic factors, generation of the best alternative strategy, implementation programme of selected strategy through budgets, and evaluation of the implemented strategy through feedback control systems.

(x) *Harvard — Boston 7s Approach* is currently gaining momentum in the Indian audit scenario. These include a critical review of superordinate goals, strategies of the corporation, structure of the organisations and systems in the company, skills of managers, staffing pattern and styles of management. The process of scenario painting by identifying these internal strengths and weaknesses, as well as external threats and opportunities, is an important element in this approach.

(y) *Human Aspect of Audit* includes the pros and cons of internal teams, external teams, team composition, generalist/specialist, frequency of audit, interview techniques, conditions of preaudit, during audit and post audit, audit working papers, and reporting responsibilities.

(z) Audit Checklist includes gathering information through interviews, developing audit questionnaire, developing audit findings based on the questionnaire, oral recommendations for improvement, discussion with the managers and preparation of the audit management report.

Role of Audit in Materials

The audit function will be of no avail, if it is performed merely as a post-mortem to check the regularity of the transactions and to see that they have had the sanction of the competent authority. Such an exercise invariably ends up as a fault-finding one and is dysfunctional to the extent that it introduces rancour and ill feeling between the auditor and the audited.

To prevent negative consequences, auditing should be an appraisal-oriented function and delve more thoroughly on efficiency and propriety criteria than mere checking of vouchers and signatures.

A TO Z OF MATERIALS AUDIT

The materials function should be audited on the basis of the following questions:

(a) How effectively has it served the consumers?
(b) What has been the service level within the investment constraints?
(c) How have losses in storage been minimised?
(d) How has obsolescence been controlled?
(e) How effectively has the purchase intelligence been used?
(f) How fruitfully has the available manpower been deployed?
(g) How smoothly does the paperwork system flow?
(h) How accurately is the information compiled?
(i) Has the materials function achieved its objective?
(j) Has the value of the cost of carrying inventory been achieved?
(k) Has extravagant expenditure been incurred in materials departments?
(l) How well have make-buy-lease criteria been applied?

Further the following aspects should be of great relevance to the auditor:

(m) All aspects have been properly accounted for.

(n) All expenditure shown in the accounts is duly authorised by competent authority.

(o) All expenditure has actually been incurred as seen from supporting vouchers and documents.

(p) *Prima facie*, there are no misappropriations or embezzlements.

(q) All receipts and expenditure are accounted for under the correct heads.

(r) They have been properly allocated between capital and revenue accounts.

(s) The appropriation and conventions are in accordance with the laws and conventions.

(t) The accounts give a true and fair picture of the state of affairs in the company.

(u) The *value* of money spent has been realised by the firm.

(v) Avoidable delays in constructions or operation have been omitted.

(w) Extravagant expenditure has not been incurred.

(x) Purchasing management is sufficiently qualified to meet only the short-term goals, but also the long-term benefits of the company.

(y) Purchase follows sound policy on source development, vendor rating, right price, right contracts and right quality at right time in order to maintain optimum inventory levels.

(z) Clear-cut policy manuals outlining procedures exist for purchasing, warehousing, inventory control, cost reduction and working capital management.

Cost Benefits of Audit

Auditing in materials management is like any other audit. It involves a lot of hard work, often unrewarding, but occasionally heart warming.

TEN POINTS TO PONDER

1. Clarify the concept of audit function.
2. Identify the need for audit in inventory and materials management.
3. Discuss the role of internal versus external auditors in the inventory and materials functions.
4. Elaborate the term *internal audit* in the context of materials function.
5. Clarify the concept of management audit in materials function.
6. Explain the significance of cost audit in materials management.
7. Describe CAG audit for the materials function in Public Sector Undertakings.
8. Examine the significance of operations audit, performance audit, safety audit, security audit, EDP audit, financial audit, social audit, and strategy audit relevant to materials management.
9. Comment on the applicability of the *7s* approach to the materials field.
10. Analyse the areas in materials management for audit purposes.

63. Human Aspects of Materials Management

MOTIVATION

It is well known that the materials function is responsible for about 60 per cent of cash outflows and a major proportion of the working capital commitment, in manufacturing organisations. Improvement of job conditions will ensure full commitment of the involved personnel. Proper procedures and well-formulated policies of selection, training, induction, promotion, and grievance-handling improve the morale of the staff and the quality of working environment. This improvement can also be brought about by financial incentives, productivity bonus, job rotation, job satisfaction, job enrichment, parity of emoluments for similar jobs, improvement in working conditions, non-financial incentives, ergonomics, capable leadership, etc. While organisational rewards may or may not be directly correlated with what the knowledge workers perceive as professional achievements, employee satisfaction is obtained through the intrinsic quality of work. The objective should be to integrate the human resources effectively in the corporate culture of the organisation. The top management has to play an important role in improving the motivation of its employees. Some organisations have introduced financial stakes for their employees by making them shareholders in order to make them more committed and involved in the work.

MASLOW'S HIERARCHY OF NEEDS

Motivation is a nebulous concept and difficult to define. In the context of motivation, Maslow's hierarchy of needs occupies an important place. At the bottom of the scale are the *basic needs* — like food, clothing and shelter — which are met by salaries and permit man to exist in his environment. *Safety need* is the next step and covers aspects such as pensions, long contracts, job security, promotion, opportunities, etc. The *sense of belonging* to a department, team, company, industry, social group or society is the third step in the hierarchy. The fourth stage, known as the *ego-status*, is motivated by a desire for recognition as an individual by means of his significant contributions, particularly new methods and knowledge. The final *self actualisation* stage is reached when a person is able to integrate personal achievement within the total organisation goal. In this stage, the best managers are found, as such a person will get the best performance of the team. These people are also known as 'Theory Y' managers and resist external pressures in order to enlarge the arena of activity by every means in their power and encourage innovation in activities. This job enrichment is a means of making work experience meaningful to the people.

PROMOTION POLICIES

Promotion policies play a crucial role in the sphere of motivation. Promoting psychophants must be avoided. Promotions can be by merit, merit cum seniority, only seniority, or reservation for some jobs. A well-laid out promotion policy acceptable to most of the peers by means of a well-tested appraisal scheme will benefit the company in the long run. The principle of self evaluation is being increasingly adopted in many progressive companies.

JOB SPECIFICATION

Matching the man to the job requirements and job to the individual is an important aspect, in the context of human assets management. This is usually done by the industrial engineering group. For this purpose, the job is studied in depth and a job analysis is carried out. Job specification consists of establishing the following parameters. (a) Title of the post (b) Superior to whom the individual has to report (c) Authority limits (d) Individual's task (e) Time constraints (f) Higher level promotion (g) Expected educational background (h) Expected knowledge/skill (i) Analytical ability (j) Counselling in times of distress (k) Preparing for advanced positions (l) On-the-job training (m) External training.

A TO Z OF RECRUITMENT CONSTRAINTS

In the context of human resource management, the significant Government regulations include the following executive orders in India:

(a) Apprentice Act 1961
(b) Contract Labour Regulations and Abolitions Act 1970
(c) Employees Family Pension Act 1971
(d) Equal Remuneration Act 1976
(e) Maternity Benefit Act 1961
(f) Payment of Bonus Act 1965
(g) Payment of Gratuity Act 1972
(h) Inter-state immigrant workmen on regulations of movement 1979
(i) Amendments to Industrial Disputes Act of 1947
(j) Restraints in recruitment through employment exchanges and pressures from top management as well as politicians
(k) Reservations for socially backward, scheduled casts/scheduled tribes
(l) Reservations for handicapped and ex-service personnel

(m) Preferences for persons displaced from land and sons of the soil
(n) Increase in employment for women

Other aspects include the following:

(o) Problems of dual career families affecting work environment
(p) Improvement in educational level of work force
(q) Rapid inflation
(r) Increasing economic and geographical mobility in workforce
(s) Rapid technological developments
(t) Use of computers in every field
(u) Growth of professionalism
(v) Increased violence, terrorism, law and order problems
(w) Increased educational opportunities in technical and managerial spheres
(x) Increased pollution
(y) Increased transportation bottlenecks
(z) Increased problems, challenges and opportunities.

MANPOWER PLANNING

In the context of human resource management, manpower planning is a very critical aspect in the materials field. Let us analyse the needs of manpower by segregating repetitive activities from the one time activities. This will help to identify the human resource requirement in materials function. The research study has indicated that only men in the age group 34–45 are employed at the decision making level of materials manager.

A TO Z OF REPETITIVE ACTIVITIES

The regularised routine activities of warehousing and materials are as follows:

(a) Receiving materials,
(b) Checking for quality,
(c) Visual inspection,
(d) Preparing goods receipt notes,
(e) Coordinating quality control for inspection,
(f) Accepting the passed materials,
(g) Preparing rejection notes for substandard materials,
(h) Completing bill formalities,
(i) Taking accepted materials into stock,
(j) Accounting for accepted items,
(k) Stocking the materials in the respective bins/locations,
(l) Application of preservatives,
(m) Preparing for issues,
(n) Accounting for material issues in the documents, and feeding relevant data to the computer,

(o) Keeping purchase department informed of regular requirements through indents,

(p) Keeping the area clean to facilitate easy handling,

(q) Checking up inventory levels,

(r) Keeping accounts updated,

(s) Raising indents,

(t) Verifying the entries,

(u) Following-up with production,

(v) Obtaining purchase requisitions,

(w) Finding out the sources either by enquiries or by tender,

(x) Placing the purchase orders and following up with suppliers,

(y) Expediting the payment of supplier's bills, and

(z) Taking necessary steps for claims and removal of rejections.

ASSESSING WORKLOAD

The workload of these routine activities can be easily computed by adopting work measurement and work sampling techniques. Three types of personnel — executive cadre for supervising, clerical cadre to look after the accounting, and semi-skilled worker cadre to do physical handling — will be needed. If heavy handling is involved, like the use of cranes, then skilled handling personnel have to be hired. Depending on the workload, the appropriate number of crane drivers and riggers has to be employed. Normally in an established concern, we can expect around 250 working days per annum from an employee, without payment of overtime. The total number of persons should take care of the normal workload and contingencies like absenteeism.

A TO Z OF NON-REPETITIVE JOBS

There are some jobs where the person concerned in the materials function has to analyse data and use his discretionary faculties. These are mainly non-routine, periodic in occurrence and include the following activities.

(a) Establishing and maintaining a codification system,

(b) Lead time analysis to locate long lead time items,

(c) Analysing the consumption pattern for long lead time items,

(d) Establishing the stock norms,

(e) Highlighting stock accumulation,

(f) Identifying abnormal consumption items,

(g) Isolating non-moving, slow-moving, surplus materials,

(h) Taking effective control on obsolete items and scrap,

(i) Periodic stock verification,

(j) Carrying out ABC/VED, MUSIC-3D analysis,

(k) Fixing EOQ levels,

(l) Fixing consumption norms,

(m) Materials budgeting,
(n) Purchase intelligence,
(o) Establishing follow-up procedures,
(p) Source development,
(q) Standardisation,
(r) Value analysis,
(s) Use of learning curve,
(t) Import substitution,
(u) Vendor rating
(v) Determining inventory carrying charges,
(w) Working out ordering/order processing costs,
(x) Steps for negotiations,
(y) Application for import licences,
(z) Development and updating of the materials management manual.

WORK MEASUREMENT

Measuring the work content of such jobs is quite difficult. Application of standard times of an organisation and methods is difficult as the performance depends on the individual. A strategy that can be adopted, is to undertake a cost-time-benefit analysis of the utilisation of such personnel — Sometimes contracting these jobs to reputed people can also be considered. Based on the general supervision standards, prevalent in the unit, the time to be spent on supervision can be approximated. The rest of the time spent can then be costed and applied to the other jobs, according to the benefits that arise from performing them. The deployment of personnel on a permanent basis, for the decision-making and analysis levels, depends on the culture of the organisation. It is desirable, however, to have a nucleus of personnel as a small care group initially. Many progressive organisations are switching over to 5 days a week, but some private organisations overwork their officers by making them work for long hours on 6 days and by converting Sunday also as unofficial working day.

DEVELOPMENT OF PERSONNEL

Under the existing Indian conditions, industries would do well to absorb diploma engineers into this cadre, and train them on the job. Trading organisations can absorb commerce graduates in the clerical cadre as numerical accounting has to be done. It is very advantageous to take in raw graduates and let them grow with the organisation by providing them with ample training and growth opportunities as this helps in motivation. The policy of treating seniors as indispensable and juniors as useless acts as a deterrent.

Direct recruits for the decision-making levels can be post-graduates, who have specialised in materials management. Such candidates are available from the Institutes of Management, Institutes of Technology, and Leading Universities that offer MBA Programmes. Direct recruits at the middle and higher level should not be perceived as an impediment to growth by the storekeepers. It is necessary that they too are given the opportunity to compete and if they are found fit, they should be preferred. This develops a healthy organisation and promotes commitment in the employees.

In the ultimate analysis, we should not forget the advantages of a regular updating of knowledge with the latest developments in the profession. The challenging task in high tech area is to ensure continuous development of manpower which is the software responsible for achieving success in effectively utilising the hardware. The need of the hour is to ensure that the prevailing restrictive work practices do not persist. Top management should develop a culture of good work ethics and team work at all levels.

TRAINING NEEDS ANALYSIS

In the process of moving from an agrarian society to an industrial economy, the workforce undergoes drastic changes. A developing economy requires more professionals, managers and skilled personnel. Professionalisation and improvement in skills can be enriched by training individuals. This is particularly significant in respect of materials function. Hence, the materials management personnel have to be educated according to the fast-changing needs of society and future trends of the economy. Product oriented practical training is necessary.

The need for training arises in the following cases: When new hands are taken for specific jobs, when promotion of certain employees with potential, has been decided upon and to meet the challenging needs of new machines, new processes and better systems. Materials Management is a neglected area in all the universities which treat this subject as a part of the production system. Only the professional organisation — Indian Institute of Materials Management, conducts part-time courses in the important sphere of materials management. The skills are required to be upgraded, improved and maintained throughout the career path of the individuals. Therefore, the training activity is a continuous process and not a one-time job.

PROGRAMME INGREDIENTS

The duration and programme contents depend upon the levels in the organisation. Detailed operational procedures are necessary at lower levels, while policy-making skills in the context of socio-economic-political changes are necessary at the higher levels. The contents should cover all aspects of purchase management, warehousing, inventory control as well as their inter-relationships with the user, finance, vendor and external regulatory agencies. External refresher courses are necessary at the higher levels, while internal programmes using the company data are adequate for the lower levels.

The pedagogical tools used in the training programme should include a judicious combination of lectures, case discussions, role playing, computerised business games, syndicate discussions, group activity, self learning, audio-visual aids, visual demonstration, exercises, visits to actual sites, and project work.

While creativity can often be attributed to one individual, most developmental attempts within the industry are built upon team effort. In this context, the following translated verse taken from Rig Veda emphasises the importance of group activity:

"Samgachdvam Samvadadwam Samvo Manamsi janataam"

This is translated as follows: "May you meet together and discuss together. May your minds comprehend alike."

The process of self development is an important aspect in training as per the dictum "God helps those who help themselves" sphere of training. Chapter 6 sloka 5 of Bhagwad Gita says as follows in this context:

"Udhdhareth Athmana Athmanam Nathmam Avasadayeth.
Athmaiva Hyathmano bandhurathmaiva Ripurathmanaha."

Let the man raise himself in the learning process; Let him not lower himself. He alone is his friend in this process of self development. He alone is his enemy!

It is difficult to assess the investment and returns in the human resources. But attempts are being made to compute human resource accounting, human resource value accounting and human resource cost accounting in some firms.

TEN POINTS TO PONDER

1. Comment on the criticality of human resource management in the field of materials function.
2. Describe the motivation process that can be adopted in a company, particularly in the area of materials management.
3. Analyse the suitability of application of Maslow's Hierarchy of needs to Indian conditions.
4. Discuss the merit versus seniority policy for promotion purposes defining merit in materials management activities.
5. Define job specification and its parameters.
6. Examine the A to Z external constraints in the recruitment of personnel in India.
7. Identify repetitive activities in materials function.
8. Workout the procedure for manpower planning for repetitive jobs.
9. Identify non-repetitive jobs in materials function.
10. Describe the process of developing and training human resources in materials function.

64. Communication and Materials Management

BENEFITS OF COMMUNICATION

Researches in India and abroad have shown that 75 per cent of the manager's time is spent in communication. This is particularly true in the case of materials manager. He has to communicate with his subordinates for the functions of source development, purchasing, follow-up, inventory control, warehousing, materials intelligence, and cost reduction. Outside his department, he has to constantly coordinate and collaborate with the user and finance functions. He has to be always in touch with the suppliers, vendors, carriers, transporters, excise, customs, imports, licensing, clearance and other agencies. Hence the materials manager has to improve his communication skills.

The materials manager is always in search of a system which provides him with the necessary information in order to taking decisions, planning and controlling all spheres of activity. The materials manager depends upon the input of information from internal and external communication.

BARRIERS TO COMMUNICATION

There are several barriers to the communication process as identified by the experts. It is necessary to locate these barriers and reduce them in order to improve the process of communication. The barriers are due to differences in perception, status difference in the hierarchy, lack of credibility, message overloading, wrong expectation, about the receiver's state of mind, timing of communication, noise and external disturbances, misunderstanding and distortion.

COMMUNICATION MEDIA

Information is distributed within an organisation through several channels.

(a) The chain of command flowing from top management through the different functional heads to the ultimate individual. There can be briefing groups to explain the corporate philosophy to the union officers.

(b) Written instructions or circulars are put up on the notice board, fixed at vantage points in the company.

(c) Informing the union office bearers who can sound as a spring board and useful vehicle for the feedback systems.

(d) Communication can be through informal work groups, people of the same caste, language, community, etc.

(e) The existence of rumour mongering groups, known as grapevine, for communication purposes is dangerous as it may lead to disinformation.

(f) Company magazines, and newsletters, carry Chairman's appreciation or messages.

(g) During emergency or flood, accidents, earthquake and serious riots, the loudspeaker system can be effectively used in order to directly address the staff.

(h) The communication process could be in the following types:

(i) One-way, (ii) two-way, (face to face), (iii) oral, (iv) written, (v) formal, (vi) informal, (vii) vertical, (viii) horizontal, (ix) upward, (x) downward, (xi) public-speaking, (xii) gossip, (xiii) rumour, (xiv) personal, (xv) negotiation, (xvi) listening, and (xvii) telephone.

A TO Z OF INTERNAL COMMUNICATION

The internal communication essentially involves the following aspects, with the various departments in the company.

(a) Forecasts and changes in the forecast for every item,
(b) Materials intelligence from market research,
(c) Packing size, material and quantity,
(d) Production scheduling,
(e) Bill of material,
(f) Material resource planning,
(g) Capacity constraints,
(h) Purchase budget,
(i) Consumption norms,
(j) Inventory levels on buffer, safety, reserve,
(k) Costs of ordering,
(l) Inventory carrying cost,
(m) Stockout cost,
(n) Working capital norms,
(o) Obsolete items disposal,
(p) Tender committee buying,
(q) Writing off items,
(r) Charging off,
(s) Reconciliation of book balance with physical balance,
(t) Quality control rejections, specifications revision,
(u) New product development,

(v) Development of drawing,
(w) Categorisation of spare parts,
(x) Cost-criticality analysis,
(y) Capitalisation of insurance spares,
(z) Cost reduction techniques.

The above list is only illustrative and not exhaustive. Communication with other functional areas, particularly finance and user — sometimes results in confrontation and hence is the need for better communication for smooth working.

A TO Z OF EXTERNAL COMMUNICATION

Materials intelligence discussed in Chapter 27 constitutes a major portion of external communication. In particular, external communication from published sources and suppliers can constitute the following:

(a) Market conditions
(b) Source of supply
(c) Demand
(d) Supply
(e) Price of the item
(f) Product availability
(g) Capacity constraints
(h) Labour contract conditions
(i) Subcontracting
(j) Tax matters
(k) Traffic and transportation media
(l) Costs of transportation
(m) Insurance costs
(n) Claims management
(o) Rejections
(p) Vendor rating
(q) Replacement of rejected goods
(r) Scrap disposal with scrap merchants
(s) Customs/excise for MODVAT benefits
(t) Import licences and foreign exchange clearance
(u) Opening letter of credit with banks
(v) Bid evaluation criteria and contractual aspects
(w) Ethical considerations
(x) Organisation climate in the supplier's plants
(y) Technical financial material, managerial assistance to the suppliers
(z) External communications are difficult to control as the users and finance of many companies feel that they have a right to deal with the vendors directly.

A TO Z TIPS FOR IMPROVED COMMUNICATION

Art of Listening

(a) The most important aspect of communication is listening. Speaking without listening cannot be effective. Listening is a rare skill, but it is often not practised. Seek not only to be understood but also understand as the other person may have his own ideas.

(b) Listening needs concentration, the desire to understand, an awareness of what is meant and judgment on how to act.

(c) Listening is a skill, which can be learned and developed. It has been observed that the percentage of messages retained after adequate training, is much higher than before training.

(d) The philosophy should be, "silence is golden and speech is silver." When you speak you do not improve your knowledge, but when you listen you learn a lot.

(e) Listen in a concentrated manner, and clarify the points. Listen to the message and not to the words. It is advantageous to reach a consensus over important points.

(f) Prepare thoroughly before communicating with external persons. The pros and cons have to be viewed carefully.

(g) It is desirable to know about the audience/other party topic, time, duration, data, analysis, conclusions, and other relevant points to be communicated.

(h) It is preferable to restrict to only the main topic with a brief introduction and conclusion.

(i) Business communication need not be oratorical with flowery language, but should be to specific points and at the same time involve the other party.

(j) Seek to clarify your ideas before communicating. The more systematically we analyse the problem or idea to be communicated, the clearer it becomes.

(k) Examine, critically the true purpose of each communication. Before you communicate, ask yourself what you really want to accomplish with your message — obtain information, initiate action, change another person's attitude. Identify your most important goal and then adapt your language, tone, and total approach to serve that specific objective.

(l) Consider the total physical and human setting whenever you communicate. Meaning and intent are conveyed by more than words alone... Consider, for example, your sense of timing — i.e., the circumstances under which you make an announcement or render a decision; the physical setting whether you communicate in private, for example, or otherwise, the social climate that pervades work, relationships within the company or a department and sets the tone of its communications.

(m) Consult others, where appropriate, in planning communications. Such consultation often helps to lend additional insight and objectivity to your message. Moreover, those who have helped you plan your communication, will give their active support.

(n) Be mindful, while you communicate, of the overtones as well as the basic content of your message. Your tone of voice, your expression, your apparent receptiveness to the responses of others — all have tremendous impact on those you wish to reach.

(o) Take the opportunity, when it arises, to convey something of help or value to the receiver. Consideration of the other person's interests and needs — the habit of trying to look at things from his or her point of view — will frequently give opportunities to convey something of immediate benefit or long-range value to him or her.

(p) Follow up your communication, by asking questions, encouraging the receiver to express his reactions, follow-up contracts, subsequent review of performance. Make certain that every important communication has a 'feedback' so that complete understanding and appropriate action result.

(q) Communicate for tomorrow as well as today. While communications may be aimed primarily at meeting the demands of an immediate situation, they must (also) be consistent with long-range interest and goals. For example, it is not easy to communicate frankly on such matters as poor performance or the shortcomings of a loyal subordinate — but postponing disagreeable communications makes them more difficult in the long run and is actually unfair to your subordinates and your company.

(r) Be sure your actions support your communications. In the final analysis, the most persuasive kind of communication is not what you say but what you do. For every manager, this means that good supervisory practices — such as clear assignment of responsibility and authority, fair rewards for effort, and sound policy enforcement — serve to communicate more than all the gifts of oratory.

(s) In interpersonal communications, do not exhibit unnecessary authority, or superiority, which will make the other party react adversely to all the proposal.

(t) Communication should be objective, adaptable, mature, rational and unemotional in order to be effective.

(u) The philosophy of communication must be based on sharing a piece of information between two or more individuals and must not be a one-way traffic.

(v) The sender of communication starts with an intended message, and chooses an effective medium of transmission, which is passed on to the receiver without distortion.

(w) It is desirable to decide as to whether the medium should examine the following aspects: personal or impersonal, importance of feedback, accuracy, speed, reliability, credibility of source, etc.

(x) The sender must understand the receiver fully, since the receiver is influenced by his own perception and attitudes for effective communication.

(y) It is necessary to set standards on tidiness, safety, punctuality, conduct, personal appearance and good human relations for effective communication.

(z) It is essential to concentrate on the quantum of communication necessary for each level. The top management is interested in targets and exceptions while the middle level is involved with details.

TIME MANAGEMENT — THE 60 STEPS

In the context of Time Management, the following 60 steps have been suggested by Mr Alan Lekein. The materials manager must try to follow these steps:

1. I count all my time as "On-time" and try to get satisfaction (not necessarily accomplishment) out of every minute.
2. I try to enjoy whatever I am doing.
3. I am a perennial optimist.
4. I build on successes.
5. I do not waste time regretting my failures.
6. I do not waste my time feeling guilty about what I do not do.
7. I remind myself: "There is always enough time for the important things", If it is important I will make time to do it.
8. Each day I try to find a new technique which I can use to help gain time.
9. I get up at 5 A.M. during the week (and go to bed early).
10. I have a light lunch so I do not get sleepy in the afternoon.
11. I read newspapers or magazines occasionally. I do glance at the headlines at the news stands to keep up on the world.
12. I skim books quickly looking for ideas.
13. I do not own a television set.
14. I have my office close enough to my home to be able to walk to work. But when I am lazy or in a hurry, I drive.
15. I examine old habits for possible elimination or streamlining.
16. I have given up forever all "waiting time". If I have to wait I consider it a "gift of time" to relax, plan or do something I would not otherwise have done.
17. I keep my watch 3 minutes fast, to get a head start on the day.
18. I carry blank 3 × 5 index cards in my pocket to jot down notes and ideas.
19. I revise my lifetime goals at least once a month.
20. I review my lifetime goals list every day and identify activities to do each day to further my goals.
21. I put signs in my office reminding me of my goals.

22. I keep my long-term goals in mind, even while doing the smallest task.
23. I always plan first thing in the morning and set priorities for the day.
24. I keep a list of specific items to be done each day, arrange them in priority order, and then do my best to get the important ones done as soon as possible.
25. I schedule my time months in advance, in such a way that each month offers variety and balance as well as 'open time' reserved for 'hot' projects.
26. I give myself time off and special reward when I have done the important things.
27. I do first things first.
28. I work smarter rather than harder.
29. I try to do only A's, never B's and C's.
30. I have confidence in my judgment of priorities and stick to them, in spite of difficulties.
31. I ask myself, "Would anything terrible happen if I did not do this priority item?" If the answer is no, I do not do it.
32. If I seem to procrastinate I ask myself "What am I avoiding?" and then I try to confront that thing head on.
33. I always use the 80/20 rule.
34. I start, find it is not necessary to do the rest.
35. I start with the most profitable parts of large projects and often find it is not necessary to do the rest.
36. I cut off non-productive activities, as quickly as possible.
37. I have developed the ability to concentrate well for long stretches of time (sometimes with the aid of coffee).
38. I give myself enough time to concentrate on high priority items.
39. I concentrate on one thing at a time.
40. I focus my efforts on items that will have the best long-term benefits.
41. I keep pushing and am persistent when I sense I have a winner.
42. I have trained myself to go down my 'to-do' list, without skipping over the difficult items.
43. I do much of my thinking on paper.
44. I work alone creatively in the morning and use the afternoons for meeting, if necessary.
45. I set deadlines for myself and others.
46. I try to listen actively in every discussion.
47. I try not to waste other people's time (unless it is something that really matters to me).
48. I delegate everything I possibly can to others.
49. I make use of specialists to help me with special problems.
50. I have someone to screen my mail and phone calls and handle all routine matters.

51. I generate as little paperwork as possible and throw away anything I possibly can.

52. I write replies to most letters at once.

53. I keep my desk top cleared for action, and put the most important thing in the centre of my desk.

54. I have a place for every thing (so I waste as little time as possible looking for things).

55. I try not to think of work on weekends.

56. I relax and 'do nothing' rather frequently.

57. I recognise that it is inevitable that some of my time will be spent on activities outside my control and do not fret about it.

58. I keep small talk to a minimum during work hours.

59. I look for action steps to be taken not to further my goals.

60. I am continually asking myself: "What is the best use of my time right now?"

Silence is a lot more articulate than noise. Those who have achieved, often speak the language of silence. And befittingly so, as silence is the best expression of confidence. Silence is golden, speech is silver. Nevertheless you need to know and so, cutting through the cacophony comes a statement of achievement.

TEN POINTS TO PONDER

1. Identify the relevance of good communication systems for the materials manager.
2. Examine the barriers in communication and suggest steps to remove them.
3. Describe the media of communication process.
4. Elucidate the internal communication aspects.
5. Explain the necessity for external communication.
6. Discuss how important it is for the materials manager to be a good listener.
7. Explain the difference between written and oral communication.
8. Elucidate the tips for effective communication.
9. Comment: "Communication is the life blood of management and materials manager spends more than 80 per cent of his time in communication."
10. Examine the steps to improve time management.

65. Reporting of Materials Management Activities

USE OF REPORTS

A report is the life blood of business management. A report is the communication of information to some one who wants it, in the most convenient form. Collection and dissemination of data are very essential in reporting. Mere facts can reflect what has happened, whereas reports give the summary of the working of the materials/purchasing/inventory/warehousing operations. The entire organisation works on the basis of two documents — receipt and issue.

A report is a record designed to help the management understand the environment and is a formal record of the work carried out, in order to take decisions.

Information will be a failure if it does not support a regular reporting of the ways in which the materials have flowed. Selective control techniques come in handy in this regard. The movement of high value critical items must be reported more frequently while low value items can be reported once a quarter or so. Too much information is often a damper and can land to a situation of the means becoming an end in itself. The important aspects in reporting are clarity, brevity, accuracy and periodicity. The efficiency and effectiveness of management at all levels depend upon the accuracy and promptness of the reports fed to them.

MONITORING RECORDS

A point that should not be overlooked is whether the documents have been rightly authorised. The materials ledger whether in the form of a bound note book/kardex records/Kalamazoo cards/Kanban system or any other similar computerised system has an individual account for every item stocked. In this account, the transactions are entered and the figures are updated. There are columns for inputs, outputs, stock balances, value, etc., so that respective entries can be made. The time-lag between the physical transaction and the posting of account should be minimised in order to have an accurate and up to date record. Large organisations can think of separate internal audit groups, to monitor the accuracy of these records. Any information that is delayed or inaccurate is misinformation and can lead to debacles in decision-making.

SYSTEMS DESIGN

The system design for the internal paper work must be such that it minimises the number of entries and at the same time, ensures an authorised account of the transactions with the least time delay. The principles of Organisation and Methods (O & M) techniques can be used advantageously. The basic tenets of these are minimisation of entries, avoidance of repetitions and standardisation of sizes, formats and accounting methods. Filing of the formats should not be complicated. The number of copies is minimised and this reduces the filing load. Colour schemes can be used to indicate the origin or the destination of the forms. A flow chart indicating the origin, flow and destination of the format will enable a better design, questions like "What is the use of the format? Can it be eliminated? Can it be combined ? Can its flow be restricted?" will enable a better design of the system.

It behaves each materials manager to get the information he needs for making decisions, with a minimum of paper work and clerical manpower. By this process he is able to achieve minimum materials costs. He should reduce the enormous amount of clerical efforts needed in his department.

Common clerical errors like posting 100 instead of 10.0 must be avoided by reducing postings. The present state of art in recording ranges from traditional ledger accounts to modern computerised systems. Each method has its own merits and demerits, and depends upon factors such as costs of formats, costs of information, benefit of information, number of transactions, each of operations and the culture of the organisation.

SOURCE DOCUMENTS

A regularly updated record of inputs, outputs and stock level is imperative for efficient warehousing management. The information on input enables a close follow-up of the orders, while the information on outputs and stocks enables effective consumption and inventory control measures. If the material cost is to be computed accurately, then we should know how the consumption has taken place. This stresses the importance of information build-up required for the proper management of stores. This information not only becomes the basis of the feedback for control of materials, but also for forecasting the future consumption.

The prime documents that form the basis of movement of materials include the following:

(a) Goods receipt note along with the supplier challan/invoice and the purchase order for the inputs.

(b) Material requisition vouchers or indents based or issue notes, as the case may be, which record the requirements against a job or work order and form the basis for recording the issues.

(c) Scrap wastage or loss notes that record the movement of materials due to the respective reasons.

EXCEPTICN REPORTING

Monthly reports consisting of purchases, inventory, stockouts, cost reduction achieved through the materials department, etc. are commonly used in the industry. In order to take timely corrective measures, the monthly reports should be submitted during the first week of the following month. Quarıerly and annual summaries, with specific emphasis on long-term aspects of materials management such as source development, value analysis, implementation, etc. are also prepared by the materials department. If the level is higher, then the information content should be given in a nutshell, and the periodicity may be monthly or quarterly on policy aspects and cost reduction. The top management is interested in exception reporting and not excessive reporting. But the user department as well as different sections of the materials department like to have detailed reports on a daily/weekly basis on all aspects of the transactions.

It is desirable to report in short sentences and short paragraphs. The report should convey the information required in the most efficient way.

The information normally fed to the top management is an abstract from the materials ledger or the account. This gives the summary of consumption, receipts, stock on hand, quantity, or order, availability of budgets, price differences, price trends, recommendations. It is necessary to attach charts, graphs, diagrams, appendices, tables and comparative previous performance of the relevant data. "A picture is worth thousand words" and hence is the need for charts.

A TO Z CONTENTS OF REPORTS

The contents of the report vary from level to level but it is desirable to include the following aspects in the report. The introductory portion consists of the following areas:

(a) Problem statement
(b) Subject
(c) Preface
(d) Objectives
(e) Foreword
(f) Purpose of report
(g) Parts/materials tested
(h) Apparatus used
(i) Definition of the terms

The main body comprises the following:

(j) Discussion

(k) Details on costs/benefits/consequences
(l) Experimental procedures
(m) Records of tests/samples
(n) Description
(o) Data obtained

The conclusion may comprise the following:

(p) Results achieved
(q) Interpretation of data
(r) Results and conclusions

The recommendatory portion includes the following:

(s) Proposed solution to problem
(t) Suggestions
(u) Comments and opinions

The appendices may comprise the following items:

(v) Charts, graphs and illustrations
(w) Computations
(x) Exhibits and diagrams
(y) Tables and comparative statements
(z) The summary/abstract/synopsis (may be given either at the beginning or in the end).

COMPUTER REPORTS

It is possible to program the computer to provide periodic reports of each vendor's and carrier's performance on aspects such as volume of business, late deliveries, rejected shipments, transit damages, price trends, budget variances, stock levels etc. The computer can break the reports corresponding to each vendor and each item. In many organisations, purchase requisition follow-up and orders are handled by the computer. The computer can also help in preparing reports on the performance of individual buyer source development officer user department. The characteristics can include, volume purchased, outstanding orders and price paid. The computer can also generate reports pertaining to overall information needs, legal status, financial positions, obligations to stockholders, product pricing and the levels of various components of inventory.

For an effective reporting system, adequate precautions are necessary on data protection, data security, data secrecy, 'virus' attack on computers, floods, fire hazards, industrial espionage, riots, civil disturbance, etc. However, the computer offers materials executives many possibilities for developing data, to improve management decision and control.

OFFICE AUTOMATION

One of the most amazing contradictions has taken place over the last few years while automation in the factory has reduced jobs and increased productivity, automation in the office has had the opposite effect of increasing jobs and reducing productivity and output per person, particularly in materials department.

Rather than reduce the paper flowing through offices, office automation has created a vast increase. Faster and cheaper photocopiers, computer printers, and fascimile copiers, have all added to the veritable deluge of paper.

Word processors have replaced typewriters and there is little doubt that they are a very efficient and effective replacement. Unfortunately, the vast majority of the correspondence, reports, etc. produced by them is unnecessary. Many managers have discovered that one way to keep busy — and appear important — is to write lots of letters. The replies come in and have to be responded to creating more paper work.

Ink jet, matrix, daisywheel, desktop publishing and laser are all different methods of putting characters into paper. Each new model is faster, and offers more and more features. The end result is the same — masses and masses of paper, much of it joined together, so that we have to spend time separating it or trying to read it in a most unacceptable way. In spite of all attempts to control it, the 'mountain of paper' just grows and grows and the legions of people needed to handle it get bigger and bigger.

What we need to strive for, is simplicity and economy. Portable computers and portable telephones, which can be linked to transmit data, are slowly and gradually replacing the executive's travelling paper office housed in his or her briefcase.

It is paradoxical that fewer and fewer people are involved in producing things, and more and more people are involved in the process of getting the paper work processed. Administration and bureaucracy now seems to be the major employers.

CUTTING THE COST OF PAPERWORK

Recently, a materials manager located in the down town area of Bombay, requested his Director's permission to destroy the records which are more than ten years old. The director gave him the permission but asked him to take xerox copies of the old records.

The materials manager is often accused of indulging in a lot of paperwork. The problem is more acute in procedure-oriented organisations. Every time, a new director joins the company, he wants new parameters, whereas no one has the courage to take the decision to discontinue the old paper work. The paper work is built on mutual mistrust, involving concurrence of more than one agency even for small value budgeted items. An organisation and methods study is a must in the materials management department in order to make an indepth analysis of the movement of files, paper work, systems, procedures, duplication of records, etc.

In every organisation, we note the rising trend of paper work required to obtain the materials. As the paper work increases, more personnel are required, resulting in increased cost of ordering, follow-up receiving, inspection and issuing. The materials manager is aware of the high cost of paper work but he is unable to look beyond his nose and is doing the fire fighting job, of getting an urgently needed items. The cost per order is about Rs 750 and the inventory carrying charges is about 30 per cent per year, with a substantial portion being accounted by the paper work. Any means of routinising the transactions or reducing the clerical errors would be welcomed by the top management. The computer offers such a possibility in this context.

Every organisation expanding, diversifying upgrading its technological and introducing innovations, resulting in an exponential increase in the number of items and transactions. But the top management is usually averse to sanctioning additional clerical posts for the increased quantum of work. Hence is the need for the materials manager to reduce the entries, documents, transactions and paper work now, to meet the future challenges.

TEN POINTS TO PONDER

1. Emphasise the importance of reporting in materials function.
2. Enumerate the uses of materials reports.
3. Explain the relevance of system design and records monitoring in the context of reporting.
4. Examine the source documents for good reporting in the materials field.
5. Articulate on the frequency of reports.
6. Critically review the basic contents of good reports.
7. Describe the signifiçance of computer reporting system.
8. Discuss the significance of office automation.
9. Suggest steps to cut down paper costs in materials field.
10. Elucidate the parameters on which materials management report should be built.

66. Professional Ethics

ETHICAL SCENARIO

Ethics is a philosophy of life, setting standards in absolute terms. The Indian Manager — mostly the materials manager — is assumed to be corrupt unless he proves his bona fides. The major outlet of the company's money is the buyer who, wields the economic power and accounts for the bulk of the company's cash outflow and is responsible for inventory, which is the single largest asset in most industries. The buyer is the unchallenged custodian of the purse of the organisation. No doubt, this is the most fertile area for breeding corruption. Hence, whatever may be the code of conduct, or ethics evolved for guidance, the same will have. no value at all, unless the managers are men with strong character, integrity and high moral standards and are honest in the dealings with suppliers/customers.

DOUBLE STANDARDS

Corruption and dishonesty are two-way affairs and cannot be treated as one-sided. The marketing executives also have a moral responsibility to observe certain codes of conduct and display a sense of honesty in selling their products. In other words, it is the buyer-seller collusion — the greatest evil in the field of buying/selling — that which is responsible for pushing up the cost-line and creating artificial scarcity in the commodity market. It is therefore, obvious that it is not worthwhile to extricate buying from selling, or selling from buying — they are inseparable twins in the realm of corruption.

No code of ethics can succeed if the top management adopts double standards — one for buying and an other for selling. Many companies expect absolute honesty only from their buyers. But, they encourage their own marketing force to resort to unfair methods to attract potential buyers. The organisations should do their utmost to maintain and uphold fair business practices. But, if the organisations themselves begin to adopt, anti-social, unfair practices, the buyer/seller can only be the unfortunate victims.

It would be imperative to stress that revolutions have never been created by the pangs of hunger; but they have been motivated by the 'want' for bread. Fagin's empire was challenged, when Oliver Twist asked for a second helping and, Jesus was betrayed because the dissatisfied disciple, Judas, wanted his 30 pieces of silver. In this context, loyalty and commitment are often mistaken as loyalty to the individuals in the organisation.

The professional integrity of the materials manager is viewed with suscipicion because of the Indian ethical scenario, of corruption in high places, kickbacks, winding up charges, bribery, commission, cut, middleman's role, costly gifts, entertainment, considerations, etc. The final blow was struck due to large kickbacks, in the last minute in purchases of Bofors field gun, HDW submarine, Westland helicopters, Air Bus 320 instead of Boeing 757, the Rs 900 crores Dulhasti power plant to French firm in Kashmir and several other governmental purchases. The punitive measures are very small compared to taking bribes. Recently, a government official imprisoned for corruption charges told a friend that he will be out of the prison by bribing the jail authorities. In the ocean of corruption, it is difficult for the manager or anyone, to be an island of virtue. Interference in buying and promotion is the order of day. Yes men, psychophants get priority over professionals, resulting in demotivating the few conscientious officers. Pursuit of excellence is almost impossible in a sea of mediocricity in ethical standards demonstrated by land scandals, political interference, booth capturing in election, and mafia influences in all walks of life.

All business enterprises are engaged in the socially purposeful activity of continuous creation of new wealth. What is more important is whether corporate goals of these business enterprises are so determined as to serve national priorities, subserve social goals, achieve human welfare, and whether honesty, morale and good ethics play a decisive role in the conduct of their business. Hence, professional ethical integrity is the need of the hour for all disciplines, skills, departments, organisations, and industries, at the micro and macro levels.

SIGNIFICANCE OF ETHICS

Ethics represent the eternal and prevalent moral standards, personal values, corporate code of conduct, and are generally part of the cultural tradition of the country. Sir Julian Huxley in his book, *Evaluation after Darwin*, mentions — "Man's most comprehensive aim is seen — not as mere survival, not as numerical increase, not as increase in the complexity of the organisation, or increased control over his environment — but as greater fulfilment, the fuller realisation of more possibilities by human species collectively and more of its component members individually". Ethics hence influences the socio-economic environment, while the socio-economic ethos is itself determined by the interaction of cultural heritage, ethics and psycho-social factors.

If one does not care for the laws of society, he is termed unethical. Professional ethics is necessary to channelise the latent talents of individuals for productive purposes. What could be more satisfying than to be engaged in work, in which every capacity or talent one may have is needed, every lesson one may have learned is used, and every value one cares about is furthered. Ethical behaviour of a man earns him respect, social recognition, moral courage and above all inner satisfaction. Ethics is concerned with the way of one's life which he has to lead according to the circumstances in which he is asked to lead it.

Business ethics may be defined as a self-generating system of moral standards in the realm of business to which a substantial majority of business executives gives voluntary assent, which leads to industrywise acceptance of certain standards of practical conduct. Personal ethics, have their source in a person's religion and philosophy of life. As the ethics of the individuals decline, the level of group ethics also decreases and degenerates. There must be some countervailing force such as altruism to offset the coding effect of selfishness, which is inherent in competition. To preserve the ethical standards, at personal, company, industrial and social levels — the service conditions and salary structure of the professional should be such that it may not leave room for them to fall prey to temptations and allurements.

PROFESSIONALISATION

If there is such a thing as a profession, as a concept distinct from a vocation, it must consist in the ideals in which its members maintain the dignity of character which they bring to the performance of duties and the austerity of the self-imposed ethical standards. To constitute a true profession there must be ethical tradition so potent as to ennoble and elevate the members whose personal standards of conduct are at a lower level and have an elevating and enabling effect. Professionalism cannot be created by resolution or overnight. It requires many years for development and they must be years of self denial, years when success by base means is scorned, years when no results bring honour except those free from the taint of unworthy deeds.

A TO Z TIPS OF ETHICAL RESPONSIBILITIES

It is easy to formulate codes for any profession, but these must be followed. The basic requirements for an individual to become professional are as follows:

(a) Faith in his profession;

(b) Loyalty to the organisation;

(c) Justice to those with whom he deals;

(d) Sound commercial transactions;

(e) Active cooperation between the executive and the customers;

(f) Disseminate information relative to and promote the welfare of the company and the industry;

(g) Promise between understanding of the ethics and his own functional aspects;

(h) Declining any advertisement that has a tendency to mislead or that does not conform to business integrity.

Specific Aspects

(i) Considering first the interest of this organisation in all transactions and believing in its established policies;

(j) Being receptive to competent counsel from his colleagues and being guided by such counsel without impairing the dignity and responsibility, of his office;

(k) Buying/selling without prejudice, seeking to obtain the maximum ultimate value for each rupee of the company or the customer;

(l) Striving consistently for knowledge of the materials and processes of manufacture and establishing practical methods for the conduct of his office;

(m) Subscribing to and working for honesty and truth in buying and selling, denouncing all forms or manifestations of commercial bribery;

(n) According prompt and courteous reception — so far as conditions will permit to all who call on legitimate business mission;

(o) Respecting others' obligations and expecting obligations to him and his concern to be respected, consistent with good business practices;

(p) Avoiding sharp business practices or reciprocity;

(q) Counselling and assisting juniors and peers in the performance of their duties, whenever occasion permits;

(r) Cooperating with all organisations and individuals engaged in activities designed to enhance the development and standing of the profession;

(s) Buying and selling on the basis of value, recognising that value represents that combination of quality, service and price which ensures greatest economy to the ultimate user;

(t) Recognising that character is the greatest asset of a person and giving it major consideration in the selection of customers/vendors and also in recruitment of personnel;

(u) Providing or accepting no gifts or entertainment in the guise of sales expense, where the intent or effect is to unduly prejudice the recipients in favour of the donors as against legitimate competition;

(v) Giving or receiving no bribes in the form of money or otherwise in any commercial transactions;

(w) Establishing and maintaining suitable contacts with Association members;

(x) Establishing contacts with and arousing interest of prospective members in local associations and at the national level;

(y) Publishing nothing that will affect the good name of an individual, an association, a company or the concerned profession;

(z) Accepting advertising solely upon the values of the medium to the advertisement.

Even though the professional may be firm in his dealings, he has to be polite and courteous.

A considerable and thoughtful professional, dedicated to sound ethical standards can do much to raise the stature of himself, his organisation, his profession and build up the image of the nation.

BUYER BEWARE

The emphasis of any individual, department, organisation, conference, training programme and journal should be on the recognition of the professional aspects of integrity of the materials manager and the appreciation of the relevance of ethical considerations in materials. The contemporary international and national compulsions recognise the relevance of socio-economic conditions and of ethics along with commercial profits.

In the relentless persuit of accumulation of wealth, one hopes that Oliver Goldsmith's words, "Ill fares the land to hastening ills a prey; where wealth accumulates and men decay", do not come true. It is, however, a fact that despite several developments, the ancient warning of 'Caveat Emptor' or buyer-beware still operates. The words of George Herbett. "The buyer needs a hundred eyes, the seller not one" are pertinent today.

Codes have been developed by several professional Associations including the Indian Institute of Materials Management. But codes are cold words. What is required is setting up examples by deeds in the sincere adherence of the codes in practice. In order that people may be happy in their work, they must be fit for it, they must not do too much of it and they must have a sense of satisfaction. The secret of success is the consistency to pursue professionally. The professional buyer must believe in the philosophy: "To spend money is easy, but to spend it well is hard." Ethical behaviour should characterise every profession and is a mark of the professional purchasing executive. Ethics in buying introduces the concept of integrity which implies honesty in the treatment of other people and leads to the highest type of professional development. Good buying ethics are not limited to refusing gifts from the suppliers, but also involves issues such as conflicts of interest on the part of top management in private, public and Government sectors — perhaps the cruelest form of hypocrisy is that the buyer is often only an accomplice and an unwilling one at that.

PROFESSIONALISM — GUIDELINES

Professionalism must have a purpose. The materials manager should be driven by the vital urges of human sympathy and awakened social conscience and his functions should reflect these. One of the attributes of professionalism is efficiency which is the means for successful completion of a task. The Chando-gaya Upanishad beautifully prescribes the means to achieve success in one's efforts.

Yadeva Vidyya karoti, shraddhaya, upanishada
Tadeva veerya-vattaram bhavati

This is translated as "What is to be done, should be done diligently through *Vidya* or knowledge *Shraddha* — conviction, or faith in oneself and his task and *Upanishada* — or deep thinking or contemplation, if it is to be successful." The manager should therefore acquire the requisite thinking through *Vidya*, have the necessary faith in himself and commitment to achieve his purpose with *Shraddha* and with deep thinking, act with the concentration of all three energies which will make action most human and efficient. The harmonious merger of technical knowledge, personal commitment and deep awareness alone can lead to efficiency and excellence.

Arthur Koestler, in his *Yogi and the Commissar* has said that man can succeed and achieve his goals only if he combines in himself the qualities of a yogi and a commissar. Like a Yogi, the professional is one who objectively reflects and, like water on lotus leaf, in a detached manner conceptualises his work. Like a commissar he is ruthless, efficient and committed to his task. His prescription for success has its roots in the Indian tradition of a yogi with equanimity of mind — Samatam Yoga Uchyate.

PHILOSOPHY OF ETHICS

In the context of ethics, it is easy to sermonise and we have a large number of such illustrations. Mahatma Gandhiji, while dealing with trusteeship and mutual confidence, has condemned the six ills of society as follows:

(a) Politics without principles
(b) Wealth without work
(c) Pleasure without commitment
(d) Knowledge without character
(e) Science without humanity
(f) Commerce without morality.

Some Government organisations have stipulated that gifts beyond Rs 100 must not be accepted. This is circumvented by the officers by receiving gifts worth less than Rs 100, every day from each client.

Some organisations ask the following questions to improve ethical practices:

(a) Do we have in our organisations well-considered statements of ethical principles to guide our employees in specific situations that arise in our business activities?

(b) Have those statements been the fruit of discussions in which all members had an opportunity to participate?

(c) Have we given our employees at all levels sufficient motivation to apply high ethical standards in their solutions?

(d) Have we provided the employees an easily accessible means of obtaining counsel on and resolution of ethical problems?

(e) Do we know whether our employees apply in their daily activities the ethical standards we have promulgated?

In the context of professional ethics, the quotation engraved on a Tablet at the entrance of the Indian Military Academy, Dehradun is worth quoting. "The safety, honour and welfare of your country always come first. The honour, welfare and comfort of the men you command comes next. Your own comfort, safety comes last, always in the end everywhere". This is the same philosophy propounded by the late President J F Kennedy of United States, when he said "Ask not what you can get from the country, but ask what you can do to the country".

So the materials manager should be a professional of high ethics commanding, not demanding the respect of all his peers, seniors & juniors.

TEN POINTS TO PONDER

1. Define the concept of business ethics.
2. Discuss the meaning of professionalisation.
3. Clarify the term professional ethics.
4. Explain the relevance of professional ethics to materials manager.
5. Enumerate the steps to be taken by the materials manager to improve ethics.
6. Illustrate the phrase 'buyer beware' with examples.
7. Examine the impact of buyer-seller relations on professional ethics.
8. Describe the code of conduct, for materials managers, developed by the professional organisation.
9. Explain the impact of individual professional ethics on the organisation's image.
10. Elucidate the concept of double standard on professional ethics as adopted by organisations in buying and selling.

67. Evaluation of Materials Management

NEED FOR EVALUATION

"We live by sales but make our profits from materials" is the slogan adopted in many progressive organisations, and this must be the central theme behind the evaluation of the materials management function. A well-designed evaluation system, primarily aims at improving the performance of the personnel, with resultant increase in the overall benefits of the organisation. The process of evaluation of materials function becomes all the more important as materials account for 60 per cent of the cash outflow and the inventory constitutes the single largest asset in the organisation. If the organisation has to be effective, it has to evaluate its performance and introduce necessary controls. A detailed approach, to analysing the performance of the materials, inventory manager is also necessary to identify the needs for improvement and to plan for individuals development programmes.

BACKGROUND SCENARIO

The evaluation process should not degenerate into a witch hunting exercise. The common Yardstick used to measure the performance of the material function is the number of stockouts — which is a negative measure. The service rendered is usually forgotten. Such an approach will be a demotivator. The inventory in any organisation consists of four components under four different heads of departments — namely, raw material with the materials function, machinery spares with maintenance, semi-finished goods under the production function and the finished goods with the marketing manager.

For the overall performance, it is necessary to recognise this reality and evaluate each one of them and then coordinate the process by properly synthesising them.

Internal Evaluation

The evaluation of materials/inventory functions can be done either by external agencies or internally. Within the organisation, the departmental staff themselves must believe in the philosophy that improvement is a continuous process. The user departments evaluate on the basis of stockout and the finance on the basis of the funds locked up in inventories. The peers/colleagues in maintenance, marketing, production, data processing and finance discuss the performance of the inventory/warehousing function in the coordination meetings. The top management based on the reports on key result areas, tries to evaluate the materials, inventory/warehousing functions. Sometimes the vigilance and internal audit are also associated with the evaluation process. Occasionally internal committees/task forces try to evaluate the performance.

External Evaluation

Many external agencies evaluate the materials/warehousing function. Just as the buyer rates the vendor, so also the supplies/ancillary units rate the buyer in turn, with regard to negotiating capabilities, adherence to quality, promptness in bill settlement, and commitment to prior appointments. The banks evaluate the firm with regard to credit-worthiness. The clients and customers assess the image of the company by its services. The professional Associations and industry/Chambers of commerce evaluate the company on the basis of ethical practices. External consultants are also sometimes associated for the purpose of improving the working. The other external agencies include transporters, carriers, railways, customs, excise, clearing agents, insurance agents, shipping agents, inspection agents, licensing departments, etc.

METHODS OF EVALUATION

In the context of evaluation, the overall objective of materials function — ensuring continuity of supply with lowest cost of the ultimate product — plays a vital role. The process of evaluation can be done on inter-firm comparison basis, i.e. comparing the performances of similar companies in the year, or intra-firm comparison, i.e. comparing the performance of the same company over a period of duration. The criteria must include all managerial aspects of forecasting, planning, coordination, control, etc. and must be the same for all the different disciplines and functions. The process in all the cases is to formulate norms and compare with the actuals.

The characteristics of evaluation encompass subjective and objective factors. The subjective factors include source development, training, morale, motivation, promptness, courteous service, good buyer-seller relations, exception reporting, user satisfaction, manager's competence, on technical operational and judgmental aspects, attitude to modern management, adherence to procedures values, systems, principles and precedences, development of manual, updating knowledge/skills etc.

Objective Criteria

The objective criteria which can be measured include information based on quantities, reports, indexes, ratios and monetary values of the performance. These include the frequency and intensity of stockouts, various kinds of ratios on consumption, obsolescence, costs, cost reduction used for planning and control. The objective criteria must be based on accurate and uptodate knowledge. The use of computer is also resorted to in the case of objective criteria. However, certain amount of subjectivity will always be there in the evaluation process as it is difficult to eliminate biases.

MBO Approach

Within the materials function, there should be an introspective measure of its own performance. For this purpose, the approach, the Management By Objective (MBO) is a popular one. In the MBO approach, objectives/goals/targets are set after discussions between the boss and the subordinate and the performance is measured as to how the set goals have been achieved. The key results or key successful factors for the function are identified first by the materials department, and an action plan is finalised before discussing with the top management. The performance at the end of the year is reviewed to identify the reasons for success/failure, which are incorporated in the subsequent plans. In this process, SWOT (Strength, Weakness, Opportunity and Threat) analysis is helpful to analyse the environment in terms of the opportunities, that are available for achievement and the threats that have to be overcome. In this scenario, individual strengths and weaknesses are deliberated on, so that the strengths can be increased and the weaknesses reduced.

The participating style of management by objectives approach deviates slightly from the Karma theory of abdicating everything to the lord. Bhagavad Gita, Chapter 2, Sloka 47 says:

Karmanyeva adhikaraysthe, Ma phaleshu kadhachana
Ma karma phala hethur, Bhumarte sangasthava karmani

Translation of this reads as follows: "Action is thy duty, fruit is not thy concern. Without expecting fruits do your duty". Obviously in the present day world, the executive not only expects benefits, but also wants to be consulted in the decision-making process instead of leaving it in divine hands.

RATIO ANALYSIS

Often ratio analysis — of putting one value in the numerator and related value in the denominator — is used as a measure of the evaluation process. A few illustrative ratios that can be used for comparisons are now given, and these can be used for evaluating the performance of stores, purchase, inventory, spares and working capital management.

(a) *Warehousing efficiency index* is obtained as the ratio of the number of requisitions delivered on time to total number of requisitions. Efforts should be made to have this rate as near to unity as possible.

(b) *Warehousing loss index*, giving a measure of how well the stocks have been preserved, handled and issued, is obtained by dividing value of annual storage loss to the average value of stocks. Efforts should be made to reduce this index.

(c) Similarly, we can build ratios for obsolescence loss, pilferage loss, and handling loss.

(d) The *handling cost index*, obtained by dividing the total handling cost by the total value of materials, expresses the handling cost as a fraction value of the materials.

(e) We can have similar devices for capacity utilisation, time spent for inspection, etc.

Inventory Turnover Ratios

(f) The *working capital turnover ratio*, indicating sales divided by the working capital, is an index of the efficiency of the working capital management. The average figure of this important ratio is about four for the Indian economy. However, this ratio is small, around one, in the case of heavy engineering organisations and is high, about 12, for trading organisations.

(g) The *ratio of sales to total inventory* is another important ratio used in many organisations. It is also necessary to evaluate each component of the working capital in terms of number of days, such as:

(i) Raw material inventory value in terms of the number of days consumption, (ii) Total spare parts inventory value in terms of number of years consumption, (iii) Semi-finished goods inventory in terms of monthly production value, (iv) Finished goods inventory in terms of number of days sales. It should again be mentioned that these components are under four separate heads of departments.

Financial Ratios

(h) The *receivable turnover ratio* — obtained by expressing the receivables to average daily sales — shows the efficiency in collection of receivables. Tandon Committee norms can also be consulted while comparing ratios in pts. (f)–(h).

(i) The *ratio of the cost of goods* sold during a year to the inventory value, indicates the quickness with which the inventory is being turned over in the business.

(j) *Liquidity ratio*, or ratio of quick assets to current liabilities, measures the ability of the firm to meet the short-term obligations immediately.

(k) *Current ratio* or ratio of current assets to current liabilities — gives a measure of the ability of the company to meet its current liabilities in time.

(l) The *material cost index* is obtained by the material cost of a product to the total cost of the product.

Planning and Forecasting Indexes

(m) The *vendor rating quality index* is obtained as a ratio of the number of lots rejected to the total lots received. Similar ratios on delivery, price, service and other aspects of vendor can be obtained.

(n) *Forecast accuracy index* is measured by dividing the forecast price to the actual price.

(o) *Planning efficiency index* is the ratio of the value of rush orders to the total value of orders.

(p) The *insurance cost index* is the ratio of the total cost of insurance to total value of insurance.

(q) The *lead time index* gives the ratio of observed average lead time to long lead items, to the ratio of planned lead time of the same long lead items.

(r) *Manpower cost index* gives the ratio of the wage bill of the materials department to the total wage bill of the company.

(s) *Obsolescence value index* gives the value of non-moving inventory items to the total inventory value.

(t) The *freight index* gives the ratio of lowest rate bid for freight to the total rate paid.

(u) The other efficiency indexes that can be worked out include the following areas:

(i) Manpower strength, (ii) Order cost, (iii) Crash order cost, (iv) Order efficiency, (v) Clerical efficiency, (vi) Purchase efficiency, (vii) Communication efficiency, (viii) Time utilisation, (ix) Claims, (x) Negotiation efficiency, (xi) Indigenous source development, (xii) Process efficiency, (xiii) Index confirmed as per requirement, (xiv) Pending orders, (xv) Follow-up efforts, (xvi) Stockout, (xvii) Priority of importance-cost-criticality availability, (xviii) Production loss, (xix) Work-in-progress, (xx) Spare parts turnover, (xxi) Discrepancy between physical value and book value, (xxii) Storage loss, (xxiii) Scrap loss, (xxiv) Standardisation, (xxv) Codification, (xxvi) Vehicle usage, (xxvii) Goods in transit value etc.

TEN POINTS TO PONDER

1. Examine the relevance of evaluation in materials and warehousing function.
2. Identify the background conditions and difficulties in the process of evaluation.
3. Discuss the internal agencies interested in evaluation.

4. Elaborate on the interests of the different external agencies in evaluation.
5. Distinguish between inter-firm and intra-firm comparisons.
6. Enumerate the subjective criteria of evaluating materials/warehousing functions.
7. Explain the significance of MBO approach in the process of evaluation.
8. Articulate on the utility of sales to inventory turnover ratio in the process of evaluation of materials/warehousing function.
9. Discuss the utility of financial ratios in the evaluation process.
10. Describe the planning efficiency indexes used in the evaluation of the materials/warehousing function.

PART VI

INTEGRATED POLICY CASE

68. Ganapathy Ram Petrochemicals

INDIAN PETROCHEMICALS SCENARIO

Dr. Ganapathy Ram, a distinguished Chemical Engineer, who has been associated with putting up petrochemical projects, is the Chairman of Ganapathy Ram Petrochemicals. He invited the author, who is the Seniormost Faculty in the Administrative Staff College of India, the premier Indian Management Institution, to suggest solution for the managerial problems faced by the Company.

Dr. Ganapathy Ram has painted the Indian petrochemical scenario in the following manner:

Chemical products derived from petroleum sources such as plastics, synthetics, synthetic rubber and miscellaneous organic chemicals and the intermediates constitute petrochemicals. Petrochemicals have become all pervading utility products in our daily chores in the form of tooth brush, tongue cleaner, soaps, detergents, utensils, polyester fabric, pills packed in laminated foils, plastic buckets, water tanks, etc. Industries like electronics and plastics survive on the lifeline of petrochemicals. It is now possible to produce materials with unique properties for specific end uses. This has resulted in not only the technological materials superiority of the product but also a price advantage over traditional like metal, wood, glass, cotton, wool, natural rubber, leather, timber, etc. For a developing economy like ours, petrochemicals are of very great economic relevance from the point of view of supplementing materials resources as well as providing employment.

The supply-demand gap has been widening resulting in the sky rocketing of world prices of major petrochemicals in early 1990s. The main reasons touted for high cost petrochemicals are the demand-supply gap, as well as the price differential, between landed cost of imports and domestic prices, being as high as 80 per cent.

Sometime back Bombay Dyeing started taking Rs 2500 per ton of deposit on each ton of DMT — Dimethyl terepthalate — in case the Government were to rescind the duty hike on paraxylene and the Government has asked Bombay Dyeing to stop collection of deposits forthwith. Although synthetic fibre and petrochemicals plants are operating at full capacity, whatever is produced is sold within no time. The Government has cut import and excise duties on a wide variety of raw material plastics like polypropylyne, HDPE, LDPE, PVC,

Polystryne, PFY, PSF and intermediaries like DMT, caprolectum, acrylonitride and mono ethylene glycol. However, users of PSF and PFY have complained that the domestic manufacturers like Reliance industries have raised their prices before the budgets, and introduce the cut only to maintain the status quo.

The IPCL produces 70 per cent of its requirements of paraxylene from its plant at Baroda and Bongaigaon refineries. The excise duty some time ago on paraxylene was a crude attempt to harm the interests of Bombay Dyeing.

The Reliance industry also manufactures PTA from its downstream captive facility for paraxylene.

The Finance Ministry has allowed the duty cut on MEG in the context of protests by the Association of synthetic fibre industry which remained silent on the duty hike on paraxylene. It is to be noted that the polyester fibre cost is more reactive on the price changes of paraxylene than on MEG, as one ton of DMT/PTA needs 0.7 ton of paraxylene and about 0.3 times of MEG. PTA stands for purified terephthalic acid and it is used in manufacture of synthetic fibres and yarn.

PETROCOMPLEX SCENARIO

The Liner Alkyl Benzene (LAB) project of Hindustan Lever, located in Punjab and based on the two critical raw materials — N-paraffin and benzene. The annual capacity is 60,000 tons and owing to delay by Government clearance, the cost escalated to around Rs 230 crores. The functionalised biopolymers project involves the use of technology supplied by National Starch and Chemical Corporation, an affiliate of Unilever.

Manali Petrocomplex

The Manali Petrochemical complex near Madras is one of the largest in the country with the commissioning of Manali Petrochemicals, UB Petroproducts and Cetex Petrochemicals, accounting for a total outlay of nearly Rs 10,000 crores — including the refinery and fertiliser. Both MRL and UBPL are engaged in the production of propylene glycol and polyols, the former required in the manufacture of pharmaceuticals, cosmetics, food flavours and other items besides tobacco processing. Polyols is used for producing poly-urethene foams. It has also other applications in industrial insulation, refrigeration, furniture, automobile components, and footwear. Both chemicals have been imported till 1990. Manali Petrochemicals is utilising the technology of ATO Chem through Technip of France for propylene oxide and propylene glycol and that of Arco of U.S. for the polyol project. Press industria AG of Switzerland is providing the know-how to UB Petroproducts. The feed stock propylene will be supplied by Madras refineries.

The Celetex petrochemicals of RPG enterprises is a smaller project, with an outlay of Rs 20 crores. The versatile solvent methyl ethyl ketone will be produced with the technology of Edeleanace of West Germany. This petrochemical will be substituting imports and the manufactured solvent has wide application in the production of lube oil, magnetic tube, printing inks, paints and varnishes. The existing facilities are also being used to manufacture MEK oxide and MEK peroxide which are used in the production of audio and video tapes.

Kothari Chemicals is producing polybutenes with the assistance of Cosden Technology Inc. of the USA. Polybutene is used in the manufacture of lube oil, adhesives, jelly filling compounds, adhesives and electrical cable compounds. The feedstock for this will be supplied by Madras Refineries.

Tamilnadu Petro Products — a joint venture of Southern Petrochemical Industries Corporation and Tamilnadu Industrial Development Corporation — is producing linear alkhyl benezene and heavy normal paraffin, with the know-how supplied by UOP of USA. Madras Refinery is producing paraffin wax with the technology provided by Union Oil of California.

MRL, along with SPIC, are producing purified terepthalic acid — (PTA) and paraxylene and orthoxylene, MRL is producing lube additives with technical and financial collaboration with Chevron Chemicals of USA.

Haldia Petrocomplex

The Chairman of Tata Chemicals has bagged the giant Rs 3000 crores, Haldia petrochemicals project, a most politicised joint sector mega project, with West Bengal Industrial Development Corporation. The joint sector company is setting up the Naphtha Cracker plant with major downstream products such as HDPE, PP, HDPE/LDPE, styrene, polystyrene, butene — 1, butadine, and acetylene black. Haldia has initially been taken as a division of Tata Tea with 11 per cent of equity vested with WBIDC. With the primary stock market and through innovative schemes, an equity of Rs 750 crore has been raised. The debt component from financial institutions is about Rs 750 crores. The Government was convinced that fragmentation of mother cracker and the major downstream projects was not in the interest of the project. The project is also self-reliant on feedback supply. The Tata top management was also convinced that the recurring revenue expenditure for naphtha imports could be taken care of, by exporting polypropylene, which is an exportable star polymer. Since the cost of storage of feedback stock is very high, the downstream units must accept ethylene, propylene and butadine on the same day, when the cracker factory becomes operational.

The annual input of one million tonnes of naptha, 'cracked' at very high temperatures, is producing 300,000 tons of ethylene, 150,000 tons of propylene, 53,000 tons of butadiene, 65,000 tons of benezene, 100,000 tons of polyethylene, 160,000 tons of LLDPPE and 80,000 tons of styrene. This is partly used to produce 40,000 tons of polystyrene, 150,000 tons polypropylene and 4500 tons of acetylene block. The complex also produces 3700 tons of C_4 raffinate. The product mix has been optimised by choosing the high value added products since PVC ranks much lower among polymers in terms of value addition.

IPCL Complex

Table 68.1 gives the capacity of IPCL considering the new cracker plants. The demand will be only 75 per cent of installed capacity in most cases.

Table 68.1 Installed Capacity of Petrochemicals

Products	Capacity 1995 (000 tons)		
	Industry	IPCL	IPCL share (%)
Basic Chemicals			
Ethylene	2,161	830	38
Propylene	798	173	21
Butadiene	118	22	19
Plastics			
HDPE	380	130	34
LDPE	292	160	55
LLDPE	180	80	44
PP	350	150	43
PVC	864	205	24
Sub total for plastics	2,066	725	35
Fibres and Chemical Intermediates			
Acrylic fibre	84	24	29
Caprolactum	120	Nil	Nil
DMT/PTA	775	30	4
LAB	284	44	15
MEG	328	115	35

The installed capacity will go up by another million ton, if all planned projects fructify — IPCL's Maharashtra Gas Cracker, Reliance Hazira complex, NOCIL's expansion, IPCL's Gandhar Project, Auriya Project of Gas Authority, Vizag Project of UB Groups, Haldia Petrocomplex, Polyolyfines Ltd. Manali Petrocomplex etc. This may result in capital cost per ton of ethylene being Rs 60,000, 40 per cent lower than Haldia's. Birlas have shown their eagerness to enter the petrochem field in a big way. G.P. Birla group through the Orient Paper, has made a bid, for Rs 125 crores HDPE project at Vizag. M.P. Birla group through Universal Cables, has sought a licence of 65 crores polyethene project. Du Pont, Canada have signed an agreement with Reliance Industries for a world-scale polyethylene plant which is expected to go on stream a Hazira. The Ambanis are trying to put up an industry of aromatics and aromatic based products like PTA and DMT with the collaboration of MRL at Manali. The Mukund group is likely to invest over Rs 1000 crores in the petrochemical complex.

Several world renowned companies have offered collaboration proposal to set up petrochemical complexes in India. Mitsui of Japan has offered a tie-up for the Vizag project. Shell International and British Petroleum have shown interest in Hazira complex. Lenide of FRG has also offered a tie-up proposal. Besides, the world Bank has shown interest in the Hazira complex.

The expansion of IPCL Baroda, and the Maharashtra Gas Cracker complex, have become fully operational since 1990. IPCL has a product range of about 30, against the competitors' range of about 15. IPCL has also a countrywide marketing network, enabling to sell larger volumes and wider ranges. With its experience and financial strength, IPCL is expected to be much ahead of others, as it reaches the market ahead of others with lower cost of production. IPCL has already increased its DMT and xylene capacities as well as butadiene extraction and PBR production facilities.

Alpha olefins are another speciality group of chemicals made by IPCL. Their main usage will be in detergent-making as an active matter replacing LAB in terms of working capacity, biodegradability and friendliness to human skin. IPCL is entering the engineering plastics, having signed in a memorandum of understanding with General Electric USA. Polycarbonates and modified polyphenyl oxide will be produced shortly. According to the Chairman of IPCL "No one will be able to offer a better product or better terms, if free and fair market forces are allowed to operate." With new entrants like Ambanis, Oswals, Mallyas, Muthaiahs, and some refineries planning to diversify into petrochemicals in the near future, there will be pressure on established units to reduce price, and production and improve the quality of petrochemical products.

Since about Rs 10 billion are likely to be invested in the petrochemicals field, by the turn of the century, there will be corporate clashes, political overtones, pressure for review of costs by the cost accounting branch of Government, kickbacks, commissions, lobbying in corridors of powers, etc. In this context, Ganapathy Ram Petrochemicals is one of the major joint sector petrochemical organisations in India having collaboration with reputed foreign firms for different products. Dr Ganapathy Ram has also mentioned that joint sector is the catch word these days to please politicians and this acts as a hedge against interference in day-to-day working. The layout for the plants in Ganapathy Ram Petrochemicals has been so designed that the capacity can be easily increased with no large fresh capital outlay. Dr Ganapathy Ram has emphasised that his plant is contributing savings in foreign exchange by producing imported items. He also emphasised that his corporate philosophy has been as follows:

> The only way you can stay ahead is not by financial wizardry but by having expert people with character.

CORPORATE MISSION

The corporate mission of Ganapathy Ram Petrochemicals is as follows:

(a) Create a sound base for the manufacture of major petrochemicals in the country and produce petrochemicals of international standards.

(b) Assist the user industries by providing technical guidance, on optimal use of petrochemicals where required.

(c) Create a strong research base in the field of petrochemicals.

(d) Earn a reasonable return on investment.

(e) Provide prompt service and quality products at reasonable price to the consumers.

(f) Ensure prompt dealings with integrity, impartiality and courtesy to the suppliers.

Dr Ganapathy Ram is the Chairman and Managing Director of the organisation. He is assisted by 5 Vice presidents for the functions of finance, projects, operations, personnel and marketing.

PERSONNEL POLICIES

Human resources are considered as the major asset in the organisation according to the Vice President Personnel. He is incharge of manpower planning, selection, recruitment, induction, training, promotion, grievance-handling, training, career planning, discipline matters, transport, travel, security, accommodation, labour welfare and union matters. He strongly believes and advocates the principle of 'do by delegation.' The total employment is 4500 engineers and 5500 other categories of staff. His department is also incharge of the formal monthly coordination meetings at different levels.

The Company's personnel policy states as follows:

(a) Develop the individual's capabilities through the provision of appropriate opportunities for advancement.

(b) Expedite redressal of grievances by the spot supervisors.

(c) Deal fairly, but firmly with the recognised representatives of employees in pursuance of healthy trade union practices.

(d) Associate employees with management by way of participation in various joint bodies.

(e) Improve the condition of backward class as per national policies.

(f) Impart training in all vital functional areas of respective disciplines.

(g) Cast the foundation for development of managers to shoulder higher responsibilities.

(h) Bridge gap between demand and availability of managers both qualitatively and quantitatively by the training school attached to the plant or by external courses.

(i) Conduct periodic assessment, at the end of training course and at the end of the year.

A few employees, belonging to the unrecognised union wanted training to be conducted only in the manufacturer's plant abroad, as was done in the initial days, and threatened to go on strike.

TECHNOLOGY AND RESEARCH

The Company is adding downstream and sidestream projects. The Project Vice-President is incharge of technology, research, foreign collaboration, development, pilot plant, quality control, inspection, energy conservation, safety, material specification etc. He is assisted by three General Managers for the above jobs with adequate number of officers.

Ganapathy Ram Petrochemicals has a well-equipped research centre, established for undertaking research in analytical spectroscopy, organic chemistry, physical chemistry, materials science, biology, pollution control, process development, effluent treatment and environmental control.

The company recognises the need to adapt to imminent environmental changes and attaches great importance to pollution control. The states' authorities on pollution control have repeatedly drawn the company's attention for better effluent treatment and pollution control.

The directional trends in the technological front indicate almost a singular obsession with process conversions and operations, safety in operating practices, energy management, and economics of alternative manufacturing routes on the products front, advances in material sciences are opening up avenues towards performance products to meet stringent and novel user requirements. But, the increasing economics of scale and access to technology for special products or processes make entry barriers higher. The Project Vice-President is confident of tackling these issues.

The current research projects include development of a paraffin dehydrogenation catalyst, experiments with low cost polymer coating for weather protection, development of rocket fuel, etc. Close coordination has been established with various scientific and research establishments of the country for sharing knowledge.

FINANCIAL ASPECTS

The Vice-President, Finance is incharge of finance, capital expenditure, working capital, audit, stock valuation, costing, cost control, budgeting and overall corporate profitability. The gross block is about 2500 crores. The equity is Rs 200 crores reserves and surplus 600 crores, current assets Rs 900 crores, current liabilities Rs 650 crores, borrowing and deferred credits Rs 1250 crores. The profit before taxes and depreciation is about Rs 250 crores. The annual sales turnover is around Rs 1500 crores. The company expects to incur an expenditure of about Rs 5000 crores in the next two years on new projects. The company also proposes to use human resources accounting and present it in the balance sheet.

The Vice-President, Marketing is incharge of market intelligence, public relations, advertisement, and he has realised that he has to face competition from a large number of manufacturers in the future. He is preparing a detailed market strategy for this purpose. Periodic seminars on use of the company's products, machinery development assistance and use of processing aids are held. The Marketing Federations and milk dairies of various states, have been approached to use the company's end products. The use of plastics in agriculture is also being pursued by the company with the help of the Government of India, selecting 15 districts for intensive preparation of plasticulture application. Workshops are periodically held in different parts of the country to popularise the company's products.

PLANT OPERATIONS

The Vice-president' operations is incharge of the total operations of the company and controls them with the help of three General Managers. The plants are xylene, dimethyl terephthalate, napthacracker, butadiene extractor, aromatics extraction, acrylanitrile, polybutadiene, rubber, linear, alkyl benezene, polypropylene, surfactants, cable compounds, homopolymer, ethylene glycol, polyvinyl chloride, acrylates, petroleum resins, low density polyethylene, linear low density and high density polyphylene, and polypropylene. The plants have been operating after obtaining foreign collaboration with different companies/countries.

The Vice-President, operations is also incharge of overall maintenance and materials, which are looked after by two General Managers. The materials function includes purchases, stores, inventory control, import substitution, follow-up, source development, materials intelligence, and cost reduction techniques. The total number of spare parts is about two lakhs which includes 35,000 bearings. The number of chemicals and other items is about 10,000. A ten digit codification has been introduced recently in the plant. The annual materials consumption including fuel and utilities is about Rs 600 crores with the imports accounting for about Rs 50 crores. The inventory value is around Rs 310 crores which includes about 20 crores worth of non-moving machinery spares.

The ordering charges are Rs 1000 per order and the inventory carrying costs are 30 per cent on the average value of inventory carried. The stockout cost or the criticality analysis has not been done. The internal administration lead time is about four months from indent to trigger off a purchase order. The company aims to have two vendors for each item but about 25 per cent are proprietary items.

MAINTENANCE PRACTICES

Engineering Services at Ganapathy Ram Petrocomplex is an important activity under the General Manager Maintenance, and covers the following functional area of plant engineering:

(a) Maintenance/overhaul of critical machines.
(b) Trouble-shooting.
(c) Design and detailed engineering for modification schemes.
(d) Undertaking minor plant projects.
(e) Cost and time estimation.
(f) Formulation and award of works contracts.
(g) Centralised workshops.
(h) Upkeep and deployment of construction equipment.
(i) Planning and placement of engineering manpower.
(j) Training of engineers and technicians.
(k) Technical evaluation of procurement cases/equipment selection.
(l) Spare parts management.
(m) Standardisation of engineering consumables.
(n) Import substitution/Development of vendors.
(o) Streamlining maintenance practices.
(p) Corrosion and inspection services.
(q) Overhauling and turnaround activities.
(r) Technical literature.

(s) As such, the engineering services have been conceptually partly decentralised.

(t) While the above responsibilities rest with the central group, routine maintenance in each plant is carried out by a task force headed by a Maintenance Manager in each plant.

(u) The maintenance task force consists of supervisory and non-supervisory employees in all the related engineering disciplines.

(v) Utilities have been integrated with a view of optimising and achieving economy of operations. These are: Power, water, steam, nitrogen, air, refrigeration, fire fighting, effluent treatment, etc.

(w) The set-up of plant maintenance vis-a-vis ESD has been enumerated in detail in the "Engineering Services Manual" being circulated herewith. It also explains at length various systems and procedures adopted.

(x) A General Manager reporting to Vice-President Operations is incharge of the total maintenance activities. He is assisted by specialists in all engineering disciplines like mechanical, instrumentation, civil, corrosion, design, modification, electronic, workshops, etc.

(y) Spare parts constitutes 75 per cent of maintenance budget. In all, 500 engineers and other categories of staff are associated with maintenance activities.

(z) All the staff have been trained in scientific maintenance management principles like, spare parts management, condition monitoring, signature analysis, vibration analysis, reliability studies, criticality analysis, maintenance systems, preventive, productive, predictive, failure analysis, computerised maintenance, design out, and maintenance documentation. A judicious combination of preventive and breakdown maintenance is the order of the day.

OVERHAULING/TURNAROUND

All planned repairs, major repairs and capital repairs are undertaken by the central maintenance organisation, under the overall control of the Maintenance General Manager. A modern engineering workshop is available which supplies about 10 per cent of the spare part needs of the organisation. Maintenance inspection, bearings department, drawings, tool room and technical literature are under the control of the central maintenance organisation.

Planning for capital repair and overhauling is initiated by the central maintenance and discussed with the operations department. The plant management and purchase section is notified about nine months in advance, to identify the activities. The operating, maintenance and purchase staff meet for planning operations of the shut-down. The spares requirement is prepared as per the recommendations of the original equipment manufacturers supplemented with the maintenance expertise available at the plant. During the shut-down, major repairs, modifications, replacements, reconditioning, refurbishing, and technological upgradations, are carried out in order to give a new lease of life to the equipment with minimum investment.

The maintenance personnel generally insist on making all items available at least two months in advance, in view of the continuous process nature of the industry. The internal audit has pointed out that this step implies locking up about Rs 1 crore of working capital in machinery spares including compressors, rotors, etc. for two months. But the maintenance personnel feel that the loss will be one hundred crores, if this is not adopted as the overhauling will be delayed.

The author interviewed a large number of officers from the different departments in order to seek their opinion with regard to the problems, procedures, systems, suggestions, recommendations. He also perused files, documents, reports formats, systems, procedures and other relevant papers in order to understand the complexities of the company. The summary of findings is now given in the form of A to Z issues.

A TO Z PROBLEMS IN PROJECT MANAGEMENT

Ganapathy Ram Petrochemicals has been taking downstream and sidestream projects, and hence expansion activity always goes on. The another interviewed the Vice-President, Projects and other project staff to ascertain their difficulties. These are listed below:

(a) Most of the projects witness cost overruns — sometimes going up to 100 per cent and time overruns — duration extending beyond one year.

(b) The lead time in decision-making process till awarding the contract is considerable due to social, political, technological, competitive, regulatory, and environmental considerations.

(c) The final decision is usually taken by the Chairman and Financial Vice-President, even though Vice-President Projects and Vice-President Operations are associated with the decisions.

(d) The manufacturing, transporting and inspection lead time, after the award of contract till commissioning of the plant, is considerable, sometimes extending up to three years.

(e) In view of the long time, disputes arise with regard to payment, particularly because of the fluctuating foreign currency values and global escalation price structure.

(f) The supplier insists on payment after completion of each stage of manufacturing and inspection has to be done by reputed, independent, international agencies.

(g) The prequalification of foreign suppliers is a tough assignment which cannot be done without visiting them and such visits are called "foreign jaunts" and usually disallowed by the finance.

(h) The bid evaluation criteria — either single bid or two bid system basis — is not fully scientific, since only few suppliers respond to the notice inviting tenders.

(i) The make/buy/lease criterian are only a theoretical exercise since many foreign agencies are interested in quoting only on turn-key basis.

(j) Even though the contract mentions penalty, liquidity damage, fall clause, risk purchase, etc. it is difficult to practise as the adjudication process is time-consuming.

(k) A major irritant is the providing of drawing for critical spares, and invariably the original equipment manufacturer refuses to give the drawings.

(l) Another problem area is, the supplier dumping unwanted slow-moving spares, as mandatory spares. This usually increases the obsolete spares.

(m) Reliability of intricate parts is not ensured by the supplier, as he himself depends on other subcontractors.

(n) The recommended part list for first overhauling is too large for the Indian culture.

(o) The service after sales, during the warranty period is poor and the suppliers' agents do not attend to the faults immediately.

(p) Even though network diagrams have been made for all the jobs — from planning the project to commissioning the equipment — the chart serves only as wall paper since all activities consume extra time with several parts becoming vital simultaneously.

(q) The supplier does not give details on life cycle costing and the economic life of the equipment.

(r) Even though the objective of a specific project is minimum cost per ton of output, it is not possible to achieve this due to the above conditions.

(s) Due to wrong planning, the new equipment is not compatible with the existing machinery, creating operational difficulties.

(t) In view of the above difficulties, the economic and financial analysis, remain valid only on paper.

(u) The equipment arrives before a stores officer is appointed to receive the machinery.

(v) The communication and coordination between the different parties, namely the equipment manufacturer, his subcontractors, inspection, insurance, transporters, purchase, product follow-up and stores need to be improved.

(w) In spite of continuous monitoring, sometimes the equipment is damaged and corroded when it reaches the site.

(x) Sometimes instructions on safety handling and technical catalogue are not complete.

(y) The temporary staff involved in materials handling, construction, and sons of the soil where the project is located, insist on getting permanent appointments as soon as the project is commissioned.

(z) The stipulated capacity utilisation by the manufacturers and the guarantees offered by them remain theoretical.

A TO Z ISSUES IN PERSONNEL MANAGEMENT

The author interviewed the staff from different sections in order to know the reactions of the corporate personnel policies adopted in Ganapathy Ram Pharmaceuticals. These are given below:

(a) Manpower planning, selection, recruitment, induction, training, career planning, grievance-redressal mechanism, promotion policies, job rotation, motivation and basis for incentives/bonus have been criticised by a sizeable number of persons in Ganapathy Ram Pharmaceuticals.

(b) Political interference at the recruitment level is a common problem faced by the personnel department.

(c) The qualifications have been completely relaxed in the case of scheduled classes, scheduled tribes and sons of the soil.

(d) The new staff do not receive a proper briefing or induction when they join the factory. Their terms of reference, targets, reporting responsibilities have not been adequately clarified to them.

(e) Formal/informal appreciation, constructive criticism, guidance, when needed, are not provided by some departmental Heads.

(f) Junior people with higher qualifications from the Institutes of management have been perceived as a threat by some departmental Heads.

(g) Formal staff meetings to pass officially the expansion plans and corporate performance are rarely held by many departments with the result gossips, rumours, and grapevine thrive in the company.

(h) The noticeboard kept in the department is not up to date and is not used as a vehicle for communication in some departments.

(i) Some Heads of the departments feel that they do not have adequate time to know about the problems of subordinates, their feelings, reactions or ideas.

(j) Some production Heads care more for the output and ignore minor accidents and safety instructions.

(k) The filing system goes on accumulating with the old documents not going to the dead material stores. Hence it takes a long time to process any paper in some departments.

(l) The delays in work schedule, due to late delivery of material and some staff being absent, are realised by the Head of the department only in the last minute creating a crisis situation.

(m) Suggestion schemes on innovations, production operations, technological changes, improved procedures, etc. are rarely welcomed by some departmental Heads who say conforming to current methods is the best.

(n) Few foreign-trained technicians have joined in middle level positions creating difficulties for others.

(o) Few retired defence personnel have been inducted at the middle level in marketing, materials, personnel and finance departments. This has created heart burn amongst some officers who have resigned to join companies in the middle east.

(p) The promotion policy of merit cum seniority has been summed up by many officers as SWEAR BY CAT (Subordination will be encourage and rewarded but your competence will only be tolerated).

(q) Loyalty to the company implies loyalty to the individual — the departmental head or Vice President — according to some staff.

(r) Some officers have accused that promotions have been awarded on the basis of right connections and not due to the individual's competence.

(s) The local politicians, defeated/former/retired/present/members of Parliament/Assembly/Nagar-palikas influence the personnel department not only in recruitment, but also in promotion.

(t) The agricultural unskilled labourers, who have surrendered their lands to Ganapathy Ram Pharmaceuticals, are forcing the company to absorb them in permanent cadres.

(u) Poor handling of grievances has resulted in lightning strikes in some production plants.

(v) Many departments complain about the poor quantity and quality of staff saying that they are managing the work with 50 per cent staff.

(w) There have been periodic complaints about the quality of subsidised food and the union has been openly accusing the contractor of making huge profits.

(x) The few casual labourers involved in construction and expansion have been agitating for making them permanent employees of the company.

(y) The Labour Union has brought to the attention of the management the unsafe methods of operation and has insisted that every employee should be given safety kits.

(z) The Officers Association and the recognised Unions have presented lists of charter of demands, on right to consultation, better service conditions, medical facilities for relatives, education allowance for children, garden allowance, magazine allowance, free water, free electricity, subsidised cooperative stores, increased promotion avenues, group insurance schemes etc.

A TO Z INSPECTION/QUALITY/CONTROL PROBLEMS

The inspection quality control staff informed about their difficulties to the author:

(a) The quality control staff is not only responsible for inspecting the incoming material quality, but also for the inprocess and final product quality.

(b) The inspection expenses including depreciation on instruments works out to about 0.8 per cent of the total production value. The Quality Control Manager has already sent a note for increased manpower and more inspection instruments.

(c) At present precision instruments to differentiate the quality of spurious bearings from others, are not readily available.

(d) Disputes in joint inspection with quality control staff, user materials staff, with the suppliers, have been increasing, resulting in demotivation of the quality control personnel.

(e) The quality control personnel are instructed to accept substandard items, during periods of shortages and emergencies.

(f) The vendors complain against the rejection of some good items by the quality control staff, when the stores do not need the items.

(g) The inspection calibration, testing facilities and other infrastructure are inadequate for the huge petrochemical complex.

(h) The Accounts Department usually does not pass the travelling claims of quality control personnel when they visit the vendors plant for inspection who insist on a separate budget approval for sanctioning the expenditure.

(i) Even though quality control is independent of manufacturing, the users pressurise the quality control to approve the standard items from reputed suppliers.

(j) The users hold the quality control responsible for any breakdown of parts in the production process.

(k) The method of sampling and the number of items to be checked for all items are all based on ad hocism, even though acceptance quality level has been prescribed for many items by the collaborators.

(l) The tendency to pass ISI marked items as good items, without inspecting, has been criticised by the plant personnel, saying that presently the quality has deteriorated with some vendors.

(m) Even though all the inspection staff are engineers with the required specialisation, job rotation between the operating and inspection does not take place.

(n) Sometimes, motives of underhand dealings with some suppliers have been attributed to the quality control inspectors.

(o) The inspection department is yet to develop a proper scientific manual consisting procedures, systems, etc. for various classes of items.

(p) Objective procedures for formally accepting marginal cases particularly for spare parts, developed by the quality control department, is still awaiting clearance from the top management.

(q) The idea of total quality management — including zero defects, in-process inspection, outgoing quality evaluation — is yet to be developed in the company.

(r) Quality control circles for tackling shopfloor problems and procedural isues have been introduced but are not very effective.

(s) It is not uncommon to hasten the inspection process to avert shutdowns in the plant.

(t) The coordination efforts of purchase, maintenance, plant production, inspection and stores need to be improved according to the inspectors.

(u) Often, the inspectors reach the central receiving warehouse at the appointed hours, but it takes a long time for the receiving section to prepare the ground and remove the items for inspection.

(v) There is a need for up to date computerised quality management for fool-proof inspection.

(w) Many inspectors feel that the productivity bonus scheme applicable to the operating personnel must be extended to them. But this has been turned down.

(x) Organisationally, the quality control staff reporting to the Project Vice President is not liked by some inspection staff; they want the Chief of Quality Control to directly report to the Managing Director for better recognition.

(y) Defects have been found in some hardware items, which have been passed on the basis of sampling and the maintenance department has made a complaint on inspection standards.

(z) The General Manager, Quality Control admits the existence of the above problems, but hopes that they will be solved in due course of time.

A TO Z OF MAINTENANCE PROBLEMS

The maintenance engineers, mechanics and fitters confided in the author their problems. The following is an illustrative, but not exhaustive list of the difficulties of maintenance personnel.

(a) In spite of attending to high technology machines and maintaining the sophisticated petrochemical plants, the maintenance personnel have not been given proper status in the organisation. They are still called the *hathoudiwallas* or men with hammer. Only finance and marketing get recognition in the company.

(b) The maintenance staff have not been consulted during the purchase, planning, installation and commissioning of the machinery which they have to maintain. But the top management feels it is not necessary.

(c) The request of the maintenance staff to visit the collaborators plant in order to know the latest techniques has been turned down.

(d) The production personnel do not allow scheduled maintenance to be carried out or even routine inspection as per the collaborator's instructions, particularly during March, where everyone feels production is the most important.

(e) Even though every one knows that the stockout cost is more than the cost of machinery spares, the spare parts are not available when badly needed, but they are available in plenty when not needed.

(f) The original equipment manufacturer is not able to supply genuine spares in spite of sending catalogue numbers and photograph of the machine as he has changed the model of the equipment.

(g) In the absence of original spare part from the original equipment manufacturer, the company has to manage with spurious bearings and items made in USA — Ulhasnagar Sindhi Association!

(h) Since the quality of the indigenous spare is not upto the mark, the maintenance budget is often exceeded.

(i) Life cycle costing has not been done in the plants in Ganpathy Ram Petrochemicals.

(j) In view of the emphasis on tight production scheduling and increased capacity utilisation, quality maintenance work is not possible.

(k) Periodic breakdowns occur due to malfunctioning, overloading, inferior quality of spares and poor maintenance.

(l) The maintenance gets blamed whenever a machine breakdown occurs as the evaluation is always on a negative scale as there is no time to go into the reasons.

(m) The quality of maintenance is affected due to dusty, hot, humid and unhealthy, working environment. Since the temperature sometimes goes upto 50°C or 120°F, some delicate instruments fail, causing bottlenecks.

(n) Drawings have not been obtained from the original equipment manufacturers. No one has taken the initiative to develop drawings.

(o) The tendency to accept the lowest quotation often increases the number of items consumed and hence the cost of maintenance goes up.

(p) The after-sales-service of the local agents of foreign supplier, even during the guarantee period is poor.

(q) The vendors return the blame back to the operations and maintenance personnel, for not adhering to the agreed schedules of maintenance.

(r) The stores department is unable to identify the item, with the result the mechanics are forced to spend a lot of time in the prohibited storage area of the warehouse.

(s) The maintenance engineers are forced to sign huge volumes of reconciliation statements of book balance with physical balance. Once resignation letter was mischievously placed amongst the heap of papers.

(t) The total lead time of converting an indent into an order, manufacturing, transporting from abroad, and inspecting sometimes exceeds five years.

(u) The inspection norms for the quality of spares purchased locally or made in the workshop vary according to the urgency of the demand, with the result that substandard spare parts are often passed into the stores.

(v) Coordination between the central maintenance, plant maintenance, tool rooms, electrical, mechanical, electronic, instrumental, civil, etc. needs to be improved.

(w) The coordination meetings between different maintenance, personnel and between heads of different departments are converted into sparring arena, where everyone wants coordination on his terms.

(x) The technical literature particularly on latest knowledge of maintenance practices, relevant to petrochemical industries is not available in the company.

(y) Due to non-availability of good quality items, cannibalisation of some parts wherever possible has become necessary.

(z) The finance department is asking too many questions on insurance spares, obsolete items, project surplus and cost reduction in maintenance without making any meaningful contribution.

A TO Z OF INDIGENISATION PROBLEMS

The Ancillary and Source Development officer, working under the Materials Manager of Ganapathy Ram Petrochemicals gave several reasons for the tardy progress in the sphere of import substitution, which we now enumerate. The difficulties of vendors, which are innumerable, have been compressed into typical issues.

Supplier's Problems

(a) Vendors want increased price for low quality items, complain about difficulties in increased cost of manufacture with diesel generator sets due to power cut.

(b) They resort to *force majure* clauses and delay the supplies.

(c) There are cost overruns and time overruns is the order of the day.

(d) Delays in replacement for rejected material.

(e) Non-removal of rejected parts from stores, even after one year, with the hope it will be consumed during shortages.

(f) Reliability of spare parts not specified.

(g) Poor after-sales-service even in warranty period by the vendors.

(h) Vendors not interested in investing in infrastructure facilities for small quantity requirements of spare parts and do not want to spend money on educational orders.

(i) The suppliers produce good quality samples according to the given specifications, but do not adhere to the specifications of approved samples, later on.

(j) Inadequate infrastructure facilities, technical as well as entrepreneurial skills.

(k) Improper testing equipments and inspection facilities with vendors.

(l) Insisting on venture capital from Ganapathy Ram Petrochemicals with a few ancillaries reporting sick later on.

(m) Employees' morale low due to poor wages resulting in frequent strikes at vendor's premises.

(n) Increasing the price after obtaining the order by more than one hundred per cent at the time of supply.

(o) Vendors insisting on monetary advances for materials every time, or requesting for supply of raw materials with the ancillary unit reporting sick and industrialists becoming rich.

(p) Reporting to psychological pricing mechanism.

(q) Poor organisation at vendor's premises.

(r) Not applying well known cost reduction techniques like learning curve and value engineering for repeat orders, and not passing advantage of learning experience to the buyer.

(s) Too much time spent on price haggling during negotiations.

(t) Lack of professional approach for quick response, prompt repair, and least repeat failures by the vendors.

(u) Often the same parts repeatedly fail in the process.

User's Perception

(v) According to the user's design, safety, maintenance, and production executives, the local vendors cannot meet the precise and high tech nature of the demands of the sophisticated petrochemical industry.

(w) No one is interested in developing drawings. Even though technically competent, designers are unable to develop due to pressures of various kinds like imposing American super clause 301 against trade restrictions, trade secrets, etc. The drawing given to a supplier is sold to a competitor in some cases.

(x) Using inferior quality raw material in the absence of imported special steel to manufacture spare parts.

(y) Lack of technical know-how on the part of vendors particularly in the field of reliability estimates and adherence to warranty clauses.

(z) Foreign collaborators to preserve their image do not permit indigenisation due to fear of safety and quality considerations.

IMPORTANT ASPECTS OF PURCHASING PROBLEMS

The officers of the materials department of Ganapathy Ram Petrochemicals confided to the author about the difficulties faced by them in carrying out their jobs successfully. Some of the major difficulties are summarised below:

(a) The user marks most of the indents as most urgent without bothering about the lead time.

(b) The user department classifies most items as proprietary and indents by brand name.

(c) Even though independent quality control inspects the items, the user complains that items are of inferior quality.

(d) The buyer is not allowed to review the specification or buy standard items.

(e) Users prefer only imported items, even though the local source is prepared to give the same item with about the same tolerance.

(f) User is not interested in the application of cost reduction techniques like standardisation, value analysis, etc.

(g) He prefers buying only from the original equipment manufacturer or well-known sources.

(h) The user does not communicate the information in process and outgoing quality rejections, which is essential for vendor rating.

(i) The maintenance department classifies all non-moving machinery spares as insurance items and is not interested in declaring any item obsolete.

(j) The user complains that the buyers are lazy and do not interpret or understand the technical specifications.

(k) During emergency purchases, the users pay a higher rate than the usual purchasing price of the company to the same buyer.

(l) The top management considers that the buyers act only as a post box of getting the indent from the department and procuring it at suppliers' price.

(m) The top management feels that the materials department has not organised its materials intelligence properly and hence does not consult the materials department in the process of capital equipment buying.

(n) The finance department often complains about exceeding purchase budget, resulting in working capital problems, without bothering about inflation.

(o) The finance department always pin-points about violating the procedures and insists on three written quotations even for proprietary items.

(p) The audit department asks too many questions about the purchasing activities, and travel cost in following up items and emergency purchases.

(q) The bill-passing section inordinately delays the bills, forcing the suppliers to increase the price.

(r) Follow-up with finance in this aspect has proved negative.

(s) The finance department always insists that the vendor with the lowest technically acceptable quotation should be negotiated to reduce the price.

(t) The finance department complains that the buyers are unable to forecast the requirement and/or the price correctly.

(u) The buyers are sometimes accused of celebrating their birthday once a week and get costly presents from the suppliers. (t) Charges of corruption, kickbacks, cuts, etc. have been alleged against some buyers, which are being looked into by a local Lokpal vigilance committee.

(v) Marketing department of Ganapathy Ram Petrochemicals has been allowed entertainment expenditure, whereas the materials department is not entitled for any entertainment amount.

(w) This issue of entitlement creates problems during periods of shortages.

(x) The buyers are accused of wrong buying in the initial stage, excess buying to please a vendor and incorrect buying of wrong quality items.

(y) Most of the suppliers do not adhere to the stipulated and mutually agreed delivery schedule in spite of the detailed follow-up steps.

(z) They resort to *force majure* clauses, if forced or threatened about blacklisting.

A TO Z OF STORES PROBLEMS

The chief custodian of Ganapathy Ram Petrochemicals, when asked about the difficulties in his operations, narrated the following points to the author. His staff also concurred with him on most of the points.

(a) Even though stores is money, it has not been appreciated in the company, as union office-bearers, sports personnel, handicapped personnel and others unwanted in other departments, have been dumped in the stores, thinking that any one can do the stores job.

(b) Infrastructure facilities like space, lighting, bins, racks, gangway, calibration of weight scales, air conditioning for some items, etc. are totally inadequate and insufficient.

(c) The stores manager is held responsible for the direct delivered items, which he has not even seen.

(d) The department classifies 30 per cent of the items as non-stock items, making codification and identification difficult.

(e) The maintenance fitter does not bother to buy an indent or sign any document, and comes with a broken piece as a specimen for immediate replacement, as the maintenance staff considers downtime more important than the document.

(f) The indents are always marked urgent with two red tags for each indent.

(g) The indents are always incomplete without work order/department/ code and detailed specifications. Only suppliers' part number is mentioned.

(h) The material indented for the mechanical department is taken by the electrical department as soon as the latter comes to know about the arrival of the item in the stores.

(i) The entire quantity of spare parts, is taken to the department, and it is difficult to know the stock levels or to take action for an expedited order.

(j) Even though about 25 per cent of the total items have not been issued even once during the last five years, the maintenance has not declared them obsolete.

(k) Sometimes the finance department insists that the glass tumblers broken in stores should be replaced by the store staff.

(l) Whenever an item is not available in the stores, the fitter/operator shouts at the stores staff and every one watches the scene with sadistic pleasure.

(m) The consumption limits, agreed to by the departments, are always violated in most of the items resulting in stockout or triggering off emergency purchase orders.

(n) The rejected items occupy a vast area in the stores, and are not immediately removed by the suppliers, who think that it will be used in emergencies.

(o) The year-end stock verification invariably results in huge differences between book balance and physical balance. It is a Herculean task to get the signature of maintenance people who have not sent the documents on the large value of store reconciliation vouchers.

(p) Identification of the spares parts numbering more than two lakh, is always a problem particularly when the indent indicates only supplier's part numbers and not internal codification.

(q) The communication within materials department, needs improvement, when the purchase follows up items which have been received. The stores department does not get copies of purchase orders and does not know the exact status of a purchase order.

(r) The inspection department rejects good quality items, whenever the stock level is high, forcing the supplier quarrel with stores officers before lifting the materials.

(s) The stores officer does not have any control on tool stores, departmental stores and other informal areas of storage.

(t) The stores officer has to execute fidelity bond as he is the custodian of all items, even the bulk material, scrap, heavy items, direct charge items, lying in open — on which he has no control.

(u) The finance officers always complain about the flabby inventory leading to high working capital, but no one in the company cares about semi-finished items or finished goods stocks.

(v) In view of the non-technical nature of stores staff, the maintenance staff enter prohibited areas and remove spare parts without any authorisation.

(w) The project surplus, or legacy of the past, accumulates and no one wants to take any decision on unwanted material.

(x) The items which have been declared obsolete by the finance have not been disposed off as the finance department wants them to be sent to the original suppliers and insist on minimum guarantee price.

(y) The material handling equipments, handling facilities are insufficient for the expansion activities of Ganapathy Ram Petrochemicals.

(z) The quality and quantity of staff need to be strengthened in the stores to meet the challenges of the technological upgradations in Ganapathy Ram Petrochemicals.

A TO Z OF LACUNAE IN MANAGEMENT DEVELOPMENT

Ganapathy Ram Petrochemicals believes in the philosophy of human resources as the major asset and attaches great importance to internal and external management development programmes. The author interviewed a large number of senior executives, who have attended these programmes.

However, the impressions of the executives of Ganapathy Ram Petrochemicals, who have attended external refresher programmes to update their managerial skills, were not always encouraging and their remarks are summarised below:

(a) The programme design is not applicable to the peculiarities of the high-tech process industry.

(b) Incompetent and inexperienced faculty teaching irrelevant topics.

(c) Too theoretical and not practical or experience-based.

(d) Emphasis only on broad general policies on strategies, scenario painting and macro level issues to be tackled by the policy-making body at the ministerial level.

(e) Lecturers with confused thinking, inability to answer questions with convictions and with poor communication abilities, many lecturers speak only with blackboard and do not see the audience.

(f) Poor time management with sessions invariably eating into the next, even though there are lectures on time management, punctuality, etc.

(g) Speaking without depth of knowledge, dedication and commitment. The same lecturer handles research as well as inventory control, lacking in depth in both disciplines.

(h) Training not based on needs of the executives of Ganapathy Ram Petrochemicals but what could be offered by the faculty.

(i) Training only to show statistics of number trained to the Ministry and Bureau of Public Enterprises.

(j) Improper mix of training media or pedagogical methodology, like lectures, business game, case studies, computer demonstrations, syndicate discussions, video films, role playing, projects, self study, exercises and problem solving. Input from the faculty constitutes 20 per cent of the entire programme.

(k) Concentrating on trivial and inconsequential matters without going into the depth of analysis.

(l) Improper infrastructure and administrative support to facilitate smooth training in-session and post-session.

(m) Many participants leaving the class room or cross talking while the lecturer is speaking.

(n) Improper sequencing of subjects and inadequate time spent on relevant topics.

(o) Background material poorly reproduced from old American books, and not relevant to local conditions.

(p) Participants learn more from Ravi Batra's book on Great depression of 1990s than from class room sessions.

(q) Greater interest in the fourteenth century philosopher Noistre-demus, comprehensive predictions including Third World War in 1999, than the classroom sessions.

(r) Learnt more by interchanging experience with other participants than in the class room.

(s) Every one treating the management development programme as a paid holiday.

(t) Two faculty members express conflicting views on the same subject on the same day resulting in confusion.

(u) Insufficient time for computer demonstration or access to personal computer during training.

(v) Case studies not realistic and not giving typical solutions.

(w) Not accustomed to sitting in class rooms continuously from nine in the morning to five in the evening.

(x) The interest and level of participants differ widely in different subjects, thereby affecting interchange of experiences.

(y) The benefits to be accrued from the programme is not commensurate with the total cost incurred.

(z) The Marketing Manager of Ganapathy Ram Petrochemicals thought that the money can be spent better by sponsoring one day cricket match between India and Pakistan, while the Operations Executives wanted development programmes to be conducted abroad, with the help of one of the collaborators.

A TO Z CAUSES FOR HIGH INVENTORY

When interviewed, the inventory controller working under the Materials Manager of Ganapathy Ram Petrochemicals gave the following as the major reasons for the very high level of inventories, particularly in the area of machinery spares.

(a) Being a continuous process industry, nobody can afford to face the consequences of stockout.

(b) Every one knows that the stockout cost, which has not been estimated, is several times the item cost. Every one lets off steam at the materials department in case of stockouts, while no one cares for the high working capital.

(c) Forced to carry more stock of imported items as during monsoon and high tides, transportation comes to a standstill.

(d) The original equipment manufacturer has dumped a lot of slow-moving spares at the time of buying the capital machinery.

(e) Sometimes, based on previous consumption pattern, spares have been ordered for machineries, which have been phased out.

(f) The operations department tends to classify all items, as critical and wants hundred per cent service level, without intimation to inventory departments.

User Problems

(g) No one is interested in developing drawings or indigenising the imported items.

(h) There are genuine difficulties in generating data pertaining to consumption norms, failure, reliability, forecasts, as the demand for any items particularly stores are erratic.

(i) The purchase department overbuys in March to consume the total budget and foreign exchange allocation as carrying forward to the next year is not possible.

(j) Tendency to classify many (about 25%) spare parts as proprietary, leading to long lead time and increased stock.

(k) The powers of delegation have not been revised during the last five years, whereas inflation has resulted in increased lead time.

(l) No initiative for standardising the equipment at the plant level, industry level or national level or for maintaining spares bank, even though every one gives one day lecture on it.

(m) Supplier at times sending more material than ordered for his economic lot size.

(n) Spares have not been segregated for into categories overhauling, proprietary routine, maintenance, rotable, capital, insurance, etc.

(o) Non-availability of scientific methods and computer software packages to the peculiarities of local conditions.

(p) Improper advance information on long shut-down, major overhauling, minor repairs, etc. necessitating high spare parts inventory.

(q) There is no motivation to reduce the stock level as the senior managers want to avoid stockout at any cost.

(r) Inventory in Ganapathy Ram implies controlling only raw material and spares, whereas it is often forgotten that the working capital commitment covers work-in-progress, finished goods and sundry debtors.

(s) The inventory includes rupees five crores worth of spare parts, dumped initially, by the original equipment manufacturer, which has not moved during the last five years. Project surplus worth, rupees three crores have also been included in the inventory figures.

(t) The total lead time for imported spares parts exceeds five years in some cases.

Documentation

(u) Documentation is not complete as fitters remove the items from stores without issue vouchers.

(v) Documentation is also incomplete when the stores is forced to give temporary codification for new items and non-stock items — about 40 per cent — which are minor modifications of existing items.

(w) The plant has entered into collaboration with 20 different countries with the result that the equipments and hence spare parts are not standardised.

(x) Both FPS and metric system are prevalent in the plant; some of the motors from Russia are classified by weight instead of horse power resulting in duplication.

(y) The chief mechanical engineer insists that about Rs 50 lakh of spares must be available at least three months, before the commencement of overhauling the equipments to avoid stockout.

(z) There is a tendency to classify every non-moving item as an insurance spare part in order to preempt declaring them as obsolete items.

A TO Z OF VENDOR DIFFICULTIES

The author interviewed some ancillary suppliers of Ganapathy Ram Petrochemicals and the Marketing Directors of the vendor organisations. The socio-political, economical, technological, design, developmental, competitive, factors on servicing and selling spares, are too many and it is difficult to measure the effect of these on spares sales/servicing according to servicing contractors, spare parts vendors and the original equipment suppliers of the company. The difficulties of the suppliers are summarised below.

(a) Vendors do not want to invest the previous working capital in slow-moving spare parts with uncertain demand, while Ganapathy Ram Petrochemicals, categorise them as urgently needed, without giving adequate time for manufacturing.

(b) Communication and transportation bottlenecks with other customers located in far flung areas, is a major issue.

(c) The large range, variety, sizes, and number due to technological upgradation, force the vendors to be highly selective in stocking the spares leading to customer dissatisfaction.

(d) The morale of the servicing staff is not high, when they are being illtreated both by the vendor and the consumer, leading to poor quality of service.

(e) While service to others is a noble thought, due to changes in the value system, in the industrial scene, the quality of service is poor even according to the vendors.

(f) The maintenance schedule of equipments and operating them according to the service manuals, are not usually done by the officers of Ganapathy Ram Petrochemicals, who blame the suppliers, for all ills.

Pricing of Spares

(g) The assumptions behind conventional scientific forecasting methods like moving average, trend analysis, exponential smoothing, correlation methods, regression analysis etc. are not valid in actual practice to determine the demand pattern, as reliable feedback from consumers are not available.

(h) Consumers always complain that the margin in spare parts is more than one hundred per cent, when the vendor adopts market skimming strategy and not market penetration strategy.

(i) It is noted by the buyers in Ganapathy Ram Petrochemicals that the suppliers sometimes adopt cost plus pricing, break-even analysis, target pricing, perceived value pricing, going rate pricing, marginal pricing, sealed bid pricing, psychological pricing, discriminatory pricing, offering discounts, proprietary item pricing, etc. depending upon the competition and urgency of requirement.

(j) Supplier's bills are always delayed and advances are rarely given according to the vendors, with price increase by manufacturing with generators not allowed. Quality and reliability are being insisted upon by the buyers with the help of buy-back clauses.

(k) All the customers insist upon opening of depots near the factory, but this is not possible due to uncertain demands.

(l) The requirement of spare parts is not only indefinite but highly erratic, making forecasts and economic lot size in manufacturing difficult.

(m) Price of an item is composed of material cost, labour cost, profit, fees, rate of interest, taxes, freight, insurance, salary, wage, bribe, *riswath*, kickforward, kickback, winding up charges, graft, honorarium, gifts, commission, discounts, warehousing charges, inspection fees, presents in the form of cash and kind, etc.

After-Sales Services

(n) Vendor's after-sales-service is always perceived as 'ASS' even though the company hides its fault by not keeping failure documents. The cost of servicing is increasing with increased inflationary pressures. Customer satisfaction is only a subjective criteria, always resulting in vendor's poor image!

(o) Since the marketing concepts revolve around the fact that the consumer is the king, one has to play to the idiosyncracies of the consuming unit, which cannot be satisfied at all, in spite of the fact high level of service is their motto.

(p) Due to scarcity of capital, the equipment's life is prolonged by the users and the original equipment manufacturer is expected to supply spares indefinitely beyond the usual ten year period.

(q) Depending upon the type of contract, the seller's service department installs the equipment and commissions the plant for which separate fees are charged.

(r) After installation, the machine is covered against one year warranty, during which preventive maintenance is carried out as per contract, but, on completion of warranty, the vendors' offer of optimal all inclusive service contract is not taken by Ganapathy Ram Petrochemicals even though the vendor is called to attend to breakdowns.

(s) Sometimes trained dealers or franchises are employed as agents but their performance is poor in view of the inadequate infrastructure facilities, resulting in repeated failure of the same item.

(t) The logic of vendor rating scheme and the competitor's rating are not revealed to the supplier.

(u) Last minute alterations, extra features in specifications, postponement and cancellation of delivery schedules, add to vendor's difficulties.

(v) The customers always complain about the initial dumping of unwanted spare parts at the time of capital equipment procurement, and want the buy-back clause to be implemented without additional costs.

(w) Inability to appreciate that vendors do not have sufficient capacity allocated for spare parts manufacturing for replacement market in view of varying demand.

(x) Lead time from subcontractors and subvendors is long since special steel and other matching components have to be imported.

(y) It is difficult to compete with spurious manufacturers who sometimes use the original equipment manufacturer's labels.

(z) It is also difficult to compare with many small scale manufacturers who quote low in view of government subsidy, reduced taxation, lower wages to labourers, savings in transportation cost, etc.

IMPORTANT ASPECTS OF MATERIALS POLICY MANUAL

Ganapathy Ram Petrochemicals, has prepared a policy manual on materials management, which has been approved by the top management. The Chairman has circulated the manual to all the officers and union office-bearers of the company with a covering note as follows: This is only a summary as the original policy manual contains five hundred pages and covers all aspects of stores, purchase, inventory control, materials management, organisation and cost reduction.

> Ganapathy Ram Petrochemicals Policy Manual — Materials Management
> I have pleasure in enclosing herewith a copy of the materials management manual. The policies are general statements of guidelines to achieve results.

The policy and procedures are very essential to be implemented for achieving the corporate goals of our organisation. The manual will help you in clearly understanding the objectives, policies, procedures, systems, strategies and responsibilities of the materials management function. The coordination meeting will discuss any additional features or difficulties.

Sd/-
G. Ganapathy Ram
Chairman, Ganapathy Ram Petrochemicals

The contents of the manual listing the general rules for internal relationship and coordination with external resources.

(a) Materials Management is the only department which serves as the point of contact between the suppliers and the company.

(b) Any department which may wish to contact a supplier directly, must do so after clearance from the materials management and must keep it informed of all transactions.

(c) Source development is the exclusive responsibility of materials department.

(d) Other company's salesmen can be received only by the materials management department. Other departments, can discuss their problems after clearance from the materials department.

(e) The materials management can question the quality, quantity, and price of the materials, asked for by the user departments in the best overall interest of the company. But any change in specification has to be approved by the indentor.

(f) The responsibility for analysis of requisition, sending out enquiries, analysis of quotations, placement of purchase orders, expediting and controlling inventory levels, lies entirely with the materials department.

(g) The development of reliable vendors and sound vendor relationship, are the concerns of the materials department.

(h) The practices of reciprocity, personnel buying, etc. will be done by the materials management under the top management direction.

(i) Materials management shall keep the concerned departments informed of the delivery schedules, accounts payable, receipts, etc., so as to ensure smooth operations.

(j) Market price levels and information on new product development will be communicated to the concerned department by the materials department.

(k) In order to effectively perform its function, materials management department will be assisted by the finance department in prompt clearance of the suppliers' bills.

(l) The finance department will also render necessary assistance to the materials department in the spheres of marine insurance, inland transport insurance, fire insurance and stores insurance as well as in claim formalities.

(m) The materials department will seek the assistance of finance and other departments in customs, excise, MODVAT, sales tax and other taxation formalities.

(n) Marketing and production departments shall keep the materials management department informed of future plans, so that expeditious action can be taken in controlling inventory levels.

(o) The materials management department will assist the top management in the process of capital equipment buying by locating sources and analysing the merits of various alternative equipments.

Thus we see that materials policy manuals serves as an indispensable tool in bringing about coordination between various external resources and streamlines the internal relationships.

Further Reading

Alijan, G.W., *Purchasing Handbook*, McGraw Hill, New York, 1985.

Alnmer, D.S., *Materials Mangement*, IRW or Home Wood, Illinois, 1974.

Auebrach, Green, *Components of Manufacturing Inventories*, Bureau of Economics, Cambridge, U.K., 1980.

Baily, P. and Tavernier, *Design of Stock Control Systems and Records*, Gower Press, London, 1970.

Baily, P. and Farmer, *Managing Materials*, Gower Press, London, 1981.

Brauen, G., *Management Decisions for Production Operations*, Dryden Press, 1971, p. 683.

Buchan, J. and Koenisberg, E., *Scientific Inventory Management*, Prentice Hall, Englewood Cliffs (N.J.).

Burton and Evans, *Effective Warehousing*, Macdon & Evan, UK, 1985.

Bureau of Public Enterprises — *Committee on Inventory Control Report*, Government of India, Delhi, 1985.

Chadda, R.S., *Inventory Management in India*, Allied Publishers, New Delhi, 1971.

Challiah, S.V., *Some Modern Concepts of Effective Stock Control*, Railway Staff College, Baroda, 1972.

Crompton, H.K., *Suppliers and Materials Management*, Business Book Publications, U.K., 1980.

Donald, P.S. Molik, *Materials Requirement of Manufacturing Industries*, Von Norstand Reinhold, New York, 1983.

Gopalakrishnan, P., *Bank Credit and Inventory Norms* — ASCI Publications, Hyderabad, 1976.

Gopalakrishnan, P. et al., *Case Method and Management Development*, ASCI Publications, Hyderabad, 1971.

Gopalakrishnan, P., *Cost Reduction Handbook*, Jaico, New Delhi, 1992.

_____, *Integrated Quality Management*, S. Chand & Co., New Delhi, 1986.

_____, *Integrated Spares Management, S. Chand & Co.*, New Delhi, 1984.

_____, *Purchasing and Materials Management*, Tata McGraw Hill, New Delhi, 1991.

Gopalakrishnan, P. and Banerji, A.K., *Maintenance and Spare Parts Management*, Prentice Hall, New Delhi, 1991, p. 385.

Gopalakrishnan, P. and Narayanan, K.S., *Computers in India — A Overview*, Popular Prakasan, Bombay, 1975.

Gopalakrishnan, P. and Rama Moorthy, V.E., *Project Management Text Book*, Macmillan, New Delhi, 1992.

Gopalakrishnan, P. and Sandilya, M.S., *Inventory Management — Text and Cases*, Macmillan India, New Delhi.

_____, *Management of Quality and Inspection*, Tata McGraw Hill, New Delhi, 1979.

_____, *Purchasing Strategies — Text and Cases*, Sterling Publishers, New Delhi, 1978.

_____, *Stores Management and Logistics*, S. Chand & Co., New Delhi, 1978.

Gopalakrishnan, P. and Sarin, A.N., *Marketing in India — Text and Cases*, Orient Longman, New Delhi, 1970.

Gopalakrishnan, P. and Sundaresan, M., *Materials Management*, Prentice-Hall, New Delhi, 1991.

_____, *Spares Parts Management — Text and Cases*, Jaico, New Delhi.

Grady, P.J.O., *Just in Time*, Kogan Paper, U.K., 1988.

Hadleyn and Whitin, J.M., *Analysis of Inventory Systems*, Prentice Hall, Englewood Cliffs (N.J.), 1968.

Hobbs, J.A., *Control Over Inventory and Production*, McGraw Hill, New York, 1973.

Joseph Orlicky, *Materials Requirement Planning*, McGraw Hill, New York, 1975.

Krishnamoorthy, K. and Sastry, P.V., *Inventories in India Manufacturing*, Institute of Economic Growth, Delhi, 1970.

Magee, J.F. and Boodman, B.M., *Production Planning and Inventory Control*, McGraw Hill, New York, 1967.

Mantri, M.K., *Stock Control Maharashtra State Road Transport Corporation*, Bombay, 1975.

Naddor, F.G., *Inventory Systems*, John Wiley, New York, 1975.

Palit, A.R., *Outlines of Materials Management*, Academic Books, Bombay, 1970.

Reins, A. et al., *Industrial Inventory Control*, Gardon and Breach Science Publishers, 1972.

Schwarz, L.B., *Multi-Level Production and Inventory Systems*, Polland, 1981.

Shingo, *Study of Toyota Production System*, Japanese Management Association, 1981.

Tai Ichi Ohio and Setswo Mito, *Just in Time* — Productivity Press Inc., Cambridge, USA, 1988.

United Nations Department of Economic Social Affairs, New York, *Study of Supply Management*, 1981.

Walter E. Goddera, *Just in Time*, Oliver Wright Publications, USA, 1986.

Vandierdonock and Miller, *Uncertainty, Complexity and Materials Management Systems*, Boston, Harvard, 1981.

Wolf, T.R., *Improving the Control of Maintenance Stores and Inventories* — Industrial and Commercial Techniques, London, UK.

Zazac, J.H. and Warshan, M.R., *Evaluating the Materials Management of Inventories*, Michigan University, 1979.